Insurgency and Counterinsurgency in the Nineteenth Century

Insurgency and Counterinsurgency in the Nineteenth Century examines insurgency and counterinsurgency across the globe in the nineteenth century.

The volume includes chapters from distinguished and rising historians from Europe, North and South America and covers irregular wars in Spain, Ireland, France, Latin America, China, USA, Africa, Central Asia and Burma. The authors explore links between insurgencies and nationalism, including learning curves and emulation in counterinsurgency. With a special emphasis on non-Western warfare, this volume includes case studies such as the Katanga and White Lotus rebellions largely unknown to Western readers. The military history of the nineteenth century thus reveals much more than the symmetrical warfare of Napoleon, Grant and Moltke. This volume shows the commonalities of responses more than their differences and refracts these through themes which crop up repeatedly in different times and places. These themes include common problems and solutions: the challenge of commanding local intelligence networks; public opinion; millenarianism, magic and religion; technology; 'hearts and minds'; the legal framework of state violence; racial stereotypes and patterns of forgetting and remembering guerrilla conflicts.

The first recent study to examine Western and non-Western warfare in equal measure, stressing the prevalence of commonalities between guerrilla warfare and counterinsurgency across the globe, *Insurgency and Counterinsurgency in the Nineteenth Century* will be of great interest to scholars of military and strategic studies, as well as modern military history. It was originally published as a special issue of *Small Wars & Insurgencies*.

Mark Lawrence is Lecturer in Modern Hispanic and Military History at the University of Kent, UK, and Director of Military History at the University of Kent (2019–2020). He is author of the award-winning *Spanish Civil Wars* (2017), *Nineteenth-Century Spain* (2019) and *Insurgency, Counterinsurgency and Policing in Centre-West Mexico, 1926–1929: Fighting Cristeros* (2020).

Insurgency and Counterinsurgency in the Nineteenth Century

A Global History

Edited by
Mark Lawrence

Routledge
Taylor & Francis Group

LONDON AND NEW YORK

First published 2021
by Routledge
2 Park Square, Milton Park, Abingdon, Oxon OX14 4RN

and by Routledge
52 Vanderbilt Avenue, New York, NY 10017

Routledge is an imprint of the Taylor & Francis Group, an informa business

Chapters 1–11 and 13–16 © 2021 Taylor & Francis
Chapter 12 © 2019 Bastian Matteo Scianna. Originally published as Open Access.

British Library Cataloguing in Publication Data
A catalogue record for this book is available from the British Library

ISBN: 978-0-367-85975-6 (hbk)
ISBN: 978-0-367-61050-0 (pbk)

Typeset in Myriad Pro
by Newgen Publishing UK

Publisher's Note
The publisher accepts responsibility for any inconsistencies that may have arisen during the conversion of this book from journal articles to book chapters, namely the inclusion of journal terminology.

Disclaimer
Every effort has been made to contact copyright holders for their permission to reprint material in this book. The publishers would be grateful to hear from any copyright holder who is not here acknowledged and will undertake to rectify any errors or omissions in future editions of this book.

Contents

Citation Information

The chapters in this book were originally published in *Small Wars & Insurgencies*, volume 30, issue 4–5 (July–August 2019). When citing this material, please use the original page numbering for each article, as follows:

Chapter 1
Why a nineteenth-century study?
Mark Lawrence
Small Wars & Insurgencies, volume 30, issue 4–5 (2019), pp. 719–733

Chapter 2
The Peninsular War guerrilla and its antecedents: humiliation forgotten, disaster prefigured: the guerra fantástica *of 1762*
Charles Esdaile
Small Wars & Insurgencies, volume 30, issue 4–5 (2019), pp. 734–749

Chapter 3
Reluctant guerrillas in early nineteenth century China: the White Lotus insurgents and their suppressors
Yingcong Dai
Small Wars & Insurgencies, volume 30, issue 4–5 (2019), pp. 750–774

Chapter 4
Regular and irregular forces in conflict: nineteenth century insurgencies in South America
Alejandro M. Rabinovich and Natalia Sobrevilla Perea
Small Wars & Insurgencies, volume 30, issue 4–5 (2019), pp. 775–796

Chapter 5
The First Carlist War (1833–40), insurgency, Ramón Cabrera, and expeditionary warfare
Mark Lawrence
Small Wars & Insurgencies, volume 30, issue 4–5 (2019), pp. 797–817

For any permission-related enquiries please visit:
www.tandfonline.com/page/help/permissions

Notes on Contributors

Ian F. W. Beckett, School of History, Rutherford College, University of Kent, UK.

Timothy Bowman, School of History, Rutherford College, University of Kent, UK.

Yingcong Dai, Department of History, William Paterson University of New Jersey, USA.

Mario Draper, School of History, Rutherford College, University of Kent, UK.

Charles Esdaile, School of Histories, Languages and Cultures, University of Liverpool, UK.

Susan-Mary Grant, School of History, Classics and Archaeology, Newcastle University, UK.

Jack Hogan, Department of International History, London School of Economics and Political Science, UK.

Mark Lawrence, School of History, Rutherford College, University of Kent, UK.

Giacomo Macola, Facoltà di Lettere e Filosofia, Sapienza Università di Roma, Italy.

Bastian Matteo Scianna, Department of History, University of Potsdam, Germany.

Nathaniel Morris, Department of History, University College London, University of London, UK.

Alexander Morrison, Faculty of History, New College, University of Oxford, UK.

Alejandro M. Rabinovich, Conicet, Universidad Nacional de La Pampa, Argentina.

Richard Reid, Faculty of History, University of Oxford, UK.

Natalia Sobrevilla Perea, School of European Culture and Languages, University of Kent, UK.

Kenneth M. Swope, Dale Center for the Study of War and Society, University of Southern Mississipi, USA.

Why a nineteenth-century study?

Mark Lawrence

The term 'guerrilla' tends to evoke twentieth-century rather than nine-teenth-century connotations. The First World War witnessed insurgent challenges to imperial rule in Europe and Asia, guerrilla revolutionaries and counter-revolutionaries in Latin America. The postwar era witnessed a brutalisation of counter-insurgency doctrines consistent with totalitarian politics, especially in China and the Soviet Union. The Second World War witnessed an intensification of counter-insurgency, especially at the hands of the exterminatory Japanese and German empires. The post-1945 era ushered in perhaps the most complex and diverse insurgency environment, as anti-colonial insurgencies were reinforced by communist, nationalistic and Maoist ideologies, which were countered with mixed success by metropolitan counter-insurgency strategies. 'People's war' thus conjured up images of Mao and Che Guevara, of revolutionary warfare, far removed from the supposedly state-centric armies and strategies of the nineteenth century. Even when the ideological certainties of the Cold War fell away from 1990, insurgencies diversified along civil war, religious and technological lines.[1]

Yet insurgency is both the oldest form of warfare and the variety with the greatest opportunities for development. Insurgency, guerrilla, partisans, and 'people's war' are nuances of the universal and immemorial phenomenon of irregular combatants waging war against formally constituted power. Military historians identify three types of 'people's war': guerrilla warfare, militia warfare, and conscription armies.[2] The former two are analysed in this volume. Strategy expert, Beatrice Heuser, describes insurgent warfare as follows; 'wars fought between parties that are fundamentally unequal, one side possessing authority, a recognised claim to a monopoly of power and a state apparatus in some form, often including armed forces'.[3] Recent research has demonstrated the diversity both of the guerrilla and of counter-insurgency throughout history. In the past three years the *Small Wars and Insurgencies* journal has published two special issues on the historical origins and contemporary impact of guerrilla warfare respectively.[4] The nineteenth century offers particular opportunity for a fresh study of global insurgency and counter-insurgency. The military history of this century reveals much more than the symmetrical warfare of Napoleon, Grant and Moltke. It also reveals well-known and less well-known insurgencies, of links between guerrilla movements and nationalism, and of complex motivations and strategies driving both insurgencies and counter-insurgencies. Contemporary strategists were much more impressed by the burden and appeal of guerrilla warfare than a cursory glance at the military academies would reveal. Even the great exponent of interstate war, Carl von Clausewitz, had a complex understanding of guerrilla warfare that has often passed unremarked by scholars studying the post-Westphalian overtones in his *On War*.[5]

Thus the nineteenth-century deserves a bespoke study, on a global scale. A global history of insurgency in this century presents us with a similar paradox as nineteenth-century global history more widely. Global empires became more antagonistic to each other even though their similarities, connections and linkages proliferated. Equally, patterns of insurgency and counter-insurgency showed increasing similarities to each other as the nineteenth century progressed, as the chapters in the following study demonstrate. Thus it is insufficient to view insurgency warfare only in a local, regional or even national context. Even such continental qualifiers as 'European' or 'American' history cannot provide a diverse understanding of insurgencies. Far-flung imperial warfare reverberated back onto the metropolis, just as metropolitan preconceptions conditioned counter-insurgency strategy. A global understanding of insurgency is all the more necessary considering how such factors as ecology, epidemiology, diasporas, and the 'informal empire' phenomenon of outsiders enjoying privileges over natives, all cut across regional, national and even continental divisions.[6] And local, asymmetrical wars after the late eighteenth century tended to become 'catastrophised' by a new political climate in national and international affairs.

This edited volume sheds new light on global insurgency and counter-insurgency in the nineteenth century. Bringing together both distinguished and rising international scholars based in Europe, North and South America, this volume provides new insights into an under-researched topic. It exposes some insurgencies unknown to most scholars, explores the links between insurgencies and nationalism, and studies the extent to which we can identify evolving patterns between reactive and progressive insurgency, along with learning curves and emulation in counter-insurgency.

A bespoke study of nineteenth-century asymmetric warfare presents us with many challenges. One of the features of modernity in the advanced polities of Europe and North America was what sociologist, Max Weber, called the monopoly on 'legitimate' violence owned by the state, as the 'private' violence related to serfdom and slavery disappeared.

Administrative and judicial reforms, policing, and economic and demo-graphic recovery, all set the trend for a reduction in the use of violence in internal affairs.[7] The onset of European war in 1792 accelerated the growth of states on the one hand whilst producing wartime strains on the other hand which promoted the phenomenon of both insurgency and counter-insurgency. Moreover, irregular campaigns took place in a context in which political discourse to some extent exaggerated violence, driving a wedge between the 'legitimate' and often ideologically promoted vio-lence of the forces of the state and the victimhood of 'illegitimate' violence of rebel communities.[8]

The growing employment of state violence against 'enemies' rather than for internal law and order helped 'totalise' warfare, influencing 'total war' strategies and transforming the strategic landscape until the end of the Second World War. Thus what space is there for a study of war aims which in being asymmetrical, were also often limited rather than 'total'? Even before the era of Revolutionary and Napoleonic warfare (and some fifty years before Clausewitz's famous musings on the subject) the Comte de Guibert understood the differences between limited war and total war. The latter was made possible, Guibert thought, by the creation of mass citizens' armies.[9] Another challenge lies in the growth of nationalism in nineteenth-century Europe, the continent which set the tone for most military evolution in this century and imbued a growing culture of militarism. Ascendant militarism between the end of the French Wars and the start of the First World War was reflected in most military strategists themselves being military officers, especially after the era of Moltke the Elder. Yet despite the growth of nationalism alongside militarism, these thinkers tended to read the works of thinkers in other countries.[10] Military strategists thought globally, and this is reflected in the global reach of this volume.

These nineteenth-century strategists grew ever more aggressive in their calculations. Unlike eighteenth-century strategists, who could not reach consensus about whether offensive or defensive war was stronger (both seemed to have their role), nineteenth-century strategists were more offen-sively minded. Napoleon and his legacy up to 1914 witnessed strategists almost universally believing in the 'cult of the offensive'. Even the small wars expert, Callwell, judged pre-emptive offensives by regular forces to be the best war-winning tactic.[11] The advancing abilities of states to mobilise troops and (especially from from mid-century) implement new weapons and logistical technology led Western strategists to emphasise recent his-torical lessons. Neither Jomini (1779–1869), Clausewitz (1780–1831) nor Auguste de Marmont (1774–1852) spent significant time learning the les-sons of ancient warfare, focusing instead on Frederick the Great and, espe-cially, Napoleon, as the godfathers of military modernity. The nineteenth century thus produced an era in lessons from living memory, even if thinkers (especially Jomini) proved slow at making sense of the technological revolution in warfare evident from the mid-nineteenth century.[12]

As the nineteenth century progressed, military thinkers grew ever more convinced that the coming war would be a war of mass and movement. Attention to guerrilla warfare could not keep up with military revolution, even though the 1815–1914 era actually witnessed relatively few major wars and relatively many guerrilla wars. Even such classically symmetrical wars as the American Civil War and Franco-Prussian War involved partisan operations, and these were of a different order to the 'partisans' attached to regular armies as scouts before the Napoleonic Wars. Before 1810 irregular warfare was considered the preserve of special forces ('partisans', or 'parties') operating behind enemy lines in support of regular forces. Yet 1810 witnessed the leap from 'partisan war' to 'people's war', owing to the ironic emulation of the French Revolutionary example in both Patriot Spain and Prussia.[13] Sibylle Scheipers suggests that 'the irregular initially emerged from the state, and not in opposition to the state,' but by the time of the Napoleonic wars irregular forces would be seen as 'an intolerable challenge to the "norm" of regular warfare'.[14] The legitimate monopoly on violence owned by the forces of the state was now threatened in ways which reached far beyond the military impact of insurgency. In the words of the counterinsurgent theorist, David Galula, 'the insurgent needs so little to achieve so much whereas the counterinsurgent needs so much to achieve so little'.[15] The nineteenth century, thanks to the experience of the French and Napoleonic Wars, recast insurgencies as threats to the state, not as an adjunct to state power as during the Early Modern era. Thus irregular forces would, in the words of Sibylle Scheipers, pose 'an intolerable challenge to the "norm" of regular warfare'.[16] The theme of the following articles comprises the conflict of 'anti-state' forces waging war against formally constituted power.

Counterinsurgency strategy

The legacy of legitimate state violence, combined with army honour codes, did not commend irregular warfare for study. Despite persistent guerrilla struggles, there was a repeated refusal of major armies to recognise guerrilla warfare as 'real' war. Academy-trained officers viewed guerrilla fighters as rebels and bandits, which was completely at odds with the romanticisation of such supposedly popular struggles as the Spanish and Russian 'people' against Napoleon. Nineteenth-century Europe witnessed a marked acceleration in ruthlessness as a consequence of the French Revolutionary and Napoleonic Wars. The French response to irregular resistance was in many cases brutal and oppressive, utilising 'flying columns' developed against the insurrection in the 1790s Vendée to mount reprisal attacks against villages thought to be collaborating with guerrillas. The French branded the irregulars universally as 'brigands' in order to strip the guerrillas of their political legitimacy, And, since 'brigands' did not have any political legitimacy, but were fuelled by criminal motives, the customs of war did not apply to them. Hence, the French forces were allowed to shoot them without trial.[17] Scheipers argues that the revolutionary state's harsh response to these early risings 'helped shape the emergence of the concept of the 'brigand' as the predominant label for the irregular fighters.[18]

For what the counterinsurgency condemned as brigandage the insurgency celebrated as heroism. Robin Hood loomed large as a template to bestow upon nineteenth-century guerrillas a mystique. Su Sanniang, female Robin-Hood-style gangleader of the Taiping rebellion, acted initially to avenge the death of her husband but soon became a charismatic leader in her own right.[19] Even as the romanticised insurgent of the nineteenth century evolved into the ideological freedom-fighter of the twentieth, military academies remained remarkably unreflective. In the extreme case of the USA, only conventional and intensive warfare was seen as 'worthy' of study and doctrine. As one US general remarked in 1970 regarding the US military's failed strategy in Vietnam: 'I'll be damned if I permit the US army, its institutions, its doctrine and its traditions, to be destroyed just to win this lousy war'.[20] US officer academies today will generally not engage with military history before the Second World War, because US military doctrine is so wedded to the centrality of technology.-[21] There was an appealing 'Whiggishness' to this obsession with military modernity. Whereas Early Modern and Modern European 'state-building' wars had many positive side-effects (extension of state machine/protection, enhanced popular participation in the state, technological innovations), the twentieth-century era of 'privatised' wars, especially those after 1945, were only destructive (not even productive economically given the reliance often on imported weapons).[22] There was little emotional appeal to engage with factional and intra-state wars from a purely military perspective.

Such intransigence is remarkable considering that the USA like other Western powers developed counter-insurgency experience over the course of the nineteenth century. The American Civil War, the subject of Susan-Mary Grant's chapter, resulted in US military law for the first time codifying the treatment of captured partisans. Whereas women and children were usually spared direct reprisals by both Confederates and Unionists during the American Civil War – atrocities tending to be committed against non-whites – captured irregulars were frequently submitted to summary justice. Even though the Lieber Code stipulated that even captured partisans were entitled to military trial, in several campaigns Union commanders, often in reprisal for Confederate ruthlessness, ordered summary shootings of irregulars.[23] Nor did US counterinsurgency experience end there. During the awkward task of cutting off civilian support for enemy insurgents, the US armed forces matched other Great Powers in adopting 'reconcentration' tactics, such as in post-Spanish Cuba and post-Spanish Philippines. Similar tactics had been employed by the Spanish themselves, who during their last colonial counterinsurgency in Cuba during 1895–98 reconcentrated civilians in the loyalist west of the island, often in response to the 'deconcentration' strategy of the Cuban Liberation Army.[24] The British turned the chaotic Spanish counterinsurgency model into a 'camp' system, less bloody than the Spanish-Cuban precedent but appalling (and well publicised) all the same. Blockhouses and wire intercepted Boer guerrillas while civilian support for the insurgency was curtailed by the forced relocation of civilians into 'concentration camps'. The Americans in the Philippines used similar ruthlessness, albeit matched with a 'civic action' programme (of using the presence of military bases to 'civilise' relocated populations). French troops campaigning in Madagascar in the 1890s pre-empted 'civic action' using what would become known as the 'ink-spot' strategy.[25]

National traditions of counterinsurgency had clearly emerged by the end of the nineteenth century. Yet as this volume shows, these traditions were not hermetic. Strategists read beyond their national literatures and the counterinsurgency strategy of nations actually overlapped with and informed each other. The Belgian-officered *Force Publique* in the Congo Free State, as Mario Draper explains, borrowed strategy from French and British precedents during its 1892–94 Congo-Arab War, Belgium possessing as yet little military tradition of its own. Little-known Belgian concepts of counter-insurgency appealed to universal – and flexible – laws of war as much as such better-known theorists as Callwell and Lyautey. Alexander Morrison's study explains how Callwell's cultural preconceptions limited his understanding of Russian counter-insurgency operations in Central Asia. His information was second-hand and he was all too inclined to support interpretations which matched his own model. On the one hand Callwell was recognisable in a Victorian sense in underestimating the sophistication of Russia's 'savage' enemies, in their fortress systems, for example, even equating the conquest of unyielding natural terrain with an unyielding 'character' of non-European opposition. On the other hand he was refreshingly matter-of-fact in his acceptance of atrocities and killings of civilians, and did not stand upon this as a faux moral issues by which inaccurately to distinguish Russian imperialism from British: indeed Callwell's coverage of Russian actions in Central Asia understood the 'Great Game' to be militarily complementary.

Mobility and pursuit

Despite the absence of railways in Europe and the USA for much of the nineteenth century, and even longer in the cases of Latin America, Africa and China, regular armies accelerated their mobility. The advent of the

French *levée en masse*, perfected by Napoleon, led to greater mobility of armies in Europe. Napoleon prided himself on his mobility: marching fast and divided on multiple roads certainly eased the efficiency with which the French lived off the land.[26] One of the paradoxes of the creeping modernisation caused by railways, factories and more sophisticated weaponry, was that it led to mass armies and *less* mobility than during the pre-twentieth century times of rapid movement on threadbare logistics and living off the land. Martin van Creveld argued that Europe's Western Front from 1915 ushered in a new era of static wafare, which persisted amidst the mass armies of the Second World War, and the logistically complex successive generations of weapons technology since the Cold War.[27] Thus our period of study in many ways represented a high point in regular armies' mobility. As a consequence nineteenth-century guerrillas in turn proved at their most successful when operating in coordination with regular forces.[28]

Accelerated mobility led to the overhaul of 'cabinet warfare' norms by regular armies on the one hand, and their revitalisation by insurgents on the other. In Europe the French Revolutionary abandonment of fortifications as centres of supplies propelled more living off the land.[29] But fortifications retained their importance as bases for insurgents throughout the nineteenth century, in Iberia, southern Italy, the Tyrol, Central Asia, China and Mexico. Yingcong Dai's study of the White Lotus rebellion shows how mountaintop forts were refuges both for insurgents and civilians trying to stay out of harm's way. Fortified topography also featured in irregular warfare in Mexico, Spain, Portugal, and Central Asia, as the chapters by Nathaniel Morris, Mark Lawrence, Charles Esdaile and Alexander Morrison respectively show.

As for nineteenth-century irregular warfare, mobility was more mixed. In most instances insurgents continued the patterns of high mobility within local regions only. Natalia Sobrevilla and Alejandro Rabinovich show how 'intermittent' mobilisation by local volunteers in South America tended to be militarily superior to the 'permanent' mobilisation of the indoctrinated and expedited levies of the state. Localism is also a feature of Charles Esdaile's study of Portugal's locally-organised militia (*ordenança*) which proved effective in harassing and ambushing longer-ranged French invasion forces. 'Intermittent' and militia forces, according to Archer Jones, pursued one side of the coin of western strategy since ancient times: 'raiding'. Raiders were unable to occupy territory or populations for extended periods of time. Instead 'persistent' forces faced this task, including in counter-insurgency campaigns.[30] Mark Lawrence's study of the Carlist insurgency of the 1830s shows how raiding was dictated by longer-term political as well as shorter-term military considerations. Population centres deemed 'Carlist' were 'liberated' by raids and treated with a view to shoring up political support, whereas hostile populations were pillaged for short-term military benefit.

Raiders, militia and 'intermittent' soldiers usually resented the growing claims of states to impose militarisation. Giacomo Macola's and Luke Hogan's study shows how pre-colonial African politics were undergoing militarisation in traditional ways. The Katanga region witnessed hit-and-run tactics, and static and siege warfare fuelled by the Sanga insurgency enjoying local knowledge of terrain and support networks. Warlords (garenganze) were all-important, and not all were able to survive European penetration in the late century. The Yeke managed to survive by reinventing themselves as a rapid deployment force in order to meet the threat posed by the Sanga, eventually defeating the Sanga. The Yeke tribe survived thanks to their core of armed gunmen, and became agents of the Congo Free State, whilst the Sanga were eventually crushed. The prospects for tribal autonomy under European conquest were thus partly dictated by pre-contact factors.

Even sophisticated imperial societies offering advanced constitutional and legal safeguards were prone to draconian policing, albeit seldom on the scale witnessed during the French Revolution, mid-century China, or colonial Africa. Nineteenth-century Ireland produced a recurrent insurgent nationalism, stemming from its subordinate position to London, the grudging emancipation of the island's majority Catholic population, and the subsequent inability for any elite in Dublin to cut across political divisions by appealing to the 'constitution', as elites in neighbouring Great Britain often did with success.[31] After 1800, despite being formally integrated into a 'United' Kingdom and the British composite monarchy, Ireland was ruled by a series of special rules with colonial overtones of counterinsurgency. As Tim Bowman argues in this volume, the Irish Constabulary, whose origins dated back to 1822, was much more of a continental-style gendarmerie than the 'servant of the citizen' local police beloved of Great Britain; indeed, only the Dublin Metropolitan Police operated in conditions deemed 'safe' enough to be routinely unarmed on the classic British model. Yet the Constabulary's varied civic duties and uneven distribution seldom gave them the appearance of a colonial occupation force except in moments of insurgency crisis, such as 1848 and 1867. Even though the Ribbonmen and Fenian insurgencies of 1848 and 1867 paled into insignificance compared to the mass risings of 1798 and 1916, Bowman's study uncovers the anxieties of the Protestant Ascendancy. The threat of Fenian infiltration into the substantial Irish elements in the British Army, along with the Irish militia, dominated the security concerns of the Dublin Castle administration and led to a series of projected countermeasures (such as the demolition of houses close to Dublin Castle, plans to use infrastructure to link isolated police 'barracks' in the countryside, and the creation of flying columns besides).

As the century progressed metropolitan public opinion became a factor in opposition to colonial powers applying brutalised counter-insurgency. Ian

Beckett's study shows how British counterinsurgency efforts in the 1880s Third Burma War were dictated by topographical and climate challenges, notions of collaborating hill tribes like the Karen being 'martial races', and a concern that the excessive imbalance in casualties in the British favour be kept secret from the press as much as possible. Beckett also shows how villagers found themselves between two fires: rapacious bandits (dacoits) and foreign imperialists in equal measure, but that part of the British success was in convincing many villagers that less pillage and more protection was to be found at British hands. The British used flying columns, as the French had done in the Vendée, and a similar incremental garrisoning of villages. Despite the significant effort and casualties expended, the campaign of the 'lost footsteps' never got appropriate recognition.

Equally, irregular resistance to Russian conquest in Central Asia, as well as British expansion in Burma, were manifest in highly mobile but also highly localised armed groups, as Alexander Morrison's and Ian Becket's articles show. But whereas the French, British and Russian counterinsurgencies tended to offer cohesive fronts, other armies were more fragmented, increasing the appeal of longer-range operations for insurgents. Nathaniel Morris shows how Manuel Lozada's long-standing campaign against the Mexican state ranged across several states. Mark Lawrence shows how Carlists in the 1830s exploited the internal political disintegration of a nominally more powerful Spanish state by launching expeditions across Spain in 1836 and 1837. Most strikingly, the White Lotus insurgents at the turn of the century performed long-range marches from Hubei which eerily anticipated Mao's ideologically very different 'Long March' of the 1930s.

Episodes such as those of Poland in 1793, 1831 and 1863, as well as the struggle of the Garibaldini, demonstrated the hybrid nature of guerrilla warfare and the growing tendency of insurgents to militarise themselves formally.[32] The French call for a 'chouan' people's war during the Franco-Prussian War after their defeat at Sedan showed the limits of irregular war in the age of the revolution in military affairs. Small tactical successes aside, the war was only indirectly impacted by French partisans, and civilians proved lukewarm to support such a struggle. For a long time, episodes of brutality by the occupying Prussian-led armies in France led historians to discern a peculiarly German preponderance for atrocity which linked the Franco-Prussian War, to exterminatory colonial policing in Germany's African colonies, to the 'Rape of Belgium' in 1914, and ultimately to the unprecedented horrors of the Nazi empire in Europe.[33] But Matteo Scianna in this volume reappraises this view with his study of German counter-insurgency during the Franco-Prussian War. Scianna shows that German responses to real and imagined *francs-tireurs* were not exceptional. Combat stresses would have led to outrages being committed by any army, and soldiers found guilty of committing atrocities were consistently subjected to courts martial.

Scianna shows how ideological perceptions linking the *levée en masse* with civilian resistance in 1871 distorted what was a 'normal' symmetrical war. Equally twentieth-century perceptions of Geman brutality retrospectively shaped understanding of Prussia's nineteenth-century militarism. A more sympathetic view of Prussian history has emerged only recently.[34] Prussia after 1945 was remembered as the villain, its progressive heritage being selectively forgotten. Patterns of *how* wars were remembered, and equally how and why they were forgotten, form a major theme in this edited volume. In some instances they are of a largely operational nature, such as Charles Esdaile's chapter explaining the failure of the Napoleonic empire to learn the lessons of the 1762 'Guerra Fantástica' in relation to Portugal's fierce topography, fortresses and capacity for guerrilla warfare.

In other instances they help to address enduring misconceptions about the role played by military history in national identity. Late-imperial China has usually been viewed as an 'amilitary' or 'demilitarised' culture, largely as a consequence of centuries of Confucianism leading to an inward-looking, bureaucratic dynasty which proved all the more vulnerable to Western military encroachments. This view has now been revised.[35] The articles by Yingcong Dai and Kenneth Swope, on insurgency and counter-insurgency respectively in the Xing dynasty, revise the old orthodoxy further, complicating and enhancing our understanding in the studies of 'White Lotus' insurgency and General Zuo's counter-insurgency. Kenneth Swope's study places counterinsurgency at the heart of the Xing dynasty's survival during the internal and external threats of the mid-nineteenth century. Zuo imposed opium bans and reindustrialisation strategies as a means of winning over 'hearts and minds', managing to suppress numerous overlapping rebellions. In a different vein of remembering and forgetting, Richard Reid's article explains how post-independence African elites, themselves forged by anti-colonial guerrilla struggles, proved to be selective about how they remembered nineteenth-century pre-colonial warfare. As Africans' resistance to the European 'Scramble' for their continent offered the Africans little military glory, post-independent states preferred to remember anti-colonial independence struggles of the twentieth century. Pre-colonial insurgencies, by contrast, were selectively remembered, usually with twentieth-century calculations in mind. Reid's study of Uganda explains the thriving pre-nineteenth century traditions of African kingdoms expanding and centralising the power of kingdoms via tightened control of professional armies. Mirambo (Miyela Kasanda, 1840–1884), a military genius who militarised the Nyamwezi tribe, was selectively remembered by twentieth-century African leaders, including by Idi Amin who needed to discover legitimacy in pre-colonial African militarism as a figleaf legitimising his own coup.

Towards total warfare

This nineteenth century, at least from a Western perspective, lies at the heart of a period in history in which strategies for waging war became 'total'. From a strategic perspective, 1945 marked the end of an era since the French Revolution during which military strategy was dominated by the desire to seek the unconditional surrender of the enemy (and often of his political institutions). Such aims had proved to be high stakes indeed whenever 'hearts and minds' did not accept regime change. At the same time, even though post-1945 strategy fundamentally shifted towards limited war once more (given the mushroom-shaped cloud), there was a growing realisation that the absence of war did not necessarily mean peace. The Cold War prospect of nuclear Armageddon made warring parties more likely to settle for less than the all-out imposition of their will in the Clausewitzian sense, meaning that post-1945 history has witnessed the logical resurrection of 'limited war' strategies.[36]

The development of the rifle, and its ability to be mass-produced and disseminated throughout the armies, resulted in a tactics being reformed. The increase in rate of fire, range, and accuracy meant that the rifle eclipsed the artillery as the monster on the battlefield, and that battlefield became ever more lethal – as a result, the way troops moved and organised themselves underwent a gradual but substantial change. It became ever more fruitless and devastating to attempt a frontal assault against such firepower, and so armies began to abandon the traditional massed frontal attacks in favour of flanking manoeuvres, and mostly abandoned the bayonet in reaction to the increased range of combat.[37] Simultaneously, the new lethality meant that the way troops attacked had to be altered: the eighteenth-century organisation of massed regiments of men became inadequate, and a new system of advance had to be created. Troops began to disperse into skirmishing lines instead of standing shoulder-to-shoulder, and as a result the system of volley fire was abandoned; smaller units were created and instead a new system of fire and manoeuvre was invented by Moltke during the Franco-Prussian War of 1870-71, leading every tactical group to remain on the offensive, and to attack, manoeuvre, and pause under cover of fire from another unit.[38] Simultaneously, fighting standing erect was abandoned and earth- or stone- works were increasingly introduced after the American Civil War. Concealment was another way of surviving, and the fashionable old, bright uniforms were gradually replaced in favour of duller, camouflaging colours like feldgrau, khaki, and horizon blue. The introduction of smokeless powder near the end of the century favoured skirmishing lines instead of the columns of the eighteenth century. The ability of a man to hide himself behind an earthwork meant that troops could extend themselves in longer lines than before and still project an appearance of defensive strength – meaning that the concentration of men per metre dropped from 1 every 10 metres in the eighteenth century down to 1 every 25 in the American Civil War and 1 in 250 metres in the Great War.[39]

The mid-century counter-insurgencies addressed in this volume witnessed accelerated innovation in weapons and logistical technology, especially in the case of the American Civil War and Franco-Prussian War. Total war was accompanied by total poltical strategies, as the price of defeat – such as the abolition of slavery in the Confederacy – promised to overturn the social order. Yet even 'total' wars were experienced in an emotional manner recognisable to veterans of pre-industrial wars. Susan-Mary Grant's study of a future Supreme Court Justice's (Oliver Wendell Holmes) experiences fighting against the Confederacy applies the concepts of morale and emotion developed in John Keegan's seminal *Face of Battle*. Grant makes the case for an 'emotional revolution' in relation to counter-insurgent warfare, as Holmes's experiences fighting are a coming-of-age progression from boyhood to manhood, from civilian to military. Studying Holmes' campaign letters, Grant reveals his emotional strain, his jadededness with the progressive Unionist ideas over the course of campaigning, and ultimately how the experience of fighting insurgents forged in Holmes a new identity of a 'warrior', which never left him even as he progressed in a distinguished career later in life.

Against the context of the relentless modernisation of symmetrical warfare, the role of the insurgent seemed to be increasingly outdated. Given that nineteenth-century insurgencies tended to be reactionary responses to revolutions rather than the reverse, it is unsurprising that Engels, Marx, Lenin and Trotsky downplayed guerrilla warfare. But there were some dissenters. From the mid-nineteenth-century there were some thinkers in Poland and Italy who identified a link between guerrilla warfare and revolutions (Mazzini being the most distinguished). The Mazzinian model of achieving democratic revolution via guerrilla warfare became more radicalised by the later part of the century. From the 1880s, the revolutionary German Socialist (SPD), Johann Most, coined the term 'propaganda by deed'.[40] Jean de Bloch at the end of the nineteenth century argued in his *War of the Future* that the individualistic potential, range and precision of the rifle made guerrilla war the war of the future instead of close-order infantry combat.

But an abiding model of nineteenth-century guerrilla is the civil war revolt against modernity. Arguably, reactionary and religious-inspired guerrilla movements in the nineteenth century were most likely to start in response to sudden political change, as the 1833 onset of the First Carlist War (the subject of Mark Lawrence's chapter) and the 1890s Canudo revolt in Brazil showed. Carlists and Canudos were anti-liberal and anti-republican respectively, and their revolts had religious overtones. A second model is liberationist and anti-colonial, such as in Latin America during 1810–1824, Haiti, 1860s Mexico, Cuban independence revolts in the 1868–78 and 1895–98 periods, and also Garibaldi's Italy and Poland. A third model is in response to new colonial campaigns penetrating from the outside, such as Shamil's famous resistance to Russian campaigning in the Caucasus, Maori Land Wars, Native Americans in the US frontier wars, Zulus, Boers, and resistance to Dutch expansion in Sumatra and French expansion in North Africa, Madagascar, Tonkin, and that of the USA in the Philippines.[41] But for most of the nineteenth century huge areas of Africa were immune to European conquest. Just as in Asia in the Early Modern Era whites at African ports were seen as traders looking to barter, not posing much of a military threat to Muslim and Animist empires throughout the continent.[42] Even amidst the late-nineteenth-century 'Scramble' African insurgencies were conditioned for reasons which had very little to do with encroaching European imperialism, as the chapter by Giacomo Macola and Jack Hogan attests.

Thus this book views nineteenth-century insurgency and counter-insurgency on a global scale. It shows the commonalities of responses more than their differences, and refracts these through themes which crop up repeatedly in different times places. These themes include common problems and common solutions; the challenge of commanding local intelligence networks; public opinion; millenarianism, magic and religion; technology; 'hearts and minds'; the legal framework of state violence; racial stereotypes; and patterns of forgetting and remembering guerrilla conflicts.

Notes

1. Beckett, *Modern Insurgencies and Counter-insurgencies.*
2. Förster and Nagler (eds.), *On the Road to Total War*, 6.
3. Heuser, *The Evolution of Strategy*, 387.
4. Heuser (ed.), *Small Wars and Insurgencies in Theory and Practice*; and Marks and Rich, "Back to the Future."
5. Daase, "Clausewitz and Small Wars," 192–194.
6. Bayly, *Birth of the Modern World*, 1–3.
7. Mark Hewitson, *Absolute War*, 129.
8. David Bell makes this case in a forthright manner (Bell, *First Total War*).
9. Heuser, *Strategy before Clausewitz*, 167–72.
10. Heuser, *The Evolution of Strategy*, 120.
11. Ibid., 146–152.
12. Heuser, *Strategy before Clausewitz*, 62–63.
13. Ibid., 187–188.
14. Scheipers, *Unlawful Combatants*, 33–34.
15. Heuser, *The Evolution of Strategy*, 389.
16. See note 14 above.

17. Scheipers, *Unlawful Combatants*, 61.
18. Ibid., 54.
19. Laqueur, *Guerrilla Warfare*, 93–98.
20. Beckett, *Modern Insurgencies and Counter-insurgencies*, 24–27.
21. Black, *Rethinking Military History*, 6.
22. Muenkler, "Clausewitz and the Privatisation of War," 226–227.
23. Beckett, *Modern Insurgencies and Counter-insurgencies*, 30–31.
24. Lawrence Tone, *War and Genocide in Cuba*, 54–58.
25. Beckett, *Modern Insurgencies and Counter-insurgencies*, 30–41.
26. Esdaile, *Wars of Napoleon*, 41–42.
27. van Creveld, *Supplying War*, 17–26.
28. Porch, *Counterinsurgency*, 13; and Walter Laqueur, *Guerrilla Warfare*, 50–52.
29. Heuser, *The Evolution of Strategy*, 80.
30. Jones, *The art of war in the western world*, 662–716.
31. Gillen, "Ascendancy Ireland, 1660-1800," 48–73.
32. Laqueur, *Guerrilla Warfare*, 65–69.
33. E.g. Isabel Hull, *Absolute Destruction*; and Laqueur, *Guerrilla Warfare*, 83–88.
34. Especially, Clark, *Iron Kingdom*.
35. E.g. di Cosmo (ed.), *Military Culture in Imperial China*, 1–22.
36. Heuser, *The Evolution of Strategy*, 17–26.
37. Fuller, *The Conduct of War 1789–1961*, 104.
38. Ibid., 120.
39. Creveld, *Technology and War*, 172.
40. Beckett, *Modern Insurgencies and Counter-insurgencies*, 14–15.
41. Walter Laqueur, *Guerrilla Warfare*, 53–99.
42. Vandervort, *Wars of Imperial Conquest in Africa, 1830–1914*, 26–27.

Disclosure statement

No potential conflict of interest was reported by the author.

Bibliography

Bayly, C A. *The Birth of the Modern World, 1780–1914: Global Connections and Comparisons*. London: Wiley-Blackwell, 2004.
Beckett, Ian F. W. *Modern Insurgencies and Counter-insurgencies: Guerrillas and Their Opponents since 1750*. London: Routledge, 2001.
Bell, David Avrom. *The First Total War: Napoleon's Europe and the Birth of Modern Warfare* . London: Bloomsbury, 2008.
Black, Jeremy. *Rethinking Military History*. London: Routledge, 2004.
Bourke, Richard, and Ian McBride, eds.. *The Princeton History of Modern Ireland*. Princeton: Princeton University Press, 2016.
Clark, Christopher. *Iron Kingdom: The Rise and Downfall of Prussia, 1600–1947*. London: Penguin, 2007.
di Cosmo, Nicola, ed.. *Military Culture in Imperial China*. Harvard: Harvard University Press, 2009.
Esdaile, Charles. *The Wars of Napoleon*. London: Routledge, 1995.
Förster, Stig, and Jörg Nagler, eds.. *On the Road to Total War: The American Civil War and the German Wars of Unification*. Cambridge: Cambridge University Press, 2002.

Fuller, J F C. *The Conduct of War 1789–1961: A Study of the Impact of the French, Industrial, and Russian Revolutions on War and Its Conduct*. Brunswick: De Capo Press, 1992.

Heuser, Beatrice. *The Evolution of Strategy: Thinking War from Antiquity to the Present*. Cambridge: Cambridge University Press, 2010.

Heuser, Beatrice, ed. *Small Wars and Insurgencies in Theory and Practice, 1500–1850*. Abingdon: Routledge, 2015.

Heuser, Beatrice. *Strategy before Clausewitz: Linking Warfare and Statecraft, 1400–1830*. London: Routledge, 2017.

Hewitson, Mark. *Absolute War: Violence and Mass Warfare in the German Lands, 1792–1820*. Oxford: Oxford University Press, 2017.

Hull, Isabel. *Absolute Destruction: Military Culture and Practices of War in Imperial Germany*. Cornell: Cornell University Press, 2005.

Jones, Archer. *The Art of War in the Western World*. Urbana and Champaign: University of Illinois Press, 2001.

Laqueur, Walter. *Guerrilla Warfare: A Historical and Critical Study*. London: Transaction Publishers, 1997.

Marks, Thomas A, and Paul B Rich. "Back to the Future: People's War in the Twenty-first Century." *Small Wars and Insurgencies* 28, no. 3 (2017): 409–425. doi:10.1080/09592318.2017.1307620.

Porch, Douglas. *Counterinsurgency: Exposing the Myths of the New Way of War*. Cambridge: Cambridge University Press, 2013.

Scheipers, Sibylle. *Unlawful Combatants: A Genealogy of the Irregular Fighter*. Oxford: Oxford University Press, 2015.

Strachan, Hew, and Andreas Hergerg-Rothe, eds.. *Clausewitz in the Twenty-first Century*. Oxford: Oxford University Press, 2007.

van Creveld, Martin. *Supplying War: Logistics from Wallenstein to Patton*. London: Cambridge University Press, 1977.

van Creveld, Martin. *Technology and War: From 2,000 B.C To the Present*. New York: Cambridge University Press, 1991.

Vandervort, Bruce. *Wars of Imperial Conquest in Africa, 1830–1914*. Routledge: Abingdon, 1998.

The Peninsular War guerrilla and its antecedents: humiliation forgotten, disaster prefigured: the *guerra fantástica* of 1762

Charles Esdaile

ABSTRACT

The brief war that took place between Spain and Portugal in 1762 is one of the least known episodes in the latter's military history, whereas, thanks to Wellington's construction of the Lines of Torres Vedras, the French invasion of 1810–11 is right at the other end of the spectrum. Yet the two episodes are closely linked to one another. At the very least, they are uncannily reminiscent in terms of their details – in both cases substantial foreign armies were vanquished through a combination of irregular resistance, scorched-earth tactics and the clever use of field fortifications – and the article therefore argues that Wellington based the plan that defeated the forces of Marshal Massena on the strategy used by the Portuguese half a century earlier.

Ask any eighteenth-century historian and they will probably say that, next to Rossbach, the greatest humiliation in the history of French arms in the period 1700–1789 is the loss of Québec. Surprisingly, however, there is another candidate for this rather dubious honour, namely the Franco-Spanish attempt in the closing years of the Seven Years' War to bully Portugal into declaring war on Britain and, more importantly, close her ports to British trade. That hardly anybody in the twenty-first century should be aware of this clash of arms is hardly surprising, for it cannot but be obscured by more dramatic events elsewhere: one thinks here of successive British successes on the high seas, in India and in Canada and also of the heroic feats of Frederick the Great.[1] – however even within living memory its lessons appear to have been forgotten, and that despite the direct involvement of both British and French troops. In 1808, then, we find Sir John Moore proclaiming that Portugal could not be defended from invasion from the forces of a Napoleon who was just as certain that it could be conquered without difficulty, and yet within easy reach there was plenty of evidence to suggest that both general and emperor could not have been more wrong. In this chapter, then, we shall discover how less than fifty years before the French invasions of 1808–1810 a full-scale attempt to subjugate England's oldest ally came to grief for almost exactly the same causes as frustrated Marshal Masséna, and, in addition, reflect upon the reasons why so little attention was paid to the campaigns concerned by that most assiduous student of history, Napoleon Bonaparte.

Let us begin with a summary of events. When the Seven Years' War engulfed Europe in 1756, the Iberian states remained neutral and that despite the close relationships which Spain and Portugal enjoyed with France on the one hand and Britain on the other. However, as the war ground on, both countries found it increasingly difficult to maintain their neutrality. First to suffer in this respect was Spain: regarded by the French foreign minister, Choiseul, as an important potential ally in the on-going struggle with Britain, in 1761 she was pressured into signing a perpetual alliance with France known as the Family Compact, a move which very quickly led to a British declaration of war. However, if the British were the primary objective of this Franco-Spanish alliance, Portugal was not far behind. From the very beginning of the war, the Royal Navy had been behaving in a distinctly cavalier fashion around the coasts of Portugal, coming and going more-or-less at will. Despite frantic Portuguese efforts to show good faith, Choiseul had become increasingly irritated at what was going on – in his view, indeed, Portugal was in effect acting as a British ally whilst at the same time insisting on the benefits of neutrality – and he now decided that the only way that the situation could be resolved was by forcing Portugal into an alliance.

On 1 April 1762, then, the French and Spanish ambassadors to Lisbon handed the Portuguese foreign minister, Luis de Acunha, an ultimatum that was not the less high-handed for the fact that the term within which an answer had to be received was a mere four days. In brief, Portugal was to join the Family Compact (on the grounds that King Joseph I's queen, Mariana Victoria, was a daughter of Phillip V of Spain), declare war on Britain, close her ports to British shipping and admit Spanish garrisons to Lisbon and Porto. Given that Portugal had only regained her independence from Spain in a long and bitter war barely a century before, there was never the remotest chance of Acunha accepting such a *démarché*, and within a month Spanish troops were marching across the frontier, though war was not actually declared by France and Spain until 20 June (though the Spaniards, at least, certainly wanted war, seeing it as a chance to restore the rule of Madrid and add Brazil to the Spanish empire in America, to the last minute it was hoped that the fiction could be maintained of a peaceful operation designed to rescue Portugal from exploitation at the hands of 'perfidious Albion').[2]

Defiant though Acunha was in the face of this aggression, Portugal was in no state to fight a war, for just seven years before she had been struck by one of the greatest disasters in her history in the form of the great earthquake of 1755. With Lisbon and many smaller places in ruins, there was simply no money available for the armed forces, and the army was therefore in the most wretched condition: dressed in rags, its regiments had been reduced to a mere skeleton, whilst it had not been paid for eighteen months; officers, then, were to be seen begging in the streets of Lisbon whilst many of the troops living by outright banditry. To quote one French visitor:

> For more than a century the military had existed in the state of the most utter abasement. The kingdom could barely put 10,000 men under arms, whilst those it could muster were peasants, or rather vagabonds, without uniforms, without arms and without discipline. If a Portuguese had some act of vengeance to pursue, all he had to do was to contact one of these savage men, and, at the cost of a modest sum, he would instantly be rid of his enemy. Even the officers ... were men of no consideration, the majors, captains and lieutenants all being valets and other lackeys of the senior commanders who continued to wait on the latter at table dressed in their very uniforms.[3]

In so far as the blame for all this was concerned, contemporary observers had no doubt that the chief fault lay with the Portuguese government's neglect of the military in favour of other concerns. As a French officer of the Portuguese corps of engineers attached to the garrison of Elvas complained to an Irish visitor some ten years after the war:

> Year rolled on after year, the army neglected and still growing worse, the officers dying out and none appointed to replace them, so that many regiments came to be commanded by subalterns and non-commissioned officers. The only use which King John the Fifth made of his troops was to dig and carry the stones to build a magnificent church and convent for 300 idlers at Mafra, but, if he abused or neglected his army, in compensation he established the holy patriarchal church of Lisbon to vie with that of Saint Peter's in Rome ... Now came the new administration of King Joseph and the Marquis of Pombal from which we all expected miracles, but the army did not improve a jot by it ... The soldiers were naked, begging charity in the streets with a monstrous string of beads in one hand and a ragged hat in the other, and their arms, when they had any, were all over rusty and without locks: a pitiful law or regulations sometimes appeared, but it meant nothing and was not even executed.[4]

At first, then, there was little fighting, though in this respect it probably helped that the province initially chosen for invasion – the remote northeastern province of Tras-os-Montes – was devoid of any garrison other than that of the mediaeval castle of Miranda do Douro which was almost immediately neutralised thanks to a chance fire which ignited the castle's main powder magazine and caused an explosion so enormous that it blew two breaches in the walls.[5] Meanwhile, the Spaniards were sensible enough to avoid conflict, paying well for the supplies which they garnered from the countryside and proclaiming loudly that they had come in peace, even, indeed, as friends, in support of which position the government of King Charles III published all the official correspondence that had preceded the invasion in an attempt to throw all the blame for the situation on the Portuguese government.[6]

In consequence of all this, the chief towns of the province – Chaves, Bragança and Torre de Moncorvo – all fell without resistance. However, the fiction of a peaceful occupation could not last. Extraordinarily enough, the Spaniards seem to have believed that, despite being one of the most barren areas of the entire Iberian peninsula, Tras-os-Montes would be able to provide enough food and fodder for the invading army to live off the country, but, even with all the good will in the world, this was simply not possible, and all the more so as the Spanish commander, the Marqués de Sarria, was accompanied by no fewer than 22,000 men.[7] Very soon, then, the local population ran out of the resources needed to meet the demands of the invaders, whereupon the latter had no option but turn to forced requisitioning. At this point, of course, what had hitherto pretty much been a phoney war became something that was entirely different. Illiterate and poverty-stricken, the inhabitants of Tras-os-Montes were devoid of national feeling in the modern sense, whilst the fact that across the frontier the language spoken was either Galician (a close cousin of Portuguese), or a hybrid dialect that continues to be spoken in a few isolated Spanish border villages to this day, meant that even the most basic cultural differences between them and their neighbours were fairly limited. The fact was, then, that it was probably immaterial to the local peasants under whose sceptre they were governed.[8] If the state was irrelevant, however, the stomach was not, and, with the population faced by mass starvation, the result was a major backlash that expressed itself in the outbreak of a savage guerrilla war.[9]

At the heart of this development was the mediaeval survival known as the *ordenança*. The *ordenança* was organised into districts commanded either by the area's chief magistrate or local worthies appointed by the state, usually prominent local landowners possessed of some military experience. Divided into companies of 250 men headed by representatives of the élites nominated by the district authorities, and composed of all able-bodied males up to the age of sixty who were not either members of the clergy or enrolled in the regular army or provincial militia, this force was in effect a home-guard that was mobilised in time of invasion and made use of as an auxiliary army. Away from areas actually penetrated by enemy forces, its role was mundane in the extreme – the apprehension of deserters, for example, or the provision of convoy escorts – but in the event of the arrival of invaders matters became completely different: in brief, its members were expected to defend their home villages, harass passing columns and, above all, protect the countryside from enemy forage parties.[10] Though its only weapons were in the first instance limited to what the men enrolled in it could provide for themselves such as scythes, pitch-forks and fowling-pieces, at the orders of the governor of the province, Francisco Sarmento, this force now turned out en masse, whilst the rest of their inhabitants fled their villages for the relative safety of encampments in the hills. In consequence, the invaders soon found themselves in serious trouble in that their attempts to requisition food were met with ferocious resistance and the punitive columns they sent out to restore order confronted with no more than ghost towns. As for food, meanwhile, what could not be carried away was destroyed, the result being that, by now almost inaccessible from Spain, the army was quickly reduced to a state of semi-starvation. With large numbers of men falling sick and perishing from want – as many as 4,000 men are reputed to have died in the hospital that was established in Braga alone – the only hope was to push on to Oporto as quickly as possible, and so, in a manner exactly reminiscent of Junot in 1807, Sarría put together a small task force commanded by an Irish émigré named Alexander O'Reilly. Spurred on by the increasingly desperate nature of the situation, O'Reilly duly pushed ever deeper into Portugal and occupied the town of Vila Real. From here it was but a short march to Oporto, but just at the crucial moment news arrived that a column of 5,000 Spanish troops trying to cross the River Douro further east south of Torre de Moncorvo had been repulsed by a force of 1,500 provincial militia and peasants commanded by Charles O'Hara, an experienced officer who was the son of the British ambassador, Lord Tyrawley. In itself this was not much of a setback, but O'Reilly now appears to have succumbed to panic. To quote an introduction which the erstwhile victor of Valmy, Charles Dumouriez, wrote, for an English translation of Vertot's 1729 history of the War of the Restoration that appeared in London in 1813:

> O'Reilly ... pushed on as far as Vila Real without meeting any resistance, but there he learned that the peasantry was arming and the defiles were dangerous upon which he turned back and made a very disorderly retreat: at Villa Pouca, and as far as Chaves, the peasants harassed him exceedingly, and had the glory of driving him back with loss and disgrace, though their number did not exceed 600, nor had they a single military man with them. This feat was highly celebrated in Portugal, and the particulars of it repeated with great pride.[11]

O'Reilly's retreat precipitated a general collapse in the Spanish position. The serious epidemic that a month of famine had produced amongst the Spanish forces was continuing to wreak havoc while, more and more emboldened, the ordenanças were everywhere killing ever greater numbers of Spanish soldiers as regiment after regiment dispersed in a desperate search for food, this process seemingly being accompanied by numerous atrocities as well as wholesale pillage. Contemporary accounts of the death toll are beyond doubt exaggerated, even downright ridiculous, but it was still heavy enough, one modern estimate suggesting that Sarria had lost 8,000 men by the time he fell back across the frontier.

With the exception of the town of Chaves, by the end of June all the territory occupied by the Spaniards had been abandoned. However, the war was far from over. A division of 12,000 French troops had been sent to reinforce the remnants of Sarria's army, and this last was now taken over by the Conde de Aranda, at that time one of Spain's most up-and-coming generals. At the same time the Spanish chief minister, Esquilache, travelled to the frontier to take personal charge of providing the coalition forces with sufficient food for six months. Within a matter of weeks, then, an army of over 40,000 was once more pushing into Portugal, this time not in Tras-os-Montes, but the province of Beira, the ultimate objective being none other than the Portuguese capital. With their numbers reduced to one third by desertion prior to the arrival of the invaders, the garrison of Almeida surrendered on 25 August after a resistance of just nine days, the performance of the governor being so lack-lustre that he was imprisoned when he and his men reached the town of Viseu after having been allowed to march out with the honours of war. Central Portugal now being open to invasion, Aranda's army headed southwest towards the River Tagus, but they were harassed at every turn by the *ordenanças*, and at the same time found themselves facing a redoubtable opponent in the form of Count William of Schaumburg-Lippe. Schaumburg-Lippe was a highly experienced soldier who had made a considerable name for himself in the earlier War of the Austrian Succession and was in late July appointed commander-in-chief. Nor, meanwhile, was this all, for July also saw the arrival in Lisbon of a British force of some 7,000 troops under General Townsend.[12]

Primarily a defensive commander, Schaumburg-Lippe realised that his tiny army – setting aside the British reinforcements, he never had more than 7,000 Portuguese troops[13] – could not hope to take on Aranda in the open field, and therefore resolved to ensconce himself in the citadel that dominated the town of Abrantes. This decision was but part of a wider picture, however. Thus, on the one hand, the populace were ordered to implement a scorched earth policy in all the districts through which Aranda was likely to advance and the *ordenança* to do all it could to impede his progress, and on the other every route that led in the direction of Lisbon was studded with small forts that could not but slow down the march of an invading army (typical of these was a series of redoubts constructed along the road that led westwards from Castelo Branco through the Serra do Telhadas and the Serra do Moradal to the River Zezere and thence to Santarem and, ultimately, Lisbon).[14] Placed in Abrantes, which afforded him a bridgehead on the north bank of the River Tagus, the Portuguese commander could threaten the flank of any force that attempted to march on Lisbon, but at the same time he was extremely vulnerable to any troops that invaded the plains of the Alemtejo to the south. With another Spanish army massing on the frontier of the Alemtejo, indeed, he was in danger of being surrounded, but the arrival of Townsend's troops meant that he could now think of offensive operations of a sort that had previously been out of the question. That being the case, then, Schaumburg-Lippe, dispatching 2,000 troops, the vast majority Portuguese, under Sir John Burgoyne (the same Sir John Burgoyne who fifteen years later was forced to surrender at Saratoga) on a daring raid that was designed at the very least to postpone the threat of encirclement, if not to wreck it altogether. In this Burgoyne was highly successful, defeating a Spanish force at the border town of Valencia de Alcántara and capturing large quantities of supplies, the net result being that the Spaniards abandoned any thought of invading the Alentejo.

If there was one thing that decided the fate of the campaign, it was probably the action of Burgoyne, for this last removed the one real hope of dealing with the problem posed by Schaumburg-Lippe's defensive strategy. Arriving before Abrantes, Aranda found that the citadel had been augmented by substantial earthwork fortifications that made launching a direct attack far too costly a prospect to be entertained, and, still worse, that the only other crossings of the Tagus in the area – those at Niza and Villa Velha de Rodao – were also strongly garrisoned. Valencia de Alcántara having deprived all hope of a Spanish force taking Schaumburg-Lippe in the rear, Aranda had no option but to try to push further west but he was forced to leave so many troops behind to watch the Portuguese positions behind the Tagus that he was able to advance no further than the village of Sobreira Formosa and that only in the teeth of determined popular resistance. Still worse, meanwhile, Schaumburg-Lippe began to send forces across the Tagus to attack the troops that had been stationed to observe their positions and even strike deep into their rear. In the Franco-Spanish camp, the situation was ever more desperate, the scorched-earth policy resorted to by the Portuguese having ensured that that rations were now running very short: there were, for example, a minimum of 12,000 sick. Fearing that he could end up being surrounded, on 15 October Aranda ordered a retreat on Castelo Branco, but, with the vengeful Portuguese closing in on his rear, this decision precipitated a complete collapse of morale and discipline alike, thousands of men deserting and throwing themselves on the mercy of the Portuguese whilst the rest fled in disorder across the frontier, abandoning much of their artillery and baggage in the process. In all, losses in killed, wounded and missing came to some 15,000 men. Desperate to salvage something from the disaster, the following month Aranda attacked again south of the Tagus, only to suffer a series of minor setbacks that afforded sufficient time for Schaumburg-Lippe to rush troops southwards to deal with the threat, the thoroughly demoralised Spaniards then falling back in some disorder across the frontier; a few days later, meanwhile, the Madrid government requested an armistice.[15]

Thus ended the campaign of 1762. As the Duc du Chapelet later noted, few armies had experienced so many fatigues and undergone so many dangers in exchange for so little glory as that of the Conde de Aranda.[16] A great victory for the Portuguese – hence the nickname of 'the unbelievable war' which was ever more attached to it – it was an object lesson in just how difficult a country Portugal was to invade, let alone conquer: as Junot found in 1807, in the wilds of Tras-os Montes, Beira and the Tagus valley alike, armies of any size were hard pressed to subsist themselves even in normal times while the difficulties were compounded by poor roads and mountainous terrain. As students of the Peninsular War will observe, meanwhile, the defeat of Sarria and Aranda fore-shadowed later events in a manner that was uncanny in the extreme. Setting aside extraordinary coincidences – for example, the destruction of the castle of Miranda do Douro by the explosion of its powder magazine and the fact that Almeida surrendered on virtually the same day in 1762 as it did in 1810 – we see all the ingredients of Wellington's defeat of Masséna, namely the implementa-tion of a scorched earth policy, the mobilization of the civilian populace by means of the *ordenança*, the dispatch of small columns of regular troops to disrupt enemy operations, and the use of field fortifications to preserve the defending army and confront the invaders with an obstacle that they could only conquer at the cost of casualties likely to be so heavy as to make their position untenable. In short, from all this, Napoleon should have realised that to send Masséna into Portugal with an army of 65,000 men was likely to prove problematic even without the extraordinary measures resorted to by Wellington in the form of the Lines of Torres Vedras. Yet, well versed in the military history of the eighteenth century though he undoubtedly was, the emperor seems to have paid not the slightest heed to the lessons of 1762, and that despite the fact that an attempt to invade Portugal in 1703 in the course of the War of the Spanish Succession also had to be abandoned on account of logistical difficulties. Why, then, was this the case?

In the first place, the answer here is one that was personal to Napoleon alone in that by 1810 such were his delusions of grandeur that he had begun to lose sight of reality. Indeed, that this was the case was clear as early as 1807. As Maximilian Foy wrote of the orders that the emperor issued in respect of Junot's march on Lisbon, for example, 'The emperor did not point out what road was to be followed, but gave peremptory orders that the march of the army should not be delayed for a single day under pretence of securing subsistence. "Twenty thousand men", said he, "can live anywhere, even in a desert."'[17]

Yet, fond though the French ruler was of declaring that there was no such word as impossible, there were other factors at work besides the vagaries of his character. In the wake of the Revolution, France was caught up in the notion of herself as *la grande nation*, and this, together with the repeated triumphs of her arms, had led to a general mood of over-con-fidence. To quote Foy again:

The Duke of Berwick, general of Phillip V, was taught in the campaign of 1704 the fate that threatens an army invading Portugal by the left bank of the Zezere. The combined forces of France and Spain were palsied in the midst of their success by topographical obstacles and the want of provisions. In 1762 on the same ground the same obstacles stopped the Spanish army under the orders of the Count of Aranda and the auxiliary corps commanded by the Prince of Beauvau, and compelled them to retreat before forces inferior both in quality and in numbers. But subsequently to the Revolution, the French were accustomed to laugh at dangers and local difficulties which would have frightened their predecessors. Without laying himself open to charges of temerity, a general might, with Napoleon's worst troops, undertake strategical enterprises in which the armies of the old monarchy would have failed. This assertion will not appear too bold to those who have studied the internal management of the great armies of Louis XIV and Louis XV. The men of that period were as fit as those of the Revolution for daring attempts, but the science of making men march was not known. Little knowledge was possessed of this essential branch of the art of war which consists in moving masses of troops rapidly over a wide extent of country for the purpose of crushing an enemy on his weak point with superior forces, or striking an unexpected blow at him in the very heart of his power. The prejudices and luxurious habits of the leadership, then, palsied and stymied the good intentions of the French soldier.[18]

To personal folly, then, was added political prejudice (a prejudice, be it said, of which there is a strong echo in Foy's own words): in brief, imbued with all the advantages supposedly stemming from the Revolution, the army of Napoleon could go anywhere and do anything. However, to a certain extent, it is also possible that the emperor may have been badly misled. Four years after the end of the war, the same Dumouriez we have met before was dispatched to Portugal to compile a report on the kingdom that was evidently designed to remedy the want of topographical intelligence which had so dogged the efforts of Sarria and Aranda, and, albeit often only in passing, the resultant work had much to say on the campaign of 1762. Whilst we cannot be certain, it is extremely difficult to believe that Napoleon did not turn to its pages when he ordered Masséna to invade Portugal, and yet it has to be said that Dumouriez's remarks are less than helpful. Let us take, for example, his comments on the siege of Almeida:

Almeida is ... the strongest fortification in Portugal. It has six royal bastions in stone and the same number of ravelins; that fronting the river Coa, which runs at a distance of a mile, is ... furnished with a cavalier for the purpose of commanding the circumjacent country; there is a good ditch and covered way. Nearly in the centre of the town on a lofty mound stands a castle, famous for its strength and its magazines bomb-proof ... The siege ... of this place ... in 1762 caused the loss of a great deal of time, provisions and treasure without obtaining any important end ... for the conquest of this fortress is of no importance with regard to the real frontier of Portugal: the conqueror of Almeida is not more certain of penetrating to the heart of the kingdom than he was before he took it. An absurd inveterate prejudice urges us often to sacrifice men and money before useless ramparts merely because the ancestors of our enemies have been such systematic fools as to fortify them.[19]

To put it mildly, these remarks are rather strange, it being very difficult to see how Beira ever could be invaded without first taking Almeida. Yet the inference is very clear: in besieging the fortress Aranda only fashioned a rod for his own back. Nor, meanwhile, does the criticism end here. In the event Almeida fell after only the briefest of sieges, but then even such little advantage as this event offered was completely squandered. As Dumouriez later wrote:

> After taking that fortress the remainder of the campaign was spent in uncertain wanderings and counter-marches: the minister disregarded the advice of his general, the Count of Aranda, who being on the spot was best qualified to take a decided part and proper measures. All military men allowed the purity of his intentions ... His plan was to move against Coimbra and also seize upon Oporto ... It is clear that if Aranda could have put this project in operation, the war in Portugal would have ended in a very different manner.[20]

As for the invasion of Tras-os-Montes, this was dealt with still more harshly. Thus:

> Contrary to all appearance, this [campaign] did nothing but what was injurious to Spain itself by a great and useless consumption of men, of horses, of cattle, of grain and, above all, of money. The Marquis of Sarria, colonel of the Spanish Guards, old, bigoted and without talents, was entrusted with the command ... But besides the want of vigour and capacity in this superannuated general, the ... extraordinary ignorance of the Spanish generals, the want of discipline in their troops, the little care that was taken to secure supplies of forage and ammunition, were circumstances very favourable to the security of Portugal. The enemy entered into the country ... on the side of Tras-os-Montes, and it was then only they discovered that there was a river to pass, but they had neither pontoons nor boats, and much time was lost in constructing them.[21]

Damning as this picture of incompetence was, Dumouriez saved the worst of his scorn for O'Reilly and his failed march on Oporto. If this operation had been abandoned, it was for one reason and one reason only, namely 'the appearance of fear which he discovered'.[22] If only he had pressed on, however, all would have been well: 'Portugal was at that time without arms and planet-struck: had the army advanced rapidly upon Oporto, it must have taken it without firing a gun. Great resources would have been found there, both in money, stores and provisions ... and the Spanish troops would not have perished, as they did [from] hunger and want of accommodation.'[23]

What emerges from Dumouriez's work, then, is not the obvious conclusion that invading Portugal with a large army was never likely to be productive of success. On the contrary, providing that the commander of an invasion conducted operations in a suitable fashion, success was entirely possible. If the campaign had been a failure, it was therefore because the Spaniards had repeatedly let slip chances of victory. To quote the Frenchman once more, 'The ignorance and want of skill which the Spaniards have displayed in their attacks on Portugal ... present examples of military exertion which alternately call forth our ridicule or compassion.'[24] Meanwhile, so far as Dumouriez was concerned, pleading the effects of the *ordenanças* was just so much special pleading: 'Whatever the Spaniards may say to the contrary, this war of the peasantry is by no means important, but to ignorant and undisciplined troops. The burning of two or three villages and the hanging of as many monks or curates, or principal persons in the parish, quickly puts a stop to the indiscrete and barbarous fury of the country people.'[25] At heart, the Frenchman continued, what counted was confidence, and this was something in which the Spaniards were completely lacking. As he wrote of the long series of conflicts between Spain and Portugal stretching back to the Middle Ages. 'It is with astonishment that we read in the pages of history that the Spaniards have almost always been beat ... It seems to be an innate infatuation in the Spaniards to afford ... a certain advantage to the Portuguese.'[26]

In short, therefore, if Portugal had survived the crisis of 1762, it was because of Spanish mismanagement and cowardice. In consequence, all that was needed was to change the *dramatis personae*. We come here to the most extraordinary of all the claims made by Dumouriez: in brief, the Portuguese hated the British and the Spaniards, but towards the French their feelings were much more ambivalent:

> The Portuguese possess an innate enmity to the English and the Spaniards. The French nation is that with which they sympathise the most, of which they bear the highest apprehension, and for which they bear the highest respect. They are governed by a prejudice: that no place can bear the attack of French besiegers. This sympathy, as well as its concomitant opinion, are both beneficial and proceed from the gaiety, the vivacity, the constancy and turn of mind common to these two peoples, a relation which able politicians may employ to their mutual interest.[27]

From this the conclusion was obvious: whilst the Spaniards were incapable of conquering Portugal, aided by their superior military knowledge and their greater ability both to overawe and win over the populace, the French were more than up to the challenge.

To conclude, whilst all that we have is inference and supposition, from all this it is possible to build up some sort of explanation for why both Napoleon and Sir John Moore were so insouciant with respect to the campaign of 1762 even though this showed very clearly both that Portugal was defensible and that even medium-sized armies would have considerable problems subsisting themselves. Having read the obvious source on the campaign, namely the report put together by Dumouriez for the French government, they were led to the unavoidable conclusion that the failure of the armies commanded by Sarria and Aranda was a function of little other than the fact that the troops concerned were Spanish.[28] However, the future hero of Valmy was all too clearly deeply influenced by the cultural prejudices typical of the age (he is, in fact, little kinder in his assessment of the Portuguese). Though he, too, had doubtless read the self-same report, Wellington was not prepared to let himself be led by the ears by an individual who the British commander may already have come to the conclusion was the self-important busy-body that he was to show himself to be in a long correspondence that lasted virtually for the full length of the Peninsular War. In 1810, then, Portugal was not just defended, but also defended in much the same manner as had been the case forty-eight years before.[29]

Where, however, does the question of guerrilla warfare fit in with all this? Here, however, the conclusion towards which we are driven might be considered to be somewhat surprising. On one level, it is clear that *petite guerre* of various sorts – in this respect one sees the activities of both bands of peasant militia and flying columns of regular troops – played a prominent role in the campaign. Yet one should not go too far here: in reality, the capacities of the *ordenança*, in particular, to inflict damage on the enemy were extremely limited and they could at best hope to worsen the logistical problems faced by the French and Spanish armies. What counted was rather Schaumberg-Lippe's skilful use of field fortifications, the effect of the entrenchments constructed at Abrantes and other places being to counter the loss of such forward position as Almeida, and in this respect it is interesting to that after the war had ended the period of military reform presided over by Schaumburg-Lippe concentrated on the reconstitution of the regular army rather than the militia. Quite clearly, then, the lesson of the campaign of 1762 was that guerrilla operations could only be part of a wider strategy that brought together a variety of the elements of the art of war. On its own, it was not a war-winner, and it is notable just how little the technique was made use of by Wellington in the campaign of 1810.

Notes

1. Even in Portuguese, there is only one recent work on the subject, namely A. Barrento, *Guerra fantástica, 1762; Portugal e o Conde de Lippe e a Guerra de Siete Anhos*. Lisbon, 2006.
2. Choiseul's explanation of events is laid out in the preamble to the declaration of war, viz. *Ordonnance du roi portent declaration de guerre contre le roi de Portugal du 20 de juin de 1762*. Paris: 1762.
3. J. F. De Bourgoing, ed. *Voyage du ci-devant Duc de Chatelet en Portugal ou se trouvent des détailles intérressans sur ses colonies*. Paris, 1798. 1–2.
4. Costigan, *Sketches of Society and Manners in Portugal*, 178–9. Another observer was the Italian traveller, Giuseppe Baretti. Herewith, for example, his reaction on seeing the garrison of Estremoz: 'Military poverty ... shines forth through the ragged coats of this wretched infantry. Indeed, the poor fellows have nothing about them that may be called good except their whiskers: if they were better dressed, such bushy and curled scare-crows would have a fine effect ... I am told that the troops kept up in this kingdom amount to no more than 8,000, and if the private men are all like those which I have in Estremoz and Lisbon, there is nowhere in Europe an equal number which look so wretchedly. The greatest part of them are absolutely in rags

and patches, and in Lisbon many of them asked my charity not only in the streets, but even when they stood sentinel.' Baretti, *A Journey from London to Genoa*, 36–7.

5. This incident is, of course, highly reminiscent of the similar fate which befell Almeida in 1810. However, unlike Almeida, the town fell without a shot being fired: although the Spaniards were before the walls of Miranda de Douro when the explosion occurred, the fire broke out whilst they were still readying their guns. According to contemporary reports, meanwhile, the death toll may have been as high as 500. See *The British Magazine*, June 1762, 329–30 .

6. For this last, see *Razón de entrar en Portugal como amigos, y sin razón de recibirlos como enemigos, manifiesto reducido a las memorias presentadas parte a parte*. Madrid: 1762.

7. As later critics observed, the whole idea of the campaign was deeply flawed. Thus: 'Fortunately for the Portuguese, nature itself has taken care to protect them from invasion and give them neighbours who are supremely ill-informed of the regions in which they might have to fight. Of this last the Spaniards gave full proof in the war of 1762. So ill-informed were they ... that they thought ... to march on Oporto via the province of Tras-os-Montes or in other words to commit an entire army to roads so dreadful that they can barely be navigated even by the inhabitants of that territory.' De Bourgoing, ed. *Voyage du ci-devant Duc du Chatelet*, II, p. 17. Not the least of the problems was that at this time hardly any Spanish travellers had published accounts of travels in Portugal. See Chinchilla, "Viajeros españoles en Portugal en el siglo XVIII," 302–26. That said, April 1762 had seen the publication of Pedro Rodríguez Campomanes' *Noticia geográfíca del reino y caninos de Portugal*, and, with it, an assessment of Tras-os-Montes that should have rung many alarm bells in the ears of the Spanish commanders. Thus: 'Tras-os-Montes is the most mountainous province in Portugal, and is but little populated compared to Entre-Duero-e-Minho. Its inhabitants are robust tillers of the soil and its accent the thickest in the entire kingdom ... The terrain is arid and very rough.' Ibid., 17–18.

8. Looking back from the perspective of the Peninsular War, British writers made much of the impact of the long period of peace that Portugal enjoyed following the end of the War of the Spanish Succession on the martial traditions of the populace. As the Scottish commentator, Andrew Halliday, wrote for example: 'On the 31 July 1750, Joseph I succeeded his father on the throne of Portugal. The first twelve years of his reign offer nothing to the military reader, and ... forty-seven years' peace had in a great measure changed the very nature of the people.' Halliday, *The Present State of Portugal and the Portuguese Army*, 91. Much the same view, meanwhile, was taken by an anonymous epistlelist in a letter written in 1777. Thus: 'A long peace had entirely destroyed all military spirit and annihilated every system of discipline.' Anon, *Letters from Portugal on the Late*, 29. However, the martial traditions alluded to in such remarks were little more than invention: in the pre-industrial world, common soldiers were invariably men who had either been pressed or driven to enlist by sheer desperation. Indeed, at least one writer writing in 1759 goes so far as in effect to deny that it ever existed. Thus: 'The spirit of the Portuguese seems never to have been thoroughly roused from the lethargy under which it sunk during those years when it was a despised province of Spain. They joined with us ... against the succession of Phillip V, but they and their allies were routed at Almansa after which time they have played no active role in Europe.' Hervey, *Letters from Portugal, Spain, Italy and Germany*, 127–8.

9. It is, perhaps, a mistake to push revisionist arguments too far here. Nationalism in the modern sense was a trait that was far from strong in the popular mentality in either Spain or Portugal in the mid-eighteenth century, but the bloody Portuguese war of independence was recent enough to ensure that anti-Spanish feeling remained rife, whilst the remote nature of many towns and villages was inclined to result in strong suspicion of all outsiders whatsoever. 'Amongst the Portuguese', wrote the Duc du Chapelet, 'hatred of all those who do not come from their country is in truth an innate vice.' De Bourgoing, ed. *Voyage du ci-devant Duc du Chatelet*, II, 4.

10. See Halliday, *Observations on the Present State of the Portuguese Army*, 54–5. According to the Duc de Chatelet, the *ordenança* were 'an utterly miserable assembly', but he admitted that 'their hatred of the Spaniards caused them to take up arms with the utmost joy'. De Bourgoing, ed. *Voyage du ci-devant Duc du Chatelet*, II, 6.

11. Dumouriez, "Introductory memoir on the military topography of Portugal". Setting aside the reverse at the crossing of the Douro, the advance of the Spanish column involved had given an insight into another major problem facing the invaders, namely want of intelligence and a lack of accurate maps. Thus: 'The *corregidoria* of Torre do Moncorvo contains twenty-six boroughs and 45,000 inhabitants ... As [the Spaniards] marched to attack it, they took it for granted they were to meet with a fortified town, and it was said that a corps of 8,000 Portuguese were to defend it. The astonishment of the Spaniards equalled their ignorance when they found Moncorvo was but a sorry village that for the last hundred years had neither wall nor gate, nor had it seen a soldier stationed there during all that time.' Ibid., iv.

12. This force does not appear to have been very impressive. Chatelet, for example, describes the troops concerned as being 'more of a threat than they were a source of strength'. De Bourgoing, ed. *Voyage du ci-devant Duc du Chatelet*, II, 18.

13. The weakness of the Portuguese army was not just a matter of manpower. According to some accounts, indeed, the miserable quality of the defending forces was undermined still further by the coming of war. Thus: 'When the moment came to go on campaign in 1762, nothing was ready: even officers were lacking, the consequence being that it was necessary to look for them in every conceivable place without the slightest regard for cost, anyone who would enlist being offered double-pay. The Portuguese *chargé d'affaires* in London, M. Pinto, became the chief recruiter for his country, and in this capacity he sent over a whole crowd of individuals of every sort and every nation ... The Portuguese having much need of new recruits, they took everyone who came their way, and that despite the fact that the new arrivals included all sorts of adventurers. Ne'er-do-wells, rogues, men on the run from some fraud, were over-night transformed into officers, and that on the flimsiest of recommendations.' De Bourgoing, ed. *Voyage du ci-devant Duc du Chatelet*, II, pp. 5–6.

14. Many of the fortifications that protected the road from Castelo Branco have recently been excavated and restored by a local cultural association entitled the Asociacion do Estudios do Alto-Tejo. See < http://www.altotejo.org/noticias/default.asp?IDN=319&op=2 >, accessed 18 September 2017.

15. By far the best account of the campaign is that contained in Schaumburg-Lippe's own account of the campaign, viz. Schaumburg-Lippe, *Mémoire de la campagne de Portugal de 1762*. Also very helpful, particularly with regard to

the role of the British expeditionary force and its relations with the Portuguese is Speelman, "Strategic illusions and the Iberian war of 1762".

16. De Bourgoing, ed. *Voyage du ci-devant Duc du Chatelet*, II, 25.
17. Foy, *History of the War in the Peninsula under Napoleon*, II, 20–1.
18. Ibid.,21–2.
19. Dumouriez, *An Account of Portugal as it appeared in 1766 to Dumouriez*, 24–5.
20. Ibid., 30.
21. Ibid., 248–9.
22. Ibid., 249.
23. Ibid., 20.
24. Ibid., 53–4.
25. Ibid., 120–1.
26. Ibid., 134.
27. Ibid., 155–6. Dumouriez was not entirely wrong to claim that the British were not popular in Portugal: with the British troops left desperately short of food, there were a number of clashes with angry inhabitants that left more than twenty soldiers dead. See Speelman, 'Strategic illusions', 455–6.
28. In fairness to Moore, it seems that the campaign was not accorded much attention when it came to archiving its details. To quote Speelman, 'When the army searched for information on Portugal during the Napoleonic invasion, the War Office possessed just a few notes and letters left there by Townsend.' Ibid., 458.
29. It is worth noting here that, just as there were strong parallels in the manner in which the two campaigns were conducted, there were strong parallels in the manner in which they were remembered. Here, for example, is how an anonymous British observer looked back on the campaign of 1762: 'England sent to Portugal officers, troops, artillery, stores and money with everything that could enable Portugal to exert her own natural strength and to supply the want of it where it was deficient. The activity of the English, assisted by the Portuguese, soon drove the Spaniards back from some advantages they had gained on the frontiers ... and in one campaign put the future of the country quite out of doubt.' Anon, *Letters from Portugal*, 31.

Disclosure statement

No potential conflict of interest was reported by the author.

Bibliography

Anon. *Letters from Portugal on the Late and Present State of the Kingdom*. London: J. Almon, 1777.

Baretti, J. *A Journey from London to Genoa via England, Portugal, Spain and France*. 3rd ed. London: T. Davies, 1770.

Barrento, A. *Guerra fantastica, 1762: Portugal e o Conde de Lippe e a Guerra de Siete Anhos*. Lisbon: Tribuna de Historia, 2006.

Costigan, A.W. *Sketches of Society and Manners in Portugal in a Series of Letters from Arthur William Costigan, Esquire, Late a Captain in the Irish Brigade in the Service of Spain to His Brother in London*. London: T. Vernor, 1787.

De Bourgoing, J.F., ed. *Voyage Du Ci-devant Duc De Chatelet En Portugal Ou Se Trouvent Des Détailles Intérressans Sur Ses Colonies*. Paris: F. Buisson, 1798.

Dumouriez, C. "Introductory memoir on the military topography of Portugal." In The History of the Revolution in Portugal in 1640 or an Account of their Revolt from Spain and setting the Crown on the Head of Don John of Braganza, Father to Don Pedro and Catherine, Queen-Dowager of England, thence continued down to the Present Time. London: Sherwood, Neely and Jones, 1813.

Dumouriez, C.F. *An Account of Portugal as It Appeared in 1766 to Dumouriez, since a Celebrated General in the French Army*. London: C. Law, 1797.

Foy, M. *History of the War in the Peninsula under Napoleon, to Which Is Added a View of the Political and Military State of the Four Belligerent Powers*. London: Truechel and Wurtz, Truechel jun. and Richter, 1827.

Halliday, A. *The Present State of Portugal and the Portuguese Army with an Epitome of the Ancient History of the Kingdom, a Sketch of the Campaigns of the Marquis of Wellington for the Last Four Years and Observations on the Manners and Customs of the People, Agriculture, Commerce, Arts, Sciences and Literature*. Edinburgh: G.R. Clarke, 1812.

Halliday, A. *Observations on the Present State of the Portuguese Army as Organised by Sir William Carr Beresford, K.B., Field- Marshal and Commander-in-Chief of that Army with an Account of the Different Military Establishments and Laws of Portugal*. London: John Murray, 1813.

Hervey, C. *Letters from Portugal, Spain, Italy and Germany in the Years 1759, 1760 and 1761*. London: R. Faulder, 1785.

Ortega Chinchilla, M.J. "Viajeros Españoles En Portugal En El Siglo XVIII: Entre El Conocimiento Y La Experiencia'." *Tempo* XXII, no. 40, May (2016): 302–326. doi:10.20509/TEM-1980-542X2016v224005.

Schaumburg-Lippe, F.W.E. von. *Mémoire De La Campagne De Portugal De 1762*. n.p., n.d.

Speelman, P.J. "Strategic Illusions and the Iberian War of 1762." In *The Seven Years' War; Global Views*, edited by M.H. Danley and P.J. Speelman, 429–460. Leiden: Brill, 2012.

Reluctant guerrillas in early nineteenth century China: the White Lotus insurgents and their suppressors

Yingcong Dai

ABSTRACT

At the turn of the nineteenth century, China's Qing dynasty (1644–1912) was hit by a sectarian rebellion. Commonly considered a breakpoint marking the end of the dynasty's golden age spanning most of the eighteenth century, the war to suppress the rebels, referred to as the White Lotus War (1796–1804) in this chapter ('White Lotus' was the umbrella name used by both the authorities and some sectarians for their teaching), exposed many structural drawbacks of the Qing political and military systems and depleted the dynasty's financial resources, which had never been recovered. Reluctant in embracing guerrilla warfare in the beginning, the insurgents quickly turned themselves into master guerrillas. Shuttling in two massive mountain ranges in central China, they managed to prolong their rebellion and fought some successful battles against their suppressors. Superior in manpower, weaponry, and logistical support, the government forces had to adapt to guerrilla warfare, albeit passively and ineptly. This chapter gives a brief introduction to this little-known episode of guerrilla war at the turn of the nineteenth century in Qing China, expounds the strengths and weaknesses of both sides, and sheds light on the roots of the war's long duration and the grim consequences to the Qing state.

Guerrilla warfare was not intended and planned when sectarians plotted a rebellion against the Qing dynasty in 1795. However, shortly after they revolted early in 1796, some insurgents adopted guerrilla tactics, which was ultimately followed by all the insurgent bands that had survived the first wave of the government's suppression campaigns. In the years, 1797–1804, multiple bands of insurgents, each consisting initially of several thousand people, managed to prolong their rebellion by shuttling in two massive mountain ranges in central China, the Qinling Mountains and Daba Mountains, keeping tens of thousands of the government troops and hired militias in mobilization. Reluctantly embracing guerrilla warfare in the beginning, the insurgents quickly demonstrated their expertise in guerrilla warfare, and fought some successful battles against their suppressors. Superior in manpower, weaponry, and logistical support, the government forces had to adapt to guerrilla warfare, albeit passively and ineptly. Yet, it took the government forces several years to completely wipe out all the rebels. This article gives a brief introduction to this little-known episode of guerrilla war at the turn of the nineteenth century in Qing China, expounds the strengths and weaknesses of both sides, and sheds light on the roots of the war's long duration and the grim consequences to the Qing state.

I

Popular sectarian societies were a perennial phenomenon in China's late imperial period, spanning the Song dynasty (960–1279), the Mongol Yuan dynasty (1279–1368), the Ming dynasty (1368–1644), and the Manchu Qing dynasty. While Buddhism was the most important theological source for most sects, ingredients from other religions, such as Zoroastrianism and Manicheanism, and teachings such as Confucianism and Taoism, were liberally mixed into many sects' tenets and practices. The Ming and Qing dynasties witnessed a surge of sectarian activities in society. Numerous new sects were either created or fissured from their parent sects. Hundreds of scriptures were compiled and circulated. Some sects were more locally centered, whereas some others gained followings in an extended area encompassing several provinces. Not caring to distinguish one sect from another, the authorities often used 'White Lotus,' the name of one of those sects, as an umbrella name for all of them.[1] Although most sects concentrated on moral teaching and accumulation of good deeds for the next life, some sects adopted a millenarianism-tainted eschatology, advocating the coming of the apocalypse, and calling for their members to rise against the ruling authorities. Therefore, both the Ming and Qing states had been vigilant toward sectarian movement, officially banning all the sects, even though they knew it would be impossible to uproot. From time to time, sects instigated uprisings and proclaimed their own regimes, but their uprising attempts often failed to materialize thanks to the government's preemptive suppression.[2]

In the early 1790s, two sects in central China, the Hunyuan and the Shouyuan sects, were engaged in a keen competition for followers in several provinces, *i.e.*, Hubei, Henan, Sichuan, and Shaanxi, which helped expose themselves to the authorities.[3] A crackdown was ordered by the emperor late in 1794. In several months, hundreds of sectarians including many chiefs were arrested and subsequently executed or exiled. Threatened by the authorities' continuous manhunt, sectarian leaders planned holding multiple uprisings in different locales simultaneously in the spring of 1796. Nevertheless, one of the key plotters, who had been respected by the sectarians of both the Hunyuan and Shouyuan sects, bowed out from the action shortly after the decision of uprising, and went into hiding.[4] As no other leaders had the mettle to command all the sectarian insurgents in different locations, the rebellion would unfold in a decentralized and frag-mented manner. Owing to the leak of the uprising plan to the local autho-rities, the first uprising started in Hubei in the February of 1796, earlier than planned, which was followed by more than half a dozen uprisings.[5] In the autumn of the year, several uprisings erupted in northeastern Sichuan, which bordered Hubei. Then five uprisings were held in Shaanxi's south-eastern corner at the end of the year and beginning of 1797. Each of those uprisings drew from several thousand to over ten thousand people, includ-ing family members of those rising sectarians.

Insurgents followed basically two strategies, either taking a city or dig-ging in on a mountaintop nearby. In Hubei, the insurgents took a few towns that were county seats, but only held one city, Dangyang, for a stretch of time, whereas they held the other towns for only days. Going onto a mountain was a more common strategy. They built an intricate defensive system on the mountaintop, digging moats, building forts, and storing wood logs and rocks. In both Sichuan and Shaanxi, all the uprisings followed this mountaintop strategy. No attempt was made to hold either a city or a town (only one town was taken in Sichuan, but the rebels did not hold it because it did not have city walls). Constrained by a localized perspective of their rebellion, few thought of leaving their home area and moving to other places or provinces at the initial stage of the conflict.

As it soon turned out, neither of the strategies unfolded to the insurgents' advantage. In Dangyang, despite an ineffective siege by the Qing forces, which lasted for months, the exhaustion of supplies doomed the insurgents. In the summer of 1796, the Qing forces recaptured the city when all the people inside, both insurgents and residents, had been out of food for days.[6] Meanwhile, the insurgents' mountaintop strategy was not successful either. They struggled on a daily basis to acquire food and water and repel enemy attacks. At times, they had to endure epidemics, as they were susceptible to diseases living in excessively packed forts.[7] In destroying their strongholds, the suppressors' firearms, e.g., cannons, grenades, and fire arrows, played a key role. Once cannon shells or grenades set the forts ablaze – and the ensuing fire often burned hundreds or thousands of insurgents and families to death – it was easier for the Qing forces to overtake the mountaintops and round up the remaining insurgents.[8] By early 1797, all uprisings in southern Shaanxi were extinguished. In Sichuan, the rebels were driven from one mountaintop to another, and suffered tremendous losses. In Hubei, except for two groups of insurgents, all other uprisings were suppressed.

Of the two, the insurgents of Xiangyang prefecture, Hubei, were particu-larly distinguished from most of their peers.[9] A strategic point in central China,[10] Xiangyang had been a sectarian enclave. In 1794, the Qing state harshly persecuted the Hunyuan and Shouyuan sects in the area. Nevertheless, the survivors persevered in their faith and maintained their extensive networks. It was from among the sectarian ringleaders there that the decision to revolt was made early in 1795. The uprising in Xiangyang started in March 1796, weeks after the first wave of uprisings in other places of western Hubei. Having failed to take the twin cities, Xiangyang and Fancheng, on the opposite banks of the Han River, the insurgents adopted mobile warfare, as the quickly assembled government forces made it impossible for them to stay in their home region. Moving eastward to central Hubei and battling the heavy Qing forces along their way, the Xiangyang insurgents soon made mobility their credo of survival. Late in 1796, they managed to return to Xiangyang after an unsuccessful eastern expedition. Early in 1797, the fewer than ten thousand insurgents and families embarked on a long trek from Hubei to Sichuan, attempting to join forces with the insurgents in Sichuan, as they heard that their counterparts had been doing well in Sichuan, where resources were abundant.[11]

The journey was difficult and long; the rebels had to make a huge detour via Henan and Shaanxi provinces to get rid of their suppressors. From February to June, they first trekked eastward along Hubei's border with Henan, and then entered Henan, from which they turned westward and thrusted into Shaanxi, where they expanded their ranks by absorbing thousands of local sectarians who embraced the rebels' cause with open arms. Many brought their entire families; some burned their houses. In the early summer, the expanded rebel forces crossed the Han River and scaled the Daba Mountains that extended hundreds of miles over the long border between Shaanxi and Sichuan. When the travelling insurgents made their way to northeastern Sichuan, and met their counterparts, the latter were on the verge of being eliminated after several deadly defeats in the spring and early summer.

However, the Xiangyang rebels' arrival in Sichuan did not result in a merger between them and the local insurgents. Some Sichuan insurgents were resentful toward those outsiders treading on their hometown's soil with a superior attitude. Meanwhile, the sojourners were disappointed with the meagerness of the region. Shortly after their arrival, the main force of the Xiangyang insurgents returned to Hubei, but a portion of their men stayed in Sichuan, perhaps to check the Qing forces that had pursued them. Despite an unfruitful sojourn in Sichuan, the Xiangyang insurgents had shown to their counterparts in Sichuan that mobile warfare might be a more efficient way to survive or even thrive. Consequently, the rebels in Sichuan left their home regions and adopted guerrilla warfare, resorting to mobility to overcome their inferiority in manpower and weaponry. They extended the perimeter of their operations to other counties and prefectures, and some ultimately trekked to either Hubei or Shaanxi. By the beginning of

1799, the mountaintop tactic was completely abandoned, when the last mountaintop strongholds held by several groups of insurgents were leveled by their suppressors one after another.[12]

In several years to come, 1799–1804, the White Lotus War became essentially a guerilla war. At first four or five bigger bands, and then about a dozen smaller bands, roamed in and out of the Qinling Mountains and the Daba

Map 1. The White Lotus War in Central China, 1796–1804.

Mountains, and fought for survival by using typical guerrilla tactics such as mobility, operation in small groups, ambushes, and night raids. Despite their numbers dwindling – initially there were more than ten thousand people in one band, but only a few thousand and then a few hundred or fewer – those guerillas kept nearly 80,000 Qing regular forces and tens of thousands of militiamen in mobilization, forcing them to adopt guerilla tactics as well divide their sizable forces into small units, demounting, and de-equipping. In fact, that those insurgents did not have a centralized leadership might have made their adaptation to guerrilla warfare more natural. Throughout the war, those rebel bands operated largely independently. Only on a few occasions did several bands coalesce to form an ad hoc coalition, but cooperation was always short-lived. After 1801, the total number of insurgents was only in the thousands. Many had retreated into primal forests in the mountains and resorted to banditry for survival.

The guerrilla war waged by the White Lotus rebels at the turn of the nineteenth century would be restaged in the twentieth century. More than 130 years after the Xiangyang insurgents' long march to northern Sichuan, in the autumn of 1932, the communist Fourth Red Army led by Zhang Guotao, one of the early leaders of the Chinese communist revolution, trekked from its base area astride Hubei, Henan, and Anhui to northern Sichuan, after failing to withstand the military pressure of the Nationalists. The communists first headed westward to Zaoyang in northern Hubei, from where they penetrated southern Henan and marched along the borders of Henan. Then they returned to Hubei at its northwestern tip, and sneaked into southern Shaanxi. In Shaanxi, the communists chose to go north, reaching the northern foot of the Qinling Mountains, in order to get rid of the Nationalist forces pursuing them. This detour gave them one more obstacle to overcome, scaling the Qinling Mountains, before they reached the Han River.[13]

At the end of 1932, the Red Fourth Army, which had dwindled in size, braving the icy water in the midst of the winter, forded the Han River near Hanzhong.[14] Then they scaled the Daba Mountains and arrived in Tongjiang,

Map 2. The two long marches: The Xiangyang Rebels' March to Sichuan in 1797 and the Red Army's March to Sichuan in 1932.

where one of the White Lotus uprisings erupted in late 1796. During their two and a half years' stay in northern Sichuan, Zhang Guotao and his comrades expanded their ranks to more than 80,000 and set up a new base area encompassing Tongjiang, Bazhou, Nanjiang, Dazhou and Taiping, at its highest point, which largely overlapped the areas where the White Lotus insurgents were most active in northern Sichuan. Although the communists were ultimately hard pressed with supply shortages which seemed to be insurmountable, their survival and convalescence suggest that the rugged and meager Daba region could be turned into a home, albeit a temporary one, for the guerrillas from afar.[15] Nevertheless, in 1797, the White Lotus insurgents from Xiangyang were not able to take root in this region.

II

Despite their disparate ideologies, the memberships of the White Lotus rebels and the communist Red Army in the 1930s were similar. When the rebellion first started in 1796, the core of insurgents in most locales were farmers, craftsmen, e.g. carpenters, goldsmiths, tailors, dyers, and handy-men, small businessmen from dealers of cotton, liquor, pigs, to shop owners and peddlers, and other professionals such as teachers, doctors, and fortune tellers. Mixed in their ranks were also clerks and runners facilitating a structure of local government. Despite their extensive networks in central China, only a small portion of the sectarian members joined the rebellion. Contrary to what has been believed by some that only the poverty-stricken were motivated to rebel, most of the sectarian-cum-rebels were property owners, and those who made their living by selling their labor were not the majority.[16] In general, they had not been faced with an economic crisis, although many might not have been satisfied with their lots. In Sichuan, a province that had been inundated with waves of migrants from other parts of the country seeking better fortunes throughout the eighteenth century, outlaws such as bandits, salt smugglers, and counterfeiters were also included in the initial crowds of the insurgents. Although they had not been the target of sectarian proselytization prior to the mobilization for revolt, the sectarian agitators turned to them at the eleventh hour in order to expand their ranks.[17] Ethnically, most insurgents were Han Chinese, the ethnic majority in China. In several uprisings in southwestern Hubei, how-ever, Bizika, an ethnic minority, were numerous among the rebels, as their communities had long been sectarian bastions.[18]

Unlike the rebellions in the late Ming dynasty, in which the disgruntled soldiers were one of the driving forces,[19] military ranks were absent among the White Lotus rebels. Because many rebels brought their families with them, and did not shy from taking more women, elderly, and children along their way, able-bodied male rebels were often outnumbered by non-combatants. Even after many rebel chiefs were killed or captured, their wives, mothers, and other relatives stayed with the rebels, sometimes for years, until their own arrest or death. Although the rebels designated their able-bodied men as 'fighters' (zhanshou) or 'spearmen' (maoshou) and appointed chiefs with various titles to lead them, the combatants were not segregated from their families; two cohorts usually marched and camped together.

Nevertheless, it was not always a liability to have numerous women in their ranks. Many women joined battles and some, especially spouses of the rebel leaders, became commanders themselves. Moreover, women were used as awards to male rebels. Ranking rebels often had more than one woman as their consorts. Having developed true affections, some of those women who had been coerced into the relationships showed their allegiances to their men even after they were captured by the government forces. The suppressors had been often amazed by the fact that those women marched with their male counterparts at great speed, even in rugged mountainous areas, given that some of them, if not all, might have had bound feet. Of course, when draft animals were available, women were always given horses or mules to ride; and women and children were ferried in boats when crossing rivers, while men often waded them.

Except for the beginning of their rebellion when they were able to galvanize their fellow sectarians to embrace their cause,[20] coercion had been the primary method for the insurgents to acquire new blood. They captured civilians regardless of their gender and age and killed those who resisted. To prevent the captives from fleeing, the insurgents sometimes tattooed 'White Lotus' in character on their faces. As stipulated by one rebel band, group chiefs would be severely penalized if anyone from their groups deserted, and the deserters would be beheaded, if they were caught.[21] In battles, those captured civilian were often used as human shield, which caused their casualties to be much higher than that of the true insurgents. Nevertheless, some coerced insurgents were eventually assimilated into the 'inner group.' Having stayed with the insurgents for a stretch of time, and with no hope to escape (sometimes no home to return to: their property had been seized and sold by local officials), they accepted what they were forced into, turning themselves into willing insurgents. For some, to be conferred with a title such as 'vanguard' (xianfeng), 'banner-bearer' (qishou), 'head scout' (tanmatou), 'commander' (zongbing), or 'marshal' (yuanshuai), might have awakened or enhanced their self-worth, helping cement their identification with their own band.[22]

As mentioned above, the theaters of this guerrilla war were mainly in the Qinling Mountains that cut through the province of Shaanxi, the Daba Mountains sprawling over the long provincial border between Sichuan and Shaanxi and extending into northwestern Hubei, and the Han River valley between the two mountain ranges in southern Shaanxi. In addition, the mountainous region north of the Yangzi River in Hubei's western border with Sichuan was also frequented by the insurgents, serving as a haven for their recuperation and reorganization.[23] In those sparsely populated mountains, the conditions for survival were harsh. While it was severely cold in winters, with snow and ice covering hills and forests, it was swelteringly hot in summertime. The only shelter available to the guerrillas were the caves and cabins abandoned by mountain residents. In the latter part of the war when many residents in the war zone went to mountaintops and built forts to protect themselves and their belongings, their unoccupied homes provided occasional resting places for the rebels in the event that their suppressors were far behind.[24] Late in 1799, when a Qing force reached the outskirt of a village in southern Shaanxi, overlooking from a hill, its commander saw the rebels in the mountain valley below resting in houses or makeshift sheds, grinding rice, or washing vegetables, and cooking smoke everywhere – it was lunchtime.[25] Over the years, the guerrillas had to brave elements, insects, and other hazards while camping in the wildness. For the suppressors, the number of bonfires' remains left by the rebels was often an indicator of their band's size.

Although they assailed their enemy's logistical convoys earlier in the conflict when they were sufficiently forceful to do so, the insurgents relied chiefly upon pillage to acquire food, clothes, draft animals, and other supplies. Some rebel bands even threatened their rank-and-filers with corporal punishment for returning from looting sorties empty-handed.[26] Having no concern over their public image, the insurgents used violence liberally, leaving shambles behind and uprooting countless civilians. In the early phases of the war, the insurgents seemed to be well equipped. Their chiefs were clothed in silks, satins, leathers, and furs, and enjoyed other luxuries. The rank-and-filers also carried with them as much booties as they could. Without the burden to maintain any logistical apparatus, they had an edge over their suppressors in mobility. Frustrated, the Qing commanders constantly complained about their disadvantage to be weighed down by their cumbersome logistical corps vis-à-vis the rebels' advantage of living off land.[27]

To protect their lives and belongings, more and more local residents moved to mountaintops and fortified themselves there, which had been a time-honored way of self-protection in the times of upheaval. While the government encouraged and supported this undertaking because it helped deny supplies to the rebels, it was a far cry that the measure succeeded in accomplishing this goal.[28] Not all the residents in the war zone participated. Even for those who had moved to the mountaintops, some left their foodstuffs at home, for they were unwilling to share them with the people in the same fort and hoped that the rebels would not go to their area. More important, the insurgents often attacked those civilian forts, looting supplies if they were able to break in, or exacting for them by threatening fort dwellers. To send the insurgents away, civilians usually chose to throw foodstuffs and other supplies to them. Despite repeated admonitions by the officials and generals not to throw supplies to the rebels, fort dwellers continued doing so until near the end of the war.[29]

In the end, years of guerrilla war exhausted the insurgents, both physically and mentally. In the late stage of the conflict, they no longer had any draft animals, and many were in rags and barefoot. Their weapons were mostly spears, knives, rocks, wood logs and the like. Although they carried some rifles taken from the Qing armies, they did not have ammunition – those rifles were only useful when they disguising themselves as the government troops in cheating civilians out of food.[30] When they approached those civilian forts for supplies, they could no longer attack them, but begged for supplies in return for money or valuables. At times, the male rebels killed women and children in emergent situations. It was not uncommon for despondent rebels to commit suicide.

Nevertheless, the backbone of the insurgents, a small number of the faithful sectarians from the bastions of the two sects, Hunyuan and Shouyuan, in northwest Hubei and northeast Sichuan, had maintained their commitment even when they had only hundreds of people left in their bands. Primarily owing to those most committed insurgents, some of whom travelled from one band to another, trying to revive their dying cause, the last several bands persisted in their guerrilla war until they were finally eliminated. At times, battles were fierce. The guerillas in rags were still capable of hitting their suppressors hard, inflicting upon them considerable casualties. As observed by both the emperor and his commanders, few original sectarian rebels surrendered throughout this war. When captured, many had kept their defiance until they were slaughtered by their captors, often in extremely brutal ways.[31]

III

Being total strangers to warfare in the beginning, the insurgents were quick learners. Having realized that they were unable to confront their enemy head on given the latter's undisputable superiority in manpower, weaponry, and financial and logistical support, the insurgents adopted guerrilla tactics to remedy their own weaknesses. As mentioned above, the insurgents from Xiangyang, Hubei, had most inclined toward mobility and never holed up on a mountaintop. After they arrived in northeastern Sichuan in the early summer of 1797 following a months-long trek, they soon embarked on another journey back to Hubei – most likely due to their schism with the local insurgents and the scarcity in resources in northeastern Sichuan. Going back, they took a different route. They first went south to the north bank of the Yangzi River and then entered Hubei along the river's Three Gorges, a narrow, zigzag stretch of the river full of dangerous riffs between towering and steep cliffs on both banks.[32]

Back to Hubei early in the autumn, the Xiangyang insurgents had a mix of successes and failures. They managed to expand their ranks to more than 20,000 people and seized several thousand horses. On their way to the north, they took a county seat by leaving behind a small dispatch to check their enemy.[33] However, they failed to take the city of Xiangyang, and suffered a disastrous defeat in its outskirt, sustaining thousands of casualties, as the Qing forces had spared no effort to safeguard this strategically critical city. Moving toward the border with Shaanxi, the insurgents stunned their enemy on 18 October 1797 by killing several ranking Manchu officers, including a Manchu nobleman of highest rank, Huilun, in Yunxian. Early in the fight, the insurgents were assaulted in several directions, losing thousands of people. However, by making a retreat, the insurgents lured their enemy to their trap. Then in a brutal and close battle, they killed Huilun and other officers.[34] The deaths of those Manchu generals shocked the Qianlong emperor, who openly expressed his anger and sorrow.[35]

Having failed to cross the Han River to the north in Hubei to fulfill their plan to go to Henan – Henan had been a dreamland in the sectarian prophecy and had dense sectarian networks – the insurgents entered Shaanxi, for the second time. In the following two months, they shuttled along the Han River's south bank, with an excursion into northern Sichuan, and attempted desperately to cross the river to the north, but suffered heavy casualties in numerous battles along their way. To avert their enemy's attention, the rebels used a typical guerrilla tactic: they divided their forces into two branches and moved to two directions. This tactic succeeded; early in February 1798, one of the two regiments made its way to the north of the Han River. Greatly alarmed, 8,000 elite Qing forces rushed to the north bank, which allowed another branch, the main one, an opportunity to cross the river on March 19–20.

Once landed in the north, the main branch again split into two. While the smaller group moved north through the Qinling Mountains and aimed at Xi'an, the provincial capital, the main group marched eastward to Shaanxi's border with Henan. However, it was soon encircled by their suppressors. One month after their crossing of the Han River, the rebels were pressed onto Shaanxi's border with Hubei. After one whole day's fight, which inflicted heavy casualties on them, they infiltrated into Hubei's Yunxi county overnight. With the aid by the local troops and militiamen in Yunxi, on 21 April 1798, the government forces routed the exhausted insurgents still in many thousands in a pincer attack. Two leaders of the band, Yao Zhifu and Wang Conger, a woman rebel, as well as their entourages, jumped off cliffs. According to the commanders of the Qing forces, they had killed more than 7,000 and captured another one thousand, but they admitted that their forces did not clear off all the rebels scattering in every direction, rather they left them to be caught by the local authorities.[36]

Not surprisingly, many insurgents managed to escape. Meanwhile, the Xiangyang insurgents who had not been with this group were active elsewhere. In the years to come, the remaining Xiangyang insurgents were instrumental in the persistence of their rebellion. One of them was Zhang Hanchao. In his seventies and likely literate – he might have lived on making divinations before the uprising – Zhang commanded great respect from his followers. In the early days of the uprising in Xiangyang, Zhang was among several rebel leaders bearing the title of 'marshal.'[37] Avoiding joining their main force to return to Hubei in the autumn of 1797, Zhang led his own band consisting of 10,000 men and women and went back to Hubei in the spring of 1798 along the same route their Xiangyang comrades had taken, i.e. the north bank of the Yangzi River. In Hubei, Zhang's men kept recruiting and kidnapping civilians, swelling their ranks to 30,000 at one point, as claimed by their opponents.[38] Like his counterparts, Zhang also tried to go to Henan. However, all his many attempts to cross the Han River at different locales failed. Following a fiasco in which thousands of his men were killed, Zhang ran westward in several groups, and eventually made his way to Shaanxi's south-eastern corner in the late summer. Then he roamed on the border areas between Shaanxi and Sichuan and along the Han River's south bank, looking for a chance to cross the river. He made sorties to northern Sichuan and Hubei, but was all forced back to Shaanxi. In the extremely hilly northeastern tip of Sichuan, his men had to abandon their horses and mules, and even killed some of their women and small children in order to move faster.[39]

In December 1798, Zhang's band – a few thousand left but barely a thousand being able-bodied men – crossed the Han River to its north bank in the river's upper stream, where it was fordable in wintertime, and entered the

Map 3. Zhang Hanchao's Guerrilla War in 1797–1799.

Qinling Mountains. In the months to follow, Zhang's band trekked through the mountains several times, never settling in one locale. It made inroads into Henan twice and penetrated Gansu more than once. Owing to his age and declining energy, Zhang delegated command to several younger chiefs, while he himself was carried in a bamboo chair. Overcome by despair at one point, Zhang tried to commit suicide, but was stopped by his followers.[40] However, Zhang's men were still effective when it came to fight. In the late spring, 1799, they defeated the new governor of Shaanxi in the mountains, and almost caught him.[41]

In the autumn of 1799, once again, Zhang headed westward, and entered Gansu province. Having suffered a series of defeats, losing most of his people – in addition to battle casualties, many hanged themselves on trees – Zhang and the band's remnants were heavily besieged in a mountain near Gansu's border with Shaanxi. Leaving a small group to engage their enemy, Zhang and his last men made a miraculous escape by slipping down a cliff covered with ivies and managed to return to Shaanxi. Going into the Qinling Mountains again, they retreated into a primal forest. Following them into the forest in small groups, the Qing troops caught them on 31 October 1799, killing more than 200 and capturing scores. Hit by a spear in the melee, Zhang Hanchao jumped from a cliff and then hanged himself on a tree.[42]

Zhang Hanchao's death did not signal the beginning of the end for the rebels from Xiangyang. Hundreds of his followers were still at large in the Qinling Mountains, who drew tens of thousands of the government forces and militias to tackle them, and were not completely cleared until a couple of years after. Meanwhile, other rebels of Xiangyang regrouped, posing new challenges to the suppressors. Unlike Zhang who had only focused on either returning to Xiangyang or making his way to Henan, some of the new bands were more unpredictable in their mobile operations, daring to go to new territories. The most outstanding and effective among them is the Gao-Ma band, named after its two co-leaders' last names, Gao and Ma.[43]

At the beginning of 1800, the Gao-Ma band moved out of the Daba Mountains where it had hidden for some time to regroup. Via southwestern corner of Shaanxi, the band thrust into Gansu after crossing the upper streams of the Han River and the Jialing River. Braving inclement climate and warding off its opponents' close pursuit, the Gao-Ma band moved quickly in Gansu's southeastern corner, which was mountainous and sparsely populated. Concerned that the band's lingering in Gansu would escalate the conflict – the province's involvement in the war had been minimal prior to this – the Qing imperial court, headed now by the Jiaqing emperor, as Qianlong passed away at the beginning of 1799, endorsed its best forces to move to Gansu from Shaanxi and Sichuan to fight the Gao-Ma band.[44]

Having lost most of its members and draft animals in several defeats in Gansu, the remainder of the Gao-Ma band, consisting of only about one thousand people, retreated to the forest on the border marches between Gansu and Sichuan in the late spring. Then it penetrated northern Sichuan after having recruited new blood. In a surprise night battle, the band assailed their enemy's encampment, killed a high-ranking Qing officer, and almost captured the governor-general of Sichuan, the highest mandarin outside of the imperial capital.[45] Again showing its high mobility, the band returned to Gansu via a Tibetan district in northwestern Sichuan, and went to Shaanxi in the summer of 1800. Moving eastward along the south bank of the Han River, it assailed a Qing outpost. In the fight that lasted for hours, they routed their enemy and killed its commander, a provincial military commander, the highest officer of the Green Standard Army in a province, along with hundreds of Qing troops. For this stunning defeat, the Jiaqing emperor dubbed the band as the most dangerous one among all the insurgent bands still active at the time.[46] Until it was exterminated early in 1801, the Gao-Ma band was a constant cause of alarm for its suppressors who drew their crack forces to deal with it suppressors and draw their crack forces to deal with it.[47]

Despite being one or two steps behind their counterparts from Xiangyang in adopting guerrilla tactics, the insurgents of Sichuan quickly caught up and shone on several occasions. Ran Tianyuan, who had co-led an uprising in northern Sichuan in 1796, poised to become an outstanding rebel commander after his uncle, another leader of the uprising, was killed and their band, the Blue Band, was nearly dismantled at the beginning of 1799. Having regrouped his band, Ran first demonstrated his military acumen in a couple of battles in northern Sichuan late in 1799. Collaborating with several other bands, Ran used ambush skillfully to knock down his enemy's advantages and inflict heavy casualties on his opponents, including the death of a Manchu aristocrat bearing the highest noble rank.[48] Having perhaps tasted the benefit of cooperation among different rebel bands, Ran created a coalition consisting of five rebel bands, all of Sichuan, and launched an expedition toward western Sichuan. On 8 February 1800, in the night, Ran's coalition crossed the Jialing River that ran through Sichuan in the middle of the province, bringing the upheaval to the river's western side that had not been touched by the conflict.[49]

As the coalition rapidly expanded its ranks to tens of thousands people by absorbing and coercing both willing and unwilling candidates, Ran conducted another successful battle in a town in Xichong county, in which his men nearly annihilated his opponents and killed their commander, a high-ranking Green Standard officer. Not until the Qing reinforcements from Shaanxi reached the frontline did Ran and his coalition encounter tough military challenge. Yet, by using ambush tactics, Ran was victorious in the first battle against the Qing elite force led by Delengtai, an experienced Mongol general, forcing the latter to pause the fight and wait for reinforcements. In a battle after Delengtai received reinforcements, Ran fought valiantly, but was captured with both his legs seriously injured. He was soon executed by 'a thousand cuts' in Chengdu, Sichuan's provincial capital. His coalition collapsed weeks after his capture, though the coalition managed to cross another river and waged assaults in the fertile Chengdu plain before it was forced to withdraw to eastern Sichuan.[50]

Between late 1800 and 1804, the remainder of the insurgents in Shaanxi, Hubei, and Sichuan continued their guerrilla warfare, although the sizes of their bands kept shrinking and they were wiped out by the suppressors one after another. In this phase, two bands were most adroit in mobility, causing tremendous troubles for their suppressors. One was led by Ran Xuesheng, a veteran insurgent from Xiangyang who had been a key consul to Zhang Hanchao. Bearing a title of 'marshal' – he was nicknamed 'Black-Eye-Sockets Marshal' – Ran had assembled a band of 1,000 diehard and able fighters. In the first half of 1801, he pulled his band out of the Qinling Mountains and made two expeditions to Gansu. In the first expedition, he had a marked success in expanding his band to several thousand people and seizing hundreds of horses and mules. Returning to Shaanxi, he routed the Qing forces near the province's western border by luring the latter into his ambush range, inflicting hundreds of casualties on his opponents. Then he re-invaded Gansu, aiming at obtaining more draft animals. Only after the commander-in-chief of the Qing suppression campaign, Eldemboo, went to Gansu to engage him in person and fought a couple of bloody battles (in one of the battles Eldemboo lost his own younger brother) did Ran Xuesheng give up his plan to go further in Gansu and return to Shaanxi.[51]

Another rebel leader who created a great deal of distress for the suppressors was Gou Wenming. Having joined his brother-in-law to rise in northern Sichuan in 1796, Gou did not become a leader until his original band, the White Band, met its demise at the beginning of 1799. Having hid deep in the Daba Mountains for months, Gou and other survivors formed a new band by taking scattered insurgents whose bands had been crushed. Not until late in 1801 did Gou's band catch his suppressors' attention when it went out of the Daba Mountains and moved westward quickly. Early in 1802, it made a brief incursion into Gansu and then returned to Sichuan. Along Sichuan's mountainous northern border, it moved eastward and disappeared in the Daba Mountains. Taking advantage of its enemy's recess for the Chinese New Year, it moved north, crossed the Han River, and entered the Qinling Mountains. For more than half a year, this band of a couple of thousand people circulated in the primal forest, often in small groups, in the heart of the extensive mountains, keeping the suppressors constantly at a loss. Not until the latter also parceled out their men and entered the forest in small teams was Gou hard pressed. Plagued by food shortages, Gou Wenming disbanded his band in the autumn of 1802 shortly before he was found and killed by the suppressors.[52]

Gou Wenming's death marks the beginning of the end for the White Lotus rebels' remarkable guerrilla war in central China. Although the Qing forces spent another two and half a years to clear up all the rebel remnants, the delay in the completion of the suppression campaign was more affected by the suppressors. Hardly able to generate new sparks after all their key leaders met their demise in 1801 and 1802, the remaining insurgents were confined in limited perimeters in their last fight for survival, although they continued to move around and seized every opportunity to exert their favorite tactic, ambush.[53]

IV

The White Lotus War occurred at a critical point of the Qing dynasty. At the beginning of 1796, days before the first uprising erupted in Hubei, the Qianlong emperor (r. 1736–1795) abdicated the throne to his son, inaugurating the new Jiaqing reign (1796–1820). Nevertheless, Qianlong, who was in his eighties, did not give up the reins of government, but continued taking charge of most state affairs, big or small, until his death early in 1799.[54] This situation complicated the Qing effort to crack down on the insurgents; the new sovereign had to wait after Qianlong's death to take over the command of the war. Having waged numerous wars both on the frontiers against foreign enemies and in the heartlands to crack down on rebellions during his sixty-year reign, Qianlong did not initially take the sectarian uprisings seriously. Echoing his officials in the battleground provinces, he considered the insurgents 'ragtag,' and those uprisings merely petty nuisances.[55] Despite repeated pleas for reinforcements from the commanders and provincial officials, Qianlong refused to send to the war the elite Banner forces in Beijing and Manchuria, but directed the battleground provinces to tap into local communities for help, which helped instigate the widespread use of hired militias in this war. Not until late 1796 did Qianlong endorse the sending of 2,000 Manchu bannermen from Beijing to the campaign, as he realized that the insurgents were more dangerous than he had estimated.[56]

Not surprisingly, complaints from the Qing commanders reemerged and intensified when most rebel bands adopted guerrilla tactics starting from late 1797. In the Sichuan theater, the insurgents attacked county seats and official supply lines, creating multiple emergencies simultaneously and keeping the government forces on the run. Yimian, the commander-in-chief of the suppression campaign at the time, who had been in Sichuan since early in the year, complained bitterly about his opponent's guerrilla tactics:

The bandits [rebels] are treacherous and scattered, difficult to round up. If the official troops intercept them head on, they split and go in different directions. If the official troops follow them and chase them, they run straight ahead once they spot their enemies. [Our soldiers] have become exhausted from the long campaign, but the bandits have become numerous because they have kidnapped [civilians].[57]

For Yimian and some other Qing commanders, the only recourse was receiving reinforcements, hoping a surge of their forces would counterbalance the insurgents' guerrilla tactics. However, some other commanders were looking at other options. Also in late 1797, Mingliang and Delengtai, two decorated generals, recommended two measures when pursuing the Xiangyang insurgents back to Hubei. First, they suggested that civilians be ordered to build forts and stow their foodstuffs and belongings there, so that the insurgents would be denied supplies. Second, they proposed that a cavalry regiment be set up to outmatch the insurgents' mobility, as the regiment would carry its own supplies for ten days, which would free it from dependence on usually cumbersome logistical services. However, skeptical of their effects, the Qianlong emperor did not give his approval to either.[58] Although fortification was carried out in some places, mostly being spontaneous initiatives by local residents, the idea of a cavalry was soon left in abeyance when all the insurgents entered mountains that were often pathless, unsuitable for horses to march.

In the rest of the campaign, the suppressors tried continuously to figure out ways to deal with the rebels' guerrilla tactics. Ideas that had been put forward by frontline officials and the emperor include using more militiamen to guard passes, digging up roads to slow the rebels' movements, burning the primal forests to drive the rebels out, poisoning water sources, etc. Initially, the commanders were resistant to turning their forces into guerrillas so that they would not lose their advantages. As Yimian put it: 'it is best for the imperial troops to fight together, whereas it is to the advantage of the bandits to divide themselves but to their disadvantage to stay together.'[59] However, they had to adapt to guerrilla warfare ultimately, albeit reluctantly, as they could not find any other method that worked. In the mountains, they often divided their huge armies and went after insurgents in smaller groups, and on foot. Despite their predominant size, in the last phase of the war, only several detachments led by able officers proved to be instrumental in fighting against the rebel remnants, while most other troops were idle or serving as guards in cities or towns or at strategic points. In fact, when the commanders were determined to defeat their opponents, the Qing forces were effective, wiping out the rebel bands one after another and in great speed.[60]

Nevertheless, the commanders were not always motivated to finish off those rebel guerrillas. From early in the suppression campaign, the suppressors exhibited a strong inclination not to swiftly put down this rebellion, but often exercised delay tactics. Sometimes, they put off exterminating a rebel band or its remnants for months. Consequently, the deadlines set up by the emperor, first Qianlong and then Jiaqing, were broken numerous times. Among other things, the suppressors saw this counter insurgency campaign a great opportunity to fish for promotions and material gains. Many had boldly misappropriated the war funds that had been more than generously supplied by the central government. One of the campaign's major expenditures was allegedly hiring large numbers of civilians as militiamen to aid regular troops in actions, but it easily played into the commanders' hands as excuses to misappropriate the fund. Although the emperors changed the campaign's commander-in-chief several times and rendered dismissals and penalties frequently, seldom had anything worked to expedite the operations. After the Qianlong emperor died at the beginning of 1799, the Jiaqing emperor tried to pull the derailed campaign back on track. He revamped the war leadership and initiated a war-front reform to tackle wartime corruption. Having encountered strenuous resistance from the front, however, Jiaqing soon abandoned his reform agenda and yielded to the commanders and the local officials in the war zone. Although only fewer than ten thousand rebels were still at large on the mountainous provincial borders by the end of 1801, it took the Qing forces several years to have them cleared. In this stage, an agonizingly slow and troubled process to demobilize tens of thousands of hired militiamen had proved to be the greatest obstacle on the way to have this campaign wrapped up, as it was difficult to suddenly relinquish those battle-tested mercenaries without causing troubles. Moreover, to delay the conclusion of hostilities would give both commanders and local officials more excuses for their colossal expenditures when facing audits in the post-war period.[61]

This ill-fated campaign against the rebel guerrillas was vastly consequential to the Qing dynasty. The abundant silver reserves in the dynasty's treasury had been nearly spent, and they were never restored in the rest of the dynasty.[62] Meanwhile, the state's failure to discipline its military at a time of crisis testifies to a loose link in Qing political structure. Having never recovered from the war politically and financially, the Qing dynasty entered the period of speedy decline as the nineteenth century progressed.

Conclusion

It was accidental that this sectarian rebellion evolved into a protracted guerrilla war in central China at the turn of the nineteenth century. When plotting uprisings, the sectarians did not envisage engaging themselves in guerilla warfare. In the spur of moment after their first clash with the authorities, most rebel groups chose to go to a mountaintop and fortify themselves there. A natural choice for insurgents who could not take a city or a walled town, this tactic had been used in the past, and would continue to be used after this rebellion. Only after the insurgents in Xiangyang, Hubei, failed to take the city of Xiangyang, did they reorient their strategy, realizing that they had to resort to mobility and speed to outmaneuver their enemy that was far more superior in size, weaponry, and resources. Thanks to their trek to Sichuan, the Xiangyang insurgents passed guerrilla tactics to their counterparts in Sichuan. Starting from late 1797, the conflict became essentially a guerrilla war.

Clearly, guerrilla warfare does not belong to only the modern era during which rebels and revolutionaries increasingly came to favour it. In fact, it had long been utilized in the premodern times. More important, guerrilla wars, sometimes, were not less significant than regular interstate wars in affecting the fortune of a state or an empire. In this case, the sectarians and their recruits – their total size had never surpassed 100,000, and fluctuated under 10,000 in the last years of the war – cost the Qing dynasty nearly ten years and its fiscal vitality to put them down. Although it was the suppressors of the rebel guerrillas who had been more responsible for the war's grave consequences, those 'ragtag' insurgents had secured a position in history for their remarkable resilience and quick mastering of guerrilla warfare, which helped tilt the fortune of the once formidable Qing dynasty, ending the last golden age in China's imperial history which spanned nearly the entire eighteenth century.

Notes

1. On the sectarian movement in late imperial China, see Overmyer, *Folk Buddhist Religion*; Seiwert, in collaboration with Ma Xisha, *Popular Religious Movement; and Heterodox Sects* and Ter Haar, *The White Lotus Teachings*.
2. For the sectarian uprisings in the Qing times, see Naquin, *Shantung Rebellion,* and *Millenarian Rebellion in China*.
3. The Hunyuan sect was founded in 1774 in southeastern Henan. 'Hunyun' literally means 'origin in chaos.' Although it was cracked down by the government more than once, it survived and expanded to several provinces in eastern and central China. For its origins and evolution, see Gaustad, "Religious Sectarianism and the State in Mid-Qing China."
4. His name is Liu Zhixie. From Anhui province in east China, Liu was a dealer of cotton, and became involved in sectarian activities in the 1770s. Liu played a key role in reviving the Hunyuan sect after a crackdown. Liu was arrested in 1800 and subsequently executed. On Liu's role in the rebellion, see Xu and Lin, "Liu Zhixie zai Chuan Chu Shaan nongmin daqiyi zhong zuoyong de kaocha" and Dai, *The White Lotus War*, 40–2, 220–1.
5. Only one uprising was near Wuchang, the provincial capital, but all other uprisings occurred in either northwestern or southwestern Hubei.
6. On the siege of Dangyang, see parts of Peng Yanqing, "Dangyang binanji"; Chen Deben's and others' confessions, *Qing zhongqi wusheng Bailianjiao qiyi ziliao* (BLJZL thereafter), Vol. 5, 28–31.

7. In a mountaintop stronghold in southwestern Hubei, an epidemic claimed thousands of lives. Zhang Zhengmo's confession, *BLJZL*, Vol. 5, 36.
8. This method was used in putting down the uprising in Laifeng, Hubei, and several uprisings in Xing'an prefecture, Shaanxi.
9. The other rebel band was from Changyang, southwest Hubei. They always found a new mountaintop to hole up after they failed to hold their old stronghold until their demise at their last stronghold in the spring of 1798. Dai, *The White Lotus War*, 54–8, 110–2.
10. During the thirteenth century, the Mongol forces had besieged the twin cities, Xiangyang and Fancheng, for three years, while the armies of China's Southern Song dynasty strenuously defended them. After the Xiangyang was lost to the Mongols in the spring of 1273 (Fancheng fell shortly before), the Mongols quickly swept the Yangzi River valleys and took the capital city of the Southern Song in 1276. In the late Ming times, rebel leader Li Zicheng captured Xiangyang late in 1642, and held it until the late summer, 1643. Another rebel chief, Zhang Xianzhong, took Xiangyang in the spring of 1641, but did not stay. Parsons, *Peasant Rebellions of the Late Ming Dynasty*, 79–80,106–13.
11. On the uprising in Xiangyang and the rebels' trek to Sichuan, see Dai, *The White Lotus War*, 61–8, 90–6.
12. On the Xiangyang rebels' return to Hubei and the transformation of the war in Sichuan, see Dai, *The White Lotus War*, 96–101.
13. On the Red Fourth Army's trek to northern Sichuan, see Zhang, *The Rise of the Chinese Communist Party, 1928–1938*, 295–317. On the Red Army's base area on the border region of Hubei, Henan, and Anhui, see Rowe, *Crimson Rain*, 286–919.
14. According to Zhang Guotao, his army went from 16,000 strong men to fewer than 9,000. But Xu Xiangqian, the commander-in-chief of the army, recalled that the number decreased from 19,000 when they left their base to 14,400 after they crossed the Han River. Xu, *Lishi de huigu*, 228, 232.
15. For the Red Army's experience in northern Sichuan, see Zhang, *The Rise of the Chinese Communist Party, 1928–1938*, 317–65. In the spring of 1935, the Red Fourth Army abandoned its base in northern Sichuan, moved to western Sichuan, and met with the Red First Army led by Mao Zedong and others in the middle of the Long March – the meeting actually inaugurated a fierce power struggle between Mao and Zhang Guotao, which marks the beginning of Zhang's demise in the communist revolution.
16. Zhuang, *Zhenkongjiaxiang – Qingdai minjian mimi zongjiaoshi yanjiu*, 202–07.
17. On Sichuan's abundance of outlaws, see Dai, *The Sichuan Frontier and Tibet*, 216–25.
18. The Bizika people had lived since ancient times in the border area astride Hubei, Hunan, and Sichuan. They are labelled as the Tujia nationality in today's China.
19. On the rebellions in the late Ming times, see Parsons, *Peasant Rebellions of the Late Ming Dynasty*.
20. For example, this happened when the Xiangyang insurgents reached Shaanxi province's southeastern corner, a sectarian enclave, in the spring of 1797. Dai, *The White Lotus War*, 93.
21. Xu, "Lun Chuan Chu Shaan nongmin qiyijun de liangge gaoshi," 192.
22. Some of the titles were the same as used in sectarian uprisings in the past, e.g., Wang Lun's uprising in Shandong in 1774. Naquin, *Shantung Rebellion*, 111–4.

23. This occurred mainly in the last stage of the conflict, 1801–1803. After some devastating defeats and losing most of their key leaders, some remaining leaders called surviving rebels to go to Hubei's western corner to regroup. Zhang Shihu's confession, *BLJZL*, Vol. 5, 154–5.

24. In 1799, rebels from a band led by Zhang Hanchao were found asleep in civilians' houses after drinking in Shaanxi. *Qinding jiaoping sansheng xiefei fanglüe* (XFFL thereafter), 116/10a-b.

25. The village was Shanggaochuan (Upper-High-Valley) in Xixiang county. XFFL, 130/3b, 130/8a-b.

26. According to a proclamation issued by a rebel band in 1797, those who did not bring back any booty would be beaten with wood plank forty times. Xu, "Lun Chuan Chu Shaan nongmin qiyijun de liangge gaoshi," 192.

27. For instance, Lebao, the campaign's commander-in-chief in 1798–1799, had complained about the rebels' guerrilla tactics more than once. XFFL, 59/30b-31a, 89/5a-10b. Starting from the early eighteenth century, local civil officials were usually responsibl for Qing armies' wartime logistical needs. They set up ad hoc logistical bureaus to manage war funds and hired civilians to transport foodstuffs and matériel. Military labor forces often outnumbered troops deployed and their compensation was one of the most expensive expenditures in wars. See Dai, "Military Finance in the High Qing Period," 306–12; and Dai, *The White Lotus War*, 384–90.

28. In the past scholarship, mountaintop fortification had been considered a key factor for the Qing dynasty to defeat the rebels. I dispute against this view in *The White Lotus War*, 327–39.

29. XFFL, 144/9a-b.

30. Zhou Shihong's confession, *BLJZL*, Vol. 5, 161.

31. The suppressors habitually slaughtered captives after battles, often in the hundreds. For rebel chiefs, they executed them in the method of 'death by a thousand cuts' (*lingchi*) by cutting off all the flesh of the executed piece by piece until only the skeleton left. Perhaps originating from Inner Asia, this practice might have started during the Song times (960–1279). During the Ming and Qing dynasties, plotting rebellion was subject to this penalty. See Brook, et al., *Death by a Thousand Cuts*.

32. For the Xiangyang band's experience after it left Sichuan, see Dai, *The White Lotus War*, 112–9.

33. The county was Xingshan, which did not have city walls. The insurgents soon abandoned it and went north.

34. XFFL, 48/19b-23b, 50/4a-9b.

35. Huilun was from an eminent Manchu aristocratic family. He inherited the highest noble rank, duke (*gong*) of the first grade, from his uncle, Mingšui, after the latter's death in the Myanmar war in 1768 in the capacity of commander-in-chief of the Qing forces. Recalling Mingšui's death sentimentally and lamenting another loss in the decorated family, Qianlong granted generous favors for Huilun's funeral and his family. XFFL, 50/18a-b.

36. While Yao died instantly, Wang was fatally injured, but confirmed her identity before killed by her captors. The Qing troops sliced Yao and Wang's corpses after cutting off their heads for display in the war zone and Beijing. *BLJZL*, Vol. 1, 323–7. XFFL, 67/9a-15b, 175/12b-13a. More about Wang Conger and her death, see McCaffrey, "Living through Rebellion," 202–12.

37. There were eight such 'marshals' in total. On Zhang's background, see Xu, "Lun Chuan Chu Shaan nongmin qiyijun de liangge gaoshi," 223–4, 229–34.
38. XFFL, 68/22a-23b. On Zhang Hanchao's guerrilla war in 1798–1799, see Dai, *The White Lotus War*, 120–2, 152–6, 158–9.
39. XFFL, 74/26b-29a, 75/8a-11a, 75/13a-15a.
40. XFFL, 83/12a.
41. The governor was Yongbao, a Manchu aristocrat, who had been the first commander-in-chief in the White Lotus War, but was dismissed and prosecuted at the end of 1796 owing to his poor performance.
42. XFFL, 127/4a-10b, 128/3b-8b, 131/13b-14a; Li Chao's confession, *BLJZL*, Vol. 5, 79–80.
43. Initially, the band was called Gao Family Regime (Gaojiaying), because it was led by Gao Junde. After Gao was captured in late 1799, the band was co-led by Gao Tiansheng and Ma Chaoli.
44. On the band's campaigns in Gansu, see Dai, *The White Lotus War*, 209–13.
45. XFFL, 174/6a-11a.
46. XFFL, 192/17b-20b, 204/10b.
47. After this battle, the Gao-Ma band continued moving eastward along the Han River. Late in 1800, it crossed the river to the north and entered the Qinling Mountains. Due to food shortages, it went to the south of the Han River early in 1801, and was annihilated in Shaanxi's southwestern corner soon after. See Dai, *The White Lotus War*, 255–6.
48. The most brilliant battle by Ran occurred on the east bank of the Jialing River in Cangxi county, Sichuan. During the fight, both sides focused on taking control of two heights. Ran first sent a dispatch to attack the Qing logistical corps, which distracted his enemy. Then Ran captured one of the heights held by the Qing troops. From the height's top, the insurgents punched and routed their opponents. The white-hot battle lasted all night. When Ran led away his men in the early dawn, he had left dead more than 300 Qing soldiers and militiamen as well as thirteen officers. XFFL, 137/16b-20a.
49. That day was the Lantern Festival, the full-moon day of the first lunar month, which marked the end of celebration of the Chinese New Year. Rebel scouts swam to the other shore, killed ferrymen, and seized boats. A few militiamen on the west bank attending a lantern party instantly scattered. For three days, thousands of insurgents and the coerced people crossed the river, by boat or on horseback. Ran Tianyuan's confession, *BLJZL*, Vol. 5, 90; Wang Ying's confession, ibid., 91. Zhang Zicong's confession, ibid., 110. XFFL, 152/16a-17a, 177/12a-14a.
50. On the rebel coalition's expedition to western Sichuan in 1800, which lasted for nearly three months and was the only time the rebels went west of the Jialing River in Sichuan, see Dai, *The White Lotus War*, 191–8, 201–5.
51. In Shaanxi, Ran Xuesheng was pressed to the south of the Han River, and move eastward. After having lost all his horses and mules and had only hundreds of people left in his band, Ran went into hiding in the Daba Mountains. He was captured in the autumn of 1801. Dai, *The White Lotus War*, 257–8, 261; Ran Xuesheng's confession, *BLJZL*, Vol. 5, 129–31.
52. On Gou Wenming's last fight in the Qinling Mountains, see Dai, *The White Lotus War*, 266–70, 273; Gou Chaojiu's confession, *BLJZL*, Vol. 5, 163–4.
53. In the last phase of the war, conflicts occurred in two areas: the border area astride Shaanxi, Hubei, and Sichuan, and north of the Three Gorges in Hubei.

Both areas were mountainous and sparsely populated. On the war's last phase, see Dai, *The White Lotus War*, Chapter 5, 'Finale, 1801–1805.'

54. On the Qianlong emperor's 'retirement,' see Elliott, *Emperor Qianlong*, 160–1. On his reign and its significance in Qing history, see Woodside, "The Ch'ien-lung reign."

55. For instance, Qianlong said so when he heard of the news of two uprisings were started in northeastern Sichuan late in 1796. XFFL, 20/19a.

56. XFFL, 18/22a-24b. After this, the Qing throne sent bannermen from Manchuria to the war four times between 1797 and 1800. Altogether, 9,500 bannermen were deployed from outside of the war zone. But most of them stayed in war for only a couple of years.

57. XFFL, 51/13a-b.

58. XFFL, 52/20a-23a, 52/29a-30b; QSLJQ, 23/3a-6a.

59. XFFL, 41/30b.

60. For example, in 1800, Delengtai quashed the rebel coalition in west Sichuan in only weeks, expelling its remainder back to the east bank of the Jialing River. Then in 1801, with high pressure from the emperor, commanders eliminated several main rebel bands and killed or caught their leaders in a relatively fast pace.

61. I have detailed the Qing state's failure in controlling the unfolding and fallout of this campaign in *The White Lotus War*, especially Chapters 2–5.

62. When the rebellion started, there were nearly 70 million taels of silver in the central government's treasury in Beijing. In 1797, the reserves were down to 27.9 million taels. By 1801, the sixth year of the war, only 16.9 million taels left. On the financial aspect of the war, see Dai, *The White Lotus War*, Chapter 7, 'Cost,' and 442–9.

Disclosure statement

No potential conflict of interest was reported by the author.

Bibliography

Brook, Timothy, Jérôme Bourgon, and Gregory Blue. *Death by a Thousand Cuts.* Cambridge, MA: Harvard University Press, 2008.

Da Qing lichao shilu [Veritable records of successive reigns of the Qing dynasty]. Multiple volumes. Reprint, Tokyo: Okura Shuppan Kabushiki Kaisha, 1937–38.

Dai, Yingcong. "Military Finance in the High Qing Period: An Overview." In *Military Culture in Imperial China*, edited by Nicola Di Cosmo, 296–316, 380–82. Cambridge, MA: Harvard University Press, 2009.

Dai, Yingcong. *The Sichuan Frontier and Tibet: Imperial Strategy in the Early Qing.* Seattle: University of Washington Press, 2009.

Dai, Yingcong. *The White Lotus War: Rebellion and Suppression in Late Imperial China.* Seattle: University of Washington Press, 2019.

Elliott, Mark C. *Emperor Qianlong: Son of Heaven, Man of the World.* New York: Longman, 2009.

Gaustad, Blaine Campbell "Religious Sectarianism and the State in Mid-Qing China: Background to the White Lotus Uprising of 1796–1804." PhD diss., University of California at Berkeley, 1994. doi:10.3168/jds.S0022-0302(94)77044-2

Gaustad, Blaine Campbell. "Prophets and Pretenders: Inter-Sect Competition in Qianlong China." *Late Imperial China* 21, no. 1, June (2000): 1–40. doi:10.1353/late.2000.0004.

Guan, Wenfa. *Jiaqingdi* [A biography of the Jiaqing Emperor]. Changchun: Jilin Wenshi Chubanshe, 1993. *Gongzhongdang* (Palace memorials). Housed in the National Palace Museum, Taipei.

Jiaqingchao shangyudang [Collection of the Jiaqing emperor's edicts]. edited by Number One Historical Archives. 25 Vols. Guilin: Guangxi Normal University Press, 2008.

McCaffrey, Cecily Miriam. "Living through Rebellion: A Local History of the White Lotus Uprising in Hubei, China." PhD diss., University of California at San Diego, 2003.

McMahon, Daniel. *Rethinking the Decline of China's Qing Dynasty: Imperial Activism and Borderland Management at the Turn of the Nineteenth Century.* New York: Routledge, 2015.

Naquin, Susan. *Millenarian Rebellion in China: The Eight Trigrams Uprising of 1813.* New Haven, CT: Yale University Press, 1976.

Naquin, Susan. *Shantung Rebellion: The Wang Lun Uprising of 1774.* New Haven, CT: Yale University Press, 1981.

Naquin, Susan. "The Transmission of White Lotus Sectarianism in Late Imperial China." In *Popular Culture in Late Imperial China*, edited by David Johnson, Andrew J. Nathan, and Evelyn S. Rawski, 255–291. Berkeley: University of California Press, 1985.

Overmyer, Daniel L. *Folk Buddhist Religion: Dissenting Sects in Late Traditional China.* Cambridge, MA: Harvard University Press, 1976.

Parsons, James Bunyan. *Peasant Rebellions of the Late Ming Dynasty.* Tucson: University of Arizona Press, 1970.

Perdue, Peter C. *China Marches West: The Qing Conquest of Central Eurasia, 1600–1800.* Cambridge, MA: Harvard University Press, 2005.

Peterson, Willard J., ed. *The Cambridge History of China, Vol. 9, Part 1: The Ch'ing Empire to 1800.* London: Cambridge University Press, 2002.

Qinding jiaoping sansheng xiefei fanglüe [Imperially sanctioned chronicle of the campaign to suppress the heretical bandits in three provinces]. 69 Vols. Reprint, Taipei: Chengwen Chubanshe, 1970.

Qing zhongqi wusheng Bailianjiao qiyi ziliao [Material regarding the White Lotus uprisings in five provinces during the mid-Qing period]. edited by History Institute, Zhongguo Shehui Kexueyuan. 5 Vols. Nanjing: Jiangsu Renmin Chubanshe, 1981.

Rowe, William T. *Crimson Rain: Seven Centuries of Violence in a Chinese County.* Stanford, CA: Stanford University Press, 2007.

Seiwert, Hubert. in collaboration with Ma Xisha. *Popular Religious Movement and Heterodox Sects in Chinese History*. Leiden: Brill, 2003.

Suzuki, Chūsei. *Shinchō chūkishi kenkyū* [A study of the mid-Qing dynasty]. Toyohashi: Aichi Daigaku Kokusai Mondai Kenkyūjo, 1952.

Ter Haar, Barend J. *The White Lotus Teachings in Chinese Religious History*. Leiden: Brill, 1992.

Twitchett, Denis, and John Fairbank. *The Cambridge History of China. Vol. 10, Late Ch'ing, 1800–1911, Part I*. Cambridge: Cambridge University Press, 1978.

Waley-Cohen, Joanna. *The Culture of War in China: Empire and the Military under the Qing Dynasty*. London and New York: I. B. Tauris, 2006.

Wang, Wensheng. *White Lotus Rebels and South China Pirates: Crisis and Reform in the Qing Empire*. Cambridge, MA: Harvard University Press, 2014.

Woodside, Alexander, "The Ch'ien-lung Reigh." In *Cambridge History of China, Vol. 9, Part One: The Ch'ing Empire to 1800*, edited by Willard J. Peterson, 230–309. London: Cambridge University Press, 2002.

Xu, Xiangqian. *Lishi de huigu* [A reflection on my life]. Beijing: Jiefangjun Chubanshe, 1984.

Xu, Zengzhong. "Shilun pingjia Wang Conger de jige wenti." [On several questions with regard to the evaluation of Wang Conger]. *Qingshi luncong* 3 (1982): 164–183.

Xu, Zengzhong. "Lun Chuan Chu Shaan nongmin qiyijun de liangge gaoshi."[On two proclamations issued by the peasant uprising army of Sichuan, Hubei, and Shaanxi]. *Zhongguo nongmin zhanzheng shi luncong* 4 (1983): 191–238.

Xu, Zengzhong, and Yi Lin. "Liu Zhixie zai Chuan Chu Shaan nongmin daqiyi zhong zuoyong de kaocha." [An examination of Liu Zhixie's role in the great peasant uprising of Sichuan, Hubei, and Shaanxi]. *Qingshi luncong* 2 (1980): 175–199.

Zhang, Guotao (Chang Kuo-t'ao). *The Rise of the Chinese Communist Party, 1928-1938: The Autobiography of Chang Kuo-t'ao*, 2 Vols. Lawrence: University Press of Kansas, 1971–1972.

Zhuang, Jifa. "Zhanzheng yu dili: Yi Qingchao Jiaqing chunian Chuan Shaan Chu Bailianjiao zhiyi weili." [War and geography: the White Lotus War in Sichuan, Shaanxi, and Hubei in the early Jiaqing period of the Qing]. In *Qingshi lunji* Vol. 10 edited by Jifa Zhuang, 109–140. Taipei: Wenshizhe Chubanshe, 2002.

Zhuang, Jifa. *Zhenkongjiaxiang—Qingdai minjian mimi zongjiaoshi yanjiu* ["True emptiness": a study on popular and secret religions in Qing times]. Taipei: Wenshizhe Chubanshe, 2002.

Regular and irregular forces in conflict: nineteenth century insurgencies in South America

Alejandro M. Rabinovich and Natalia Sobrevilla Perea ⓘ

ABSTRACT

In the decades following independence from Spain, 'civil wars' ravaged the newly established polities in South America. Former vice-regal capitals inherited a larger portion of the colonial administration and had larger economic resources and a hegemonic project they were able to have permanent and professional armed forces, capable of leading the offensive and giving battle following the European rules of military art. The central hypothesis of this work is that there is a necessary relationship between the shape of these asymmetrical conflicts, their outcome and the political territorial configuration of each country in post-revolutionary Spanish America. When permanent armies took over from local militias, the capital kept the integrity of its territories and there was a tendency towards political centralization. When this did not happen and the militias managed to find a way to defeat their centralizing enemies, the local powers had an opportunity to renegotiate their participation in the political body, and sought to maintain their independence, which was manifest in federal agreements, otherwise a process of territorial fragmentation began. More than a difference between regular and irregular forces there was one between *intermittent*, and *permanent* mobilization.

In 1808, after being invaded by his former allies, the Spanish King abdicated to Napoleon's brother Joseph. One of the main consequences was a monarchical crisis that set in motion what has been described as the 'retro-version of sovereignty'. It meant that in the monarch's absence, the people were sovereign, because as it was they who had given the King the right to govern, so if he was not there, that right returned to the people. In the two Viceroyalties established the longest, Mexico and Peru, there was little change, but in many other areas elites demanded sovereignty return to the people. Based on this principle Juntas were created to govern in name of the absent King Ferdinand VII, both in the peninsula and some places in the Americas. But this was not the only consequence, as smaller cities, and indeed even small towns and villages attempted to reclaim the right to be autonomous.[1] In the following decades, 'civil wars' ravaged the newly established polities in the continent. Evidently, circumstances varied and local particularities gave each conflict its own characteristics, but in all cases we find towns (later provinces) confronting each other hoping to determine who had legitimate control of the territory.

For the most part, the former vice-regal capitals inherited a larger portion of the colonial administration, and this was especially so in the oldest viceroyalties, but was also the case in the newer ones created in the eighteenth century in New Granada and the Río de la Plata. Capitals, having larger economic resources and a hegemonic project to control their hinterland, were able to have permanent and professional armed forces, capable of leading the offensive and giving battle following the European rules of military art. In contrast, attempts to set up a bureaucracy in smaller towns and provinces were often thwarted by a lack of fiscal resources. Consequently these smaller places were forced to fight with more rudimentary militias, leading in the majority of cases to what we would now call 'asymmetrical military conflicts', where those fighting chose to use alternative tactics to the ones usually employed at the time.[2]

The central hypothesis of this work is that there is a necessary relationship between the shape of these asymmetrical conflicts, their outcome and the political territorial configuration of each country in post-revolutionary Spanish America. When permanent armies took over from local militias, the capital kept the integrity of its territories and there was a tendency towards political centralization. When this did not happen and the militias managed to find a way to defeat their centralizing enemies, the local powers had an opportunity to renegotiate their participation in the political body, and sought to maintain their independence, which was manifest in federal agreements, otherwise a process of territorial fragmentation began. All territories had to navigate this tension between provincial autonomy and the centralizing ambitions of capitals, and some proved more successful than others.

The case of the former viceroyalty of the Rio de la Plata, where the current republics of Argentina, Bolivia, Paraguay and Uruguay are located, is extreme, because in the nearly half a century between the start of the revolution for independence in 1810 and the period known, as the 'national reorganization' at mid-century, it experienced permanent unstoppable warfare.[3] The terms war 'of independence', 'international' or 'against the Indian' are just different names given to a complete state of unending war that took many shapes and was made up of a series of conflicts, military campaigns and political confrontations that are difficult to unravel. From the strictly military point of view, the rupture of the colonial order engendered an impressive profusion of fighting forces. This included the armies created in the image of the contemporaneous European ones, financed from Buenos Aires, but also militias of all kinds, including the independent guerrillas of the countryside, more organized corps, volunteer companies and the Indian and *mestizo* (mixed race) combat units.[4] These constantly mobilized forces made war in their own way and in their own time and particularly with their own specific interests. They evolved due to their circumstances: the line squadrons could become militias, they could transform into revolutionary guerrillas or vice-versa. The destabilizing effect of all this diverse activity led to the severe unsettling of the economic structure of the country and generated unprecedented levels of everyday violence.[5]

In contrast, in the former viceroyalty of Peru the process of independence had two distinct moments. During the first, between 1809 and 1819, it became the bulwark of the loyalist forces in South America, fighting against all those who sought to organize Juntas and creating a strong disciplined army from the existing militias in the southern Andes.[6] The second period that lasted from 1820 to 1826 comprised war and invasion from powerful liberating armies that came after fighting long campaigns in Chile and Colombia, during which they had become professionalised. This period witnessed the entrenchment of the loyalist army in the south, which reached up to 20,000 men, as well as the development of guerrilla warfare in the Andean region, waged mostly by indigenous people, sometimes to aid independence and on other occasions to support the loyalists.[7] Upon final capitulation at the end of 1824, some of these continued to fight in the name of the King as guerrilla forces for several years, in specific pockets.[8] But this was an anomaly as the newly created Peruvian Republic was able, for the most part, to keep control of the recently created independent army, including those who surrendered. The civil wars that ensued tended to be between factions of the army and some militias and did not follow the same pattern of total war found in the Rio de la Plata. Guerrilla fighting had its place, but it did not play a central one.

In this chapter we compare the Rio de la Plata and Peru in an attempt to situate the insurgency experience in the larger region during the nineteenth century. We consider that to be able to understand irregular war it is imperative to gain some understanding of the general conditions of warfare and the consequences of the colonial experience, as well as the way in which the wars for independence tended to create the environment for the conflicts that followed during the nineteenth century. To do so we begin with an examination of the colonial defence system, the changes brought to the region by the wars of independence and how these had long-lasting effects. A second section reviews the types of mobilization and cohesion, in particular in the Rio de la Plata where war became entrenched. A final section looks at how irregular fighting happened in the Rio de la Plata where it became more entrenched.

From colonial militias to armies of independence

Militias organized by local populations had traditionally defended Spanish American colonies, with only a small number of soldiers sent from the peninsula to guard major ports and presidios.[9] After the disastrous loss of Havana during the Seven Year war, the Bourbons set out to thoroughly reform the system by developing what they described as disciplined militias, which meant that these ones would undergo a much more rigorous training, and would be commanded by professional soldiers.[10] Rolled out first in Cuba and then in Peru, Mexico and beyond the plan achieved mixed results. In many cases those who joined did not really train or fight, as the main incentive to become members were privileges, such as wearing uniforms, tax exemptions, access to their own corporate courts and a safe place to gain official favour for promotion. One of the places where the system was most severely tested was in the Andes where the Tupac Amaru's revolt of 1780 led to the swift professionalization of the militia. This was because a swift response to the threat the insurgents posed was needed.[11] The provinces affected by the uprising developed a highly trained fighting force, using the militias, while local elites invested in sending their offspring to obtain further training in the peninsula.[12]

In contrast the military forces in the Rio de la Plata were much less developed.[13] The cost of the revolutionary war in Europe and the weakness of the Spanish fleet, made it increasingly difficult to bring reinforcements from the peninsula. In 1802 only half of the positions were filled in the two regiments of the Rio de la Plata, and just about 2000 troops guarded the viceroyalty.[14] The organization of disciplined militias, which was well advanced in Peru, was nothing but a project. In June 1806 when a British expedition arrived in Buenos Aires, the authorities capitulated and the King's army was disarmed. The local population reacted en-masse chasing out the invaders and in 1807 they organized volunteer militias to stop a second invasion. The new army did not resemble other colonial troops: 8000-men strong, it trained every day, voted for their officers and after the crisis of 1808, answered to American notables.[15]

Militarization in the southern Andes grew in 1809 when two Juntas were set up in the cities of Chuquisaca and La Paz in what today is Bolivia. Appealing to the idea of the return of sovereignty to the locality, both these cities attempted to set up a government that would not depend on either Lima or Buenos Aires.[16] An immediate military response was sent from the southern provinces of Peru, led by a member of the local elite, who had trained for many years in the military in the peninsula, and who had in fact been sent by the Junta of Seville to share the news of the political situation in his local region. José Manuel de Goyeneche worked with the militias in his region, and with the support of men like him from Cuzco and Arequipa trained them into becoming more like a regular army. He was also supported by Indian notables, some of whom were veterans of the fight against Tupac Amaru.[17] A much smaller force came from Buenos Aires, and both Juntas were quickly dismantled, and one of the consequences was the creation of a strong loyalist army in southern Peru, trained by professional soldiers.

In 1810 when the Buenos Aires elite decided to depose the viceroy and set up their own Junta, unlike most other viceroyalties they had their own military force ready to fight. Positioned at the very south of the continent, once the capital city had control over the region it sought to take over the rest of the viceroyalty, with the aim to 'export' the revolution to its neighbours, principally to the Upper provinces that had recently attempted to create Juntas. There they met the newly organized army of Upper Peru, which was made up of local militias under the command of members of the local elite. The spirit of the army sent from Buenos Aires was of a purely offensive revolutionary war.[18] The new patriot armies aimed to attack, so they were forced to operate thousands of kilometres from the capital for months and years without any lines of communication facing considerable well-trained loyalist forces, that after 1814 were reinforced from Spain. The revolutionaries were convinced only modern regular armies could face this enemy. The volunteer militia regiments were therefore transformed into regular units using Spanish ordinances and the most recent French regulations, which they had translated. In the wake of the Napoleonic Wars, dozens of unemployed European officers were hired to command.

Organizing and training these regular forces was so taxing that the government in the Rio de la Plata found itself close to financial collapse. There was also much worry because popular sectors resisted forced recruitment.[19] The way the permanent armies were organized took a toll, straining the inhabitants in the theatres of operations. The fronts were situated in agrarian societies where there was little economic surplus and there was not much in terms of population.[20] In that context, the draft of many thousands of men, who ipso facto became economically unproductive, had a profoundly destabilizing effect. As these armies wrenched men from their homes, they requisitioned horses and committed so many excesses, not dissimilar to the ones that characterized the enemy.[21] These forces became in the eyes of the local population an instrument and the expression of a new form of tyranny, that of the central government of Buenos Aires.[22]

This regular militarization promoted by the Junta was overwhelmed everywhere in the first days of conflict. In all fronts the temptation to mobilize local rural population and Indians as auxiliaries was great.[23] But these populations only rose spontaneously to support certain causes. So while the armies remained in the front, the militias, the country folk, and the mobilized Indians played a relatively secondary role. But after the first defeats against the highly trained loyalists from the much more densely inhabited Andean regions and regular forces retreated, territorial defence returned exclusively to the local population, who were forced to maintain a desperate fight to escape the reprisals of the new occupier. In Upper Peru, particularly at Salta and at the Banda Oriental, where this kind of dynamic was mostly repeated, the local forces gained experience. Through a relentless war of resources they would resort to chasing the enemy from their lands, and this progressively instilled in the population the idea that regular armies were not indispensable to save the new patria.

Just as the spirit of the governmental armies was offensive, local armies were defensive.[24] But the greatest incentive that led to the raising of the first rural forces was to prevent attacks by the invaders on producers and their villages. The popular volunteer militias initially took up arms to guard their houses, their livestock and their possessions. They operated in a terrain they knew perfectly and were supported by a social fabric they were defending. They fiercely attacked armies that approached their territories, but once they had moved away from their province, the defenders' ardour quickly faded. On most occasions, after some days of pursuit, each returned home.[25] In contrast the troops raised in the southern Peruvian viceroyalty rarely had to move too far away from their homes, and their commanders understood that to maintain the loyalty of their Indian recruits they had to treat them more like a militia than a regular army, so they allowed them to travel with their women, and to abandon the camps when it was time to carry out agricultural tasks.[26] The men who joined were promised a piece of land as a prize for participating in the army.[27]

In spite of this success, in the areas where the Andes meet the Amazon basin in present day Bolivia, small guerrillas controlled valleys that had never been very integrated into the colonial state. These so-called 'little republics' or *republiquetas* remained undefeated by the loyalist forces and they carried out guerrilla attacks, supported by the local population and aided by geography. During the decade of war that followed the 1810 revolution in Buenos Aires, each province of the former viceroyalty developed a militia force with strong links to the land.[28] In 1814 in the Banda Oriental, in 1815 in Salta and a little later elsewhere in the littoral, these local forces rose against permanent armies sent from the capital, just as they had done with the loyalists. They were all successful. This new kind of mobilization became a show of force and a revolt against the government. Each local power would enter into war and effectively defend their locality, resisting regular armies, and refusing obedience to the government. In 1820, central administration dissolved in the Rio de la Plata. Each old or new province continued independently, participating in campaigns against their neighbours and joining and fighting in coalitions with rival leagues and federations.

A large portion of the professional army had been exported in 1817 over the Andes to aid the independence of Chile. Led by José de San Martín, very few of these men who had been recruited in Buenos Aires and the border province of Cuyo ever returned. After the campaigns in Chile they travelled on to Peru in 1820, when San Martín decided not to intervene in the internal problems in the Rio de la Plata and instead became the head of 'The Army of the Andes' that was no longer anchored in a particular state and whose legitimacy came from its own officers.[29] In Peru, the war was prolonged and difficult as the loyalist forces clung to the Andean region with a highly trained army that had been mobilized for more than a decade, and had the advantage of having leaders who had ample experience of guerrilla wars gained in the Peninsular War.[30] The militia in the north of the country had no experience fighting, and the guerrillas who rose to support independence did not travel very far away from their localities, following the logic of self-defence. Independence in Peru was finally achieved in 1824 when the highly trained army of Simón Bolívar entered the fray from Colombia.

In the Rio de la Plata the defensive forces that came to being between 1810 and 1820 were at the basis of an extraordinary territorial and political fragmentation. The revolutions and the battles in the territory of the former viceroyalty left the country in pieces: what had previously been one unified sovereignty became more than a dozen local governments – republics, provinces, and diverse coalitions – retaining their right to war and their own military force.[31] Below this administrative structure, the fragmentation was even more extreme. The commander of a militia squadron, a judge of the peace, a property owner, a well-known brigand if they had the capacity to mobilize some dozens of men, could become military or political leaders. Power, authority, force were radically atomized. A new question came up, however: how had the local militias defeated regular armies that were better trained and had much better armament? How had defence helped at the time of attack?[32]

In contrast to this experience, in the newly created republics of Bolivia and Peru it was possible to maintain cohesion, in spite of the large regional and provincial differences. In Peru conflict between the north and south of Peru was long-lasting, leading to several civil wars, whilst in Bolivia, even though there was severe competition between the provinces over the location of the capital and the role of the central and provincial authorities, it did not descend into the fractious fighting seen in the Rio de la Plata where a centralized state proved elusive. This was because war was waged with a combination of armies and local militias that were highly professionalised like in southern Peru, or that in fact barely were involved in the fighting like in the north. Peru and Bolivia attempted a brief union in a Confederation between 1836 and 1839, but this failed due to opposition from Chile, and from the Argentine Confederation, that was formed from a loose union of provinces in the Rio de la Plata.[33] In the aftermath what characterized fighting in both Peru and Bolivia were army factions, who claimed to be sovereign. In both places military men took control of the state apparatus, but were keen to always have constitutional and parliamentarian systems to legitimize their rule. Guerrillas and insurgencies tended to be very rare and army factions fighting for power dominated the confrontation.

Types of mobilization and cohesion

In the Rio de la Plata warfare became entrenched and there was a variety of combat units. To understand how they functioned and illuminate how conflict worked in the period we will carry out a general, yet simplified, review of the types of combat units. We propose that the study of the forces using a dichotomy between regular-irregular forces is often misleading.[34] Instead we find it much more useful in this context, to privilege a distinction deeply anchored in the local practices of war: that of the *intermittent*, as opposed to *permanent* mobilization. This way we will be able to measure the level of organization of the total universe of forces using a functional criterion, devoid of a value judgment that can be easily verified case by case.

During permanent mobilization, recruits were incorporated in the military forces exclusively for a predetermined number of years.[35] Even in times of peace, they would remain at the barracks, at the camps or at the garrison; they lived a wholly militarized life. The result from this type of organization was the professional soldier. Because permanent mobilization is the kind most used today, it was also the best known. But let us stop for a moment on the question of how that system could work in South America at the start of the nineteenth century. As a social technique, permanent mobilization presented specific inconveniences. Because the recruits were dedicated solely to war, this meant that the rest of society had to be capable of producing an economic surplus to cover their needs. Also, because the men incorporated were not considered the best kind in their original social milieu, it was necessary to attract them or to have a severe form of coercion. Permanent mobilization tended to create a class of men in arms that could easily become dangerous to the other social groups.[36]

If the feeble revolutionary government of the Rio de la Plata decided to put up with the cost of this system it was because the purely military advantages of permanent mobilization were perceived as being decisive. They could be surmised in one word: cohesion. Due to an enormous disciplinary effort, permanent mobilization was reputed to create combat units with a great degree of cohesion that elevated their possibilities of victory, especially when facing a less solidly put together force. They favoured turning individual combatants into tactical units reliant on something larger than themselves.[37] The target of military discipline was the body of the soldier.[38] The aim was to separate the new recruit from his previous way of life, his external links, and his social background. The men selected were enclosed in confined spaces like barracks and military camps. They were no longer subject to regular justice and they were excluded from economic activities. Their time was strictly controlled and they no longer operated with the usual rhythm of the village or the countryside. The objective was to produce the symbolic social death of the recruit and his rebirth in a new order, that of a soldier, who had esprit de corps, had been rechristened with a nom de guerre and had exchanged his former family for his brothers in arms.[39]

Then the disciplinary system began. Corporal training, where the ins and outs of the movements were regulated, became simple automatic responses that could be reproduced in unison with the rest of the unit. By indefatigable repetition of exercises and marches, it became possible to remove all corporal spontaneity, all initiative, to make them lose their instinctive reflex for fear. The men learnt to battle in a rigorous order, formed in many lines, from the tallest to the shortest, keeping exact distances between the corps. Regular armies employed drill in battle, and by spending time at the barracks permanent units gained base cohesion, so that the fragile corps, in the flesh of the individual recruit, would create an invincible giant in the company, the battalion and the regiment.

Intermittent mobilization was completely different. It did not seek to create war specialists, instead the idea was to make war an integral part of each man's life. Instead the aim was to turn it into a normal activity that was adjusted to the citizen's other roles, productive, familial, and political.[40] Instead of snatching fathers from their homes or those who produced what was needed by communities for subsistence, this type of militarization trained and mobilized men in their own social context. This system's underlying logic began with two premises that differentiated it from permanent mobilization. 1- it had to be a certain type of war, which took place episodically, with each episode not lasting more than some days, all in all lasting just some weeks. 2- war activities were compatible or even interchangeable with peaceful civilian activities.

This dynamic did not produce a soldier or military man, instead, what emerged were warriors and militiamen. Or better still, as they were fond of saying in the Rio de la Plata: 'warrior-citizens'.[41] Therefore, just as the regular regiments were the typical case of permanent mobilization, the militias were the typical case of intermittent mobilization. In all the provinces of the Rio de la Plata, local militias were made up of all the adult men (between 16–45) with a stable and legal address. From the last census of the village or department, the men were divided into units of active and passive militias, with the obligation of serving for a number of years. In contrast to the regular army, militia members were not required to abandon their habitual way of life. During peacetime, service did not extend beyond training (one or two days a month, generally on Sundays or feasts days) and some active service that tended not to last too long: guards, garrisons, patrols, convoying of prisoners and herds. In times of war, militias could be mobilized to face the enemy, respond to a call or go on campaign. But in some cases their services could not legally be for more than six months a year, and they generally adopted a rotation system that did not impose on any individual more than two months of service a year.[42]

These militias only had basic training, and learning some simplified tactic for cavalry and infantry. They were not expected to keep rigorous formation in battle, or to use their armament in specific ways. In spite of this, they were considered to be tactically comparable to regular forces, although they tended to be more poorly armed, dressed and trained. The specificity of their forces was not found in the tactics or ordinances used, but in how they imbricated war and the social milieu. Instead of taking men away from their environment, militia organizations tried to operate as close as possible to the social structure of the countryside and the village. The heads of the large properties in the Pampas the *estancias*, the rural producers of good standing, the proprietors of shops were food and drink were sold were placed at the head of the companies, in charge of commanding their peons, their protégés, their clients.[43]

More importantly the vertical social rapport was reinforced by a very closed horizontal space: in a same company, fathers would meet their sons, their neighbours, their *compadres* and colleagues. The men who worked the land and provided sustenance at the end of the month were the same ones who were assembled to go into battle. This did not mean there was a perfect continuity between normal economic life and the one within the militia, because the latter retained its own logic that could very quickly enter into conflict with day-to-day life. But in a warring society, such as the one of post-revolutionary Rio de la Plata, serving in the military and supporting the local economy became the two main facets of society. Both were indispensable parts of a way of life adapted to survive in times of war and revolution.

Where economic, familial and proximity links predominated more cohesive base groups and units emerged. In the Rio de la Plata they were called *montones*, which were gatherings of haphazardly assembled men that did not necessarily respond to the official division of men into squadrons or territorial companies. They were the base of all the spontaneous resistance movements and were at the heart of militia mobilization. The central figure was what their contemporaries called the *caudillejo*, the small cacique, or minor captain. Their influence and prestige were narrowly circumscribed to the local area, where they could possibly control dozens of men.[44] They could come from any ethnic background, or social condition; the only indispensable precondition was that he could resemble a commander of a base group. These *montones* would rarely be more than 60 or 80 men, and the optimal one was between 30 and 50. The type of cohesion that kept these units together was not traditional military discipline.[45] Although corporal punishment did indeed exist, and could be very cruel and perfunctory, as war advanced experienced chiefs sought to provide their militiamen with incremental liberties. Even during campaigns, men would leave to return home to visit or work. A good militia captain, if he was well entrenched in the region, did not whip up a scandal. He knew that he had no interest in treating these men as deserters. They would return some days later, and at least the commander could always count on them to join up again once they were called to take up arms.

These troops were rarely paid and lived off what they could extract from the countryside It was therefore only natural that the men, and women who looked after the camps would turn to pillage. Militiamen knew their rights and privileges very well, and these depended much more on usage than on the law. During certain campaigns they did not even have to have a salary, but the commander who engaged them would provide free food from the local *estancias*, or share part of the bounty taken from the enemy. It was also the case that a commander that did not demobilize in time for crucial rural activities like harvest or planting risked a revolt that could be the end of his forces.[46] These companies were made up of a complex web of favours, debts, coercion, persuasion, protection and solidarity. Their structure largely surpassed the classic military one as it paid close attention to the social, and for this reason cohesion worked in a different, more efficient manner. Men were regularly mobilized legally in this way, and when they responded to the call of the legitimate government – this military organization of society became a militia. When men were mobilized illegally – meaning without an official governmental sanction, even when the commander called the militia units – they became known as a *montonera*[47]: a broad autonomous rural guerrilla, which sometimes was revolutionary.

These intermittent groups of militiamen, or *montoneros*, essentially had the same structure: a small company that was capable of independent action, adapted to live on the terrain that worked with ease with a dynamic of gathering/dispersion. Their ability to recruit men and keep them together was strongly determined by seasonal factors: the climate, the calendar of agricultural work, and the availability of pasture.[48] These companies trained in the keeping of armament, but not in a systematic manner. Often their equipment was what each of the militiamen had been able to obtain from the enemy, and general provision resulted in a lot of variety. Even if they were not well adapted to battle in a coordinated manner, which came from long hours of training, they still developed ways of coordinating their fire.

These weaknesses, however, were compensated by the intrinsic ability brought to the militias and *montoneras* by the local men and which were derived from the rural population's way of life. Instead of respecting drill and discipline, like solders in the regular army, they spent their lives on horseback. The country folk of the Rio de la Plata had an incomparable equestrian ability typical of the people who come from prairie areas.[49] Their dexterity in the handling of the long knife, the lasso and *boleadoras*, which was a local adapted lasso with two balls they threw at enemies or animals, was unsurpassed. In their long expeditions they practiced hunting and rodeo, they knew how to read animal traces, withstand the most difficult weather and take advantage of the natural resources. They could not be good soldiers in a regular army, but they instinctively knew manoeuvres such as overflow, encirclement and pursuit, which they used in their quotidian work with the battle troops.

In the context of post-revolutionary political fragmentation the structural advantages of intermittent mobilization over regular forces were many. The men mobilized in militias were only paid for the days of actual service. The cost of training and arming them was much lower than that of a permanent army. In periods of crisis for the central administration the institutional effort needed to set them up was equally low: the militias demanded less bureaucracy, fewer shops, barracks, schools, offices, and workshops. In contrast to the regular army units, intermittent mobilization did not wrench arms from productive work. The men who were forced to serve were not displeased in the same measure. This type of mobilization was typical of the prairie areas and did not work in the same way in the Andean region. There, however, the armies tended to develop a system of intermittent mobilization more similar to the militia than to the permanent kind, at least in the first half of the nineteenth century.

Irregular wars, irregular forces

Intermittent forces could be felt even before they came into contact with their enemy.[50] As they advanced, they prepared the terrain. Cattle was hidden, water points poisoned, inhabitants whose loyalty was in doubt were imprisoned or intimidated. The enemy's advance was made as painful as possible. Without local guides, or informers, it could be impossible to find remount horses or fresh produce. It could become very dangerous to separate from patrols: local intermittent forces permanently prowled, at some distance, ready to jump at any opportunity. Defensive tactics were the norm in this war of resources, harassment and exhaustion.[51] In the face of invasion local units had a vast defensive repertoire. They organized ambushes near the best pastures. They set the vegetation on fire as enemy troops passed. At night, they knew how to create havoc with the horses at their enemy's camp, with devastating effect. Using the lasso, they captured and carried away guards, officers, and scouts. Deploying ruse as strategy, they excelled in taking the invaders' horses. So, if everything went well for the defenders, an attacker who had penetrated a province with a very strong presence could return home some weeks later – without having engaged in combat – demoralised, decimated and, what was worse of all in the Pampa: on foot.

In post-revolutionary South America, excluding the campaigns against the Indian, most of the confrontations were civil wars between opposing factions who spoke the same language, shared the same ethnic origin, and historical and social characteristics. Modern national sentiment was still embryonic: men were not yet linked to parties, and would change sides based on convenience.[52] Under these conditions, as the *montonera* approached, it was normal for the leaders of the regular army to tremble. They did not necessarily fear military defeat. But they could feel plotting brewing among their subaltern officers; they noticed that deserters did not return, joining the enemy instead, they saw how officers who had lost control of their men feared being stabbed in the back. It was as if there was a law: the force of attraction always benefitted the most irregular troops.[53] This had to do with the fact that the soldiers in the permanent army were mostly recruited by force, subjected to the laws of discipline to subdue their instincts, deprived of the chance to pillage, and rarely saw their salary.

As a result, in the Rio de la Plata the turncoat became one of the most important figures of the military equation. To 'protect' recently integrated men, intermittent forces detached a thin line of sharpshooters, who behaved in a manner very similar to that of a regular company of mounted hunters. These men advanced in pairs and were deployed with long intervals between each, approaching to shooting distance. Then one of the men of each pair would remain in the back with both horses held from the bridle, while their companion dismounted and opened fire with a carbine aiming to harass the enemy and provoke a reaction.[54] If the commander attacking in this way was not sure of his troops, he had no option but to attack en masse. The sources speak of a less prudent commander who sent his own sharpshooters, only to see them attach a piece of white cloth on the tip of their lances and eagerly change sides to the *montonera* while those at the other side taunted them. The effect of contagion could be formidable: in certain battles, squadrons stopped their charge at the last moment to take a place on enemy lines.[55]

In contrast, if the commander of the permanent forces had enthusiastic, well-paid and well-trained troops, he could detach some cavalry companies to attack the sharpshooters. In an instant they would remount and flee. Those on foot were not worried that their companion would leave them, as these pairs were made up by brothers, *compadres*, or men linked through symbolic ties such as having godchildren in common, and members of the same family so they would back each other till death. At the same time, these pursuits were very dangerous for those attacking. Usually those in flight did nothing but shoot their persecutors, often leading them to already prepared ambushes. They used depressions in the terrain, in wooded areas or buildings to trap entire squadrons; they were always ready to pursue disorganized troops.

If there was no ambush, regular squadrons would re-join at the principal line of the enemy formation. This line varied a lot depending on the degree of regularization of the intermittent forces. In the first half of the nineteenth century, a great number of the militia units managed to achieve a degree of cohesion similar to that of a permanent force; they could, if they needed to, go into battle following such conventions as having cavalry formed in echelons at the flanks, with the infantry and the cavalry at the centre and reserve at the back. If this band was made up of more irregular forces, the main line tended to resemble sharpshooters, described earlier. The infantry, that was either small in number or inexistent, was always formed at the centre, at a fixed point; their role was limited to serve as a rally point for the cavalry if they became dispersed. The cavalry would either be two thirds of those in battle, or even their totality.[56] They preferred to place themselves on the wings, formed on a single line, which they elongated substantially as they approached the enemy. If they saw an artillery attack the intervals between horsemen could increase by several meters.

This is when they would use the tactics known as 'horseshoe' or 'crescent moon'.[57] The most mobile and light horsemen, preferably Indians, would take position at the flanks. Upon seeing such a small grouping, the least experienced regular forces would try to charge upon the centre in an attempt to separate them. Alternatively, those at the centre would hide while the intermittent line forces would bend with ease, and elongate the arms along the flanks behind the attackers, encircling them. In properly executed militia charges against the cavalry those at the centre would disband in all directions. Not encountering resistance, the attacking soldiers would cry victory, occupy the adversary's position and marvel at the cowardice of their enemy. It was only then, at second glance, that they noticed they had paid scant attention to their back and that all their baggage, their combat supplies and their reserve horses had been taken, their retreat line had been compromised; and they could not fight as their munitions had been taken.

At this point, in a few minutes intermittent forces would reform and attack. These were still the *montones*, base groups that profited from their capacity for dispersion and reassembly. It was at this crucial moment of battle that their specific type of cohesion showed its value. The militia member, the *montonero* could easily leave, but he would always return to his captain and his men. A militia army or *montonera* was very weakly welded in general formation and could easily break in their unit bases, but what it lacked in organic articulation it made up in its ability to regroup, and their strength therefore resided in that each of its component parts could continue battling independently. At this stage if the permanent army still had enough cavalry, they could repeat the operation and charge each of these new regroupings, but they would be further dispersed by following them. In all probability, the base groups would regroup again, and only if the army pressed matters would they eventually retire definitively. The regular army would thus have won the day, left to start again the next day. The danger for permanent forces really appeared when signs of fatigue became clear, or when a squadron moved considerably away from the line. Because it would be a mistake to believe that the intermittent forces would systematically refuse contact, on account of being so light. On the contrary their dodging tactics were nothing but preparation for the most furious corps à corps fight. A small mistake was enough, the smallest space amongst the ranks of the regular troops for the militias and the Indian to forge into a powerful charge of foragers.

In the Rio de la Plata, the clash that became decisive was known as the *entrevero*. Here each side had to prove their cohesion to the last consequences, on a brutal man-to-man combat that would decide victory. It was at this point that the intermittent cavalry soldiers were most appreciated. The militia and the *montonero* lived for this moment. Once the *entrevero* began the men on both sides would fight mostly with bladed weapons, all formation and order would be lost: this was every man for himself. Instead of a rigid body to body combat, such as a Greek style phalanx, in rectangular formation, in the *entrevero* the dynamic was spinning, men and horses intertwining in a movement close to circular. Here, neither the lance nor the rifle was of great use. The fight was the realm of the knife, the sabre and even pistols. It is hard to describe the chaos that followed as hundreds of men and their horses began to intertwine in this way. Visibility was reduced due to the dust, and any notion of securing the front, the flanks, the back were pointless; bodies arrived from all sides. To make matters worse, the militia troops rarely wore uniforms; they distinguished themselves with a simple coloured ribbon or feather in their cap. With the visibility conditions that characterized the *entrevero*, it is evident that these markings did not amount to much and the militiamen attacked and killed indiscriminately, without really knowing who was falling under the knife.[58] While the horse caracoled, the blows fell on the right and the left without mercy. Those who fell from their horses had very few chances of coming out alive: they were trampled to death.

The *entrevero* at the battle of Malabrigo of 1845 described by Prudencio Arnold lasted a half an hour at most. One of the longest known was the combat of *Concepción del Río Cuarto*, on 8 July 1821: this infernal spectacle lasted more than three quarters of an hour. The physical and psychological resistance of those in combat was tested to the limit during this veritable eternity.[59] The advantage in this decisive instance was clearly on the side of the intermittent forces. Due to their training, regular soldiers were conditioned to fight in a certain order, precise distances, following mechanical movements with their comrades to the right, the left and the back.[60] In the whirlwind of the *entrevero*, all this was irredeemably lost. The soldier found himself in an incomprehensible place where their training was useless when compared with instinctive reaction, and blind fury was the only asset. But the more trained the regular army was, the more difficult it was to pass from one mode to another. Under these conditions, terrorized soldiers descended into panic which destroyed all vestige of cohesion.[61] In contrast the militia man of *montonero*, could continue to battle to his advantage as long as he kept his horse and his dagger.

As a consequence, if the intermittent warring forces found themselves giving a more traditional battle, they always had the possibility to provoke a moment that would descend into the *entrevero*. On the contrary the key for the regular forces was to prevent this kind of fight at all cost. As long as the infantry would keep their cohesion, they could return to formation and attempt to reach the closest village. In the depopulated Pampa this risked being very far away, but some troops were saved this way. But if their structure was broken and an *entrevero* ensued what chance did they have to escape? The gaucho, but even more so the Indian, were experts in pursuit and could wreck a troop in flight, in the same way they pursued a herd of beasts to finish the game.

At this moment of rupture, the difference between the type of cohesion typical of permanent armies and that of intermittent ones could be appreciated in all clarity. The militia unit was easy to break, but it was just as easily reassembled, above all dispersion could be a good thing for it. Unlike the cohesion of regular forces, that tended to be rigid, militia cohesion was flexible instead. When the militia squadron disappeared, the men could reassemble as long as they continued to be welded by relationships that went beyond the purely military. When their unity was broken they did not form a mass of isolated individuals: they continued to be part of the same kinship network of neighbours, with reciprocity and dependency mechanisms; they would all return home together. Conversely the regular soldier's ties with the army were all broken, once its unity was shattered in battle. He would throw away his gun, in a heretic gesture he would defy all military ethos, of all the appearance of the military family. A wounded beast, tracked by the enemy, by his own army and by the government, he would dispose of the uniform, run to survive, hide in the fields or with some peasants who provided hospitality. In any case they were completely lost for the army.

Concluding remarks

One of the most important consequences of the wars of independence in the South American region was that the process led to a prolonged war which brought different types of mobilization where intermittent forces became more useful than permanent ones. This process was much more acute in the Rio de la Plata where after 1820 there was no centralised state until the 1860s. In this half-decade the wars could not really be described as insurgencies, but instead they were characterised as the confrontation between intermittent and permanent forces, where the former tended to be more successful than the latter. In the Andes the process was not so drastic as the armies that emerged from the wars of independence took over the organisation of the new states of Bolivia and Peru, and fighting was between factions of these armies. These armies, however, tended also to be more intermittent than permanent, although they did not have the same kinds of structures as the *montones* found in the Pampas.

Notes

1. On this point see Lempérière, "Revolución, guerra civil, guerra de independencia."
2. Arreguin-Toft, *How the Weak Win Wars.*
3. Verdo, *L'indépendance argentine.*
4. Demelas, "De la 'petite guerre' à la guerre populaire."
5. Fradkin, "'Facinerosos' contra 'Cajetillas'?"; Slatta, "Rural Criminality and Social Conflict"; and Salvatore, "The Breakdown of Social Discipline."
6. Sobrevilla Perea, "Luchando por 'la patria'."
7. Rabinovich, "La máquina de la guerra y el Estado."
8. Méndez, *The Plebeian Republic.*
9. There are many works on the colonial army, one of the first ones is by McAllister, *The "Fuero Militar" in New Spain.*
10. See for instance Campbell, *The Military and Society in Colonial Peru*; Kuethe, *Military Reform and Society in New Granada*; and Archer, *The Army in Bourbon Mexico.*
11. For the most recent study see Walker, *The Tupac Amaru Rebellion.*
12. The work of Leon Campbell looks at this issue in detail *The Military and Society in Colonial Peru.*
13. Fernandez, *El Ejército de América antes de la independencia.*
14. See "Informe del Subinspector General sobre el deficiente estado de preparación militar del virreinato, 1802", in Juan Beverina, *El Virreinato de las*

Provincias Unidas del Río de la Plata, su organización militar, Buenos Aires, Círculo Militar, 1992, 437–443.

15. Rabinovich, "The Making of Warriors."
16. Roca, *Ni con Lima ni con Buenos Aires*.
17. Sobrevilla Perea, "Luchando por 'la patria'."
18. The offensive vocation of the revolution was evident after its first move. In the popular petition of 25 May 1810, that demanded the creation of a new government, the revolutionaries demanded the immediate departure of an armed expedition of 500 men on to the interior provinces. See Anon, "Acta capitular del día 25 de Mayo de 1810."
19. With regards to the issues of putting together regular armies and the creation of new States, see Storrs, *The Fiscal-Military State in Eighteenth-Century Europe*; and Tilly, *The Formation of National States in Western Europe*.
20. Garavaglia, *Les hommes de la Pampa*.
21. Fradkin, "Las formas de hacer la guerra," 167–214.
22. For example "Copia del oficio dirigido al cabildo de Buenos Aires por el Gobernador de Santa Fe, Brigadier General Estanislao López el 14 de Septiembre 1820", *Revista de la Junta de Estudios Históricos de Mendoza*, Segunda Época, n°3, 1966, 337–344.
23. On the Indian war culture see Beccara, *Guerre et ethnogenèse Mapuche dans le Chili colonial*.
24. The definition of the military concept of defense and its relationship to attack can be found in von Clausewitz, *De la guerre*, 399–425.
25. For the first wars of 1814, see Lopez "La guerra de independencia en Salta," 113–35; Manuel Otero, "Informe sobre los servicios del Coronel Don Luis Burela de Salta en la Guerra de la Independencia", and "Informe sobre los servicios del general Don Pablo de la Torre", in *Memorias: de Güemes a Rosas*, Buenos Aires, Ediciones Argentinas, 1946, 33–116. Yabén, *Los capitanes de Güemes*,17–19 .
26. For a detailed description of this see Joaquín de la Pezuela, *Compendio de los sucesos ocurridos en el ejército del Perú y sus provincias (1813–1816)* Santiago de Chile: Bicentenario, 2011.
27. Archivo General de Indias (AGI) Lima,1014A, Proclama de José de Abascal, Lima Abril 25 1812.
28. Rabinovich, "La militarización del río de la plata, 11–42.
29. Rabinovich, "La máquina de la guerra y el Estado".
30. Sobrevilla Perea, "From Europe to the Andes and Back,"472–88.
31. Verdo, "La guerre constituante,"246.
32. These are the main questions that guide current research on asymmetrical warfare. Baud, *La guerre asymétrique ou la défaite du vainqueur*.
33. Sobrevilla Perea, *The caudillo of the Andes*.
34. For the classic study of the problem of "regularity" in the military, see, Schmitt, *Théorie du Partisan*.
35. A comparative study of different models of militarization can be found in Keegan, *A History of Warfare;* 221–234, Corvisier, *La Guerre*, 210–7; and *Armées et sociétés en Europe de 1494 à 1789*, Paris, PUF, 1976.
36. This idea of this danger is deeply seated in Indo-European mythology see Dumezil, *Heur et malheur du guerrier*.
37. Muir, *Tactics and the experience of battle in the age of Napoleon*, 68–76.

38. The best modern studies on discipline and control is Foucault, *Surveiller et punir*.

39. Gresle, "La 'société militaire,"777–98.

40. For a similar experience in Venezuela see Hebrard, «Cités en guerre et sociabilité au Venezuela (1812–1830), »123–48.

41. See for example the proclamation of the Cabildo of Buenos Aires 24 sept.1807, Biblioteca Nacional Argentina, http://www.bibnal.edu.ar/webpub/digital.asp; also the proclamation of Supreme Director Pueyrredón to the people of Salta, le 18 October 1816; Gümes, *Güemes documentado*, 67.

42. Cansanello, *De súbditos a ciudadanos*.

43. Garavaglia, "Ejército y milicia,"153–87.

44. Just in the case of Upper Peru we know the names of 132 little *caudillos*. See Bidondo, *La guerra de la independencia en el Alto Perú*, 180.

45. For a detailed analysis see Rabinovich, *La société guerrière*, 200–15.

46. There are many examples of militia mobilization in the archive; a good example is that of the greatest militia commander of Caillet-Bois (dir.), *Archivo del Brigadier General Juan Facundo Quiroga*, 13, 284, 289, 303, 304.

47. *Montonera*: from the Spanish word *montón*, was used to describe those who had similar elements. Name given in South America to a type of irregular war in the countryside. Definitions vary with time see Real Academia Española, *Diccionario de la Lengua Española*, 1869, 1884 and 1899 editions. Available online: http://www.rae.es/rae.html.

48. Fradkin, *La historia de una montonera*, 39.

49. On the people of great planes see Lebedynsky, *Armes et guerriers barbares au temps des grandes invasions ; et Les Scythes. La civilisation des steppes (VIIe-IIIe siècles av. J.-C.)*, Paris, Ed. Errance, 2001.

50. On indirect tactics see Gérard Challiand, *Stratégies de la guérilla*, Paris, Payot, 1994.

51. These tactics are explained in detail ins "Orden de Miguel de Güemes a Vicente Torino, 6 junio 1820" ; Güemes, *Güemes documentado*, vol.8, 48–52.

52. Rabinovich,"El fenómeno de la deserción en las guerras de la revolución e independencia del Río de la Plata," 33–56.

53. The examples are numerous, see Baron de Holmberg's report of his defeat in Espinillo, *Partes de batalla de las Guerras Civiles 1814–1821*, vol.1, Buenos Aires, Academia Nacional de la Historia, 1973, pp.7–16 .

54. The complete description of this « montonera » tactic is provided by the expert hand of José María Paz. See the notes of his study of combat de la Herradura in his *Memorias Póstumas*, vol.1, Buenos Aires, Ed. Emecé, 2000, pp.271–289.

55. This is exactly what happened to commander Manuel Dorrego at the combat of Guayabos. See *Partes de batalla de las Guerras Civiles 1814–1821*, vol.1, pp.66–70.

56. Recent approaches to the history of cavalry in *historique des armées: Le cheval dans l'histoire militaire*, n°249, 2007.

57. The prairie horsemen in Venezuela known as the *llaneros* used a similar tactic; see Clément Thibaud, *Républiques en Armes*, 284–287.

58. Prudencio Arnold, *Un soldado argentino*, Buenos Aires, EUDEBA, 1970, 66–72.

59. Manuel A. Pueyrredón, *Escritos históricos*, Buenos Aires, Julio Suárez ed., 1929, 35–40. Cf. Damián Hudson, *Recuerdos Históricos sobre la Provincia de Cuyo*, vol.2, 1898, 413–414.

60. Stéphane Audoin-Rouzeau, "Vers une anthropologie historique de la violence de combat au XIXe siècle: relire Ardant du Picq ?", Revue d'histoire du XIXe siècle, 2005, n°30.
61. On panic in combat, see Corvisier, "Le moral des combattants, panique et enthousiasme: Malplaquet, 11 septembre 1709". Jean Chagniot, "Une panique: les Gardes françaises à Dettingen (23 juin 1743)".

Acknowledgments

Some sections of this paper were published in "Milices et guérillas paysannes face a la armée régulier: le combat asymétrique au Rio de la Plata et la fragmentation territoriale (1810-1852)", Hispania Nova, Revista de Historia Contemporánea, num. 13, 2015, pp. 164-187. Translated by Natalia Sobrevilla Perea. We thank the Leverhulme Trust for making the writing of this joint article possible.

Disclosure statement

No potential conflict of interest was reported by the authors.

ORCID

Natalia Sobrevilla Perea (iD) http://orcid.org/0000-0001-9592-7551

Bibliography

Anon. "Acta Capitular Del Día 25 De Mayo De 1810." In *Partes oficiales y documentos relativos a la Guerra de la Independencia*, Vol. 1, 11–14. Buenos Aires: Archivo General de la Nación, 1900.

Archer, Christon. *The Army in Bourbon Mexico, 1760–1810*. Albuquerque: University of New Mexico Press, 1977.

Arreguin-Toft, Ivan. *How the Weak Win Wars: A Theory of Asymmetric Conflict*. Cambridge University Press: New York, 2005.

Beccara, Gillaume. *Guerre Et Ethnogenèse Mapuche Dans Le Chili Colonial. L'invention Du Soi*. L'Harmattan: Paris, 1998.

Campbell, Leon G. *The Military and Society in Colonial Peru 1750–1810*. Philadelphia: American Philosophical Society, 1978.

Chagniot, Jean. "Une Panique: Les Gardes Françaises À Dettingen (23 Juin 1743)." *Revue d'Histoire Moderne Et Contemporaine* 24 (1977): 78–95. doi:10.3406/rhmc.1977.967.

Corvisier, Andre. "Le Moral Des Combattants, Panique Et Enthousiasme: Malplaquet, 11 Septembre 1709." *Revue Historique Des Armées* 3 (1977): 7–32.

Demelas, Marie-Danielle. "De La "petite Guerre" À La Guerre Populaire: Genèse De La Guérilla Comme Valeur En Amérique Du Sud." *Cahier Des Amériques Latines* 36 (2008): 17–36.

Fernandez, Juan Marchena. *El Ejército De América Antes De La Independencia: Ejército Regular Y Milicias Americanas, 1750–1815*. Madrid: MAPFRE, 2005.

Fradkin, Raul. "Facinerosos" Contra "cajetillas"? La Conflictividad Social Rural En Buenos Aires Durante La Década De 1820 Y Las Montoneras Federales." *Illes I Imperis* 5 (2001): 5–33.

Fradkin, Raúl O. *La Historia De Una Montonera. Bandolerismo Y Caudillismo En Buenos Aires, 1826*, 39. Buenos Aires: Ed. Siglo XXI, 2006.

Fradkin, Raúl O. "Las Formas De Hacer La Guerra En El Litoral Rioplatense." In *La Historia Económica Y Los Procesos De Independencia En La América Hispana*, edited by Susana Bandieri, 167–214. Buenos Aires: AAHE/Prometeo Libros, 2009.

Garavaglia, Juan Carlos. *Les Hommes De La Pampa. Une Histoire Agraire De La Campagne De Buenos Aires (1700–1830)*. Paris: Ed. EHESS et MSH, 2000.

Gresle, François. "La "société Militaire": Son Devenir À La Lumière De La Professionnalisation." *Revue Française De Sociologie* 44, no. 4 (2003): 777–798.

Kuethe, Allan J. *Military Reform and Society in New Granada, i773-I808*. Gainesville: University of Florida Press, 1978.

Lemperiere, Annick. "Revolución, Guerra Civil, Guerra De Independencia En El Mundo Hispánico, 1808–1825." *Ayer* 55 (2004): 15–36.

Lopez, Sara Mata de. "La Guerra De Independencia En Salta U La Emergencia De Nuevas Relaciones De Poder." *Andes. Antropología E Historia* 13 (2002): 113–135.

McAllister, Lyle N. *The "fuero Militar" in New Spain: 1764–1800*. Gainesville: University of Florida Press, 1957.

Méndez, Cecilia. *The Plebeian Republic: The Huanta Rebellion and the Making of the Peruvian State, 1820–1850*. Durham: Duke University Press, 2005.

Muir, Rory. *Tactics and the Experience of Battle in the Age of Napoleon*. Yale University Press: Londres, 1998.

Paz, José María. "Notes of His Study of Combat De La Herradura." In *Memorias Póstumas*, edited by Ireneo Rebollo, 271–289. Vol. 1.Buenos Aires: Ed. Emecé, 2000.

Rabinovich, Alejandro M. "The Making of Warriors: The Militarization of the Rio De La Plata, 1806–1807." In *War, Empire and Slavery, 1770–1830*, edited by Richard Bessel, 81–98. Basingstoke: Palgrave Macmillan, 2010.

Rabinovich, Alejandro M. "El Fenómeno De La Deserción En Las Guerras De La Revolución E Independencia Del Río De La Plata. Elementos Cuantitativos Y Cualitativos Para Un Análisis. 1810–1829." *Estudios Interdisciplinarios De América Latina Y El Caribe* 22, no. 1 (2011): 33–56.

Rabinovich, Alejandro M. "La Máquina De La Guerra Y El Estado: El Ejército De Los Andes Tras La Caída Del Estado Central Del Río De La Plata En 1820." In *Las Fuerzas De Guerra Y La Construcción Del Estado: América Latina*, edited by Juan Carlos Garavaglia, Juan Pro, and Eduardo Zimmermann, 205–240. Rosario: ProHistoria, 2012.

Rabinovich, Alejandro M. "La Militarización Del Río De La Plata, 1810–1820. Elementos Cuantitativos Y Conceptuales Para Un Análisis." *Boletín del Instituto de historia argentina y americana "Dr. Emilio Ravignani"*, 3º Serie, nº 37, segundo semestre de 2012, 11–42.

Roca, José Luis. *Ni Con Lima Ni Con Buenos Aires: La Formación De Un Estado Nacional En Charcas*. Lima: Instituto Francés de Estudios Andinos, 2011.

Salvatore, Ricardo D. "The Breakdown of Social Discipline in the Banda Oriental and the Littoral, 1790–1820." In *Revolution and Restoration. The Rearrangement of Power in Argentina, 1776–1860*, edited by Mark Szuchman and Jonathan Brown, 74–102. University of Nebraska Press, 1994.

Slatta, Richard W. "Rural Criminality and Social Conflict in Nineteenth-Century Buenos Aires Province." *The Hispanic American Historical Review* 60, no. 3 (1980): 450–472. doi:10.2307/2513269.

Sobrevilla Perea, Natalia Sobrevilla. "From Europe to the Andes and Back: Becoming 'los Ayacuchos'." *European Historical Quarterly* 41, no. 3 (2011): 472–488. doi:10.1177/0265691411405296.

Sobrevilla Perea, Natalia Sobrevilla. "Luchando Por 'la Patria' En Los Andes 1808–1815." *Revista Andina*, Cuzco n.52, 2012a. doi:10.1094/PDIS-11-11-0999-PDN

Storrs, Christopher, ed. *The Fiscal-Military State in Eighteenth-Century Europe*. Ashgate: Farnham, 2009.

Tilly, Charles. *The Formation of National States in Western Europe*. Princeton: Princeton University Press, 1975.

Verdo, Geneviève. *L'indépendance Argentine: Entre Cités Et Nation (1808–1821)*. Paris: Publications de la Sorbonne, 2006.

von Clausewitz, Carl. *De la guerre*. Paris: Les éditions de Minuit, 1955, 399–425.

Walker, Charles F. *The Tupac Amaru Rebellion*. Cambridge: Harvard University Press, 2014.

Yabén, Juan. *Los Capitanes De Güemes*. Buenos Aires: éd. Lito, 1971.

The First Carlist War (1833–40), insurgency, Ramón Cabrera, and expeditionary warfare

Mark Lawrence

ABSTRACT

The period 1833 to 1840 witnessed a brutal civil war in Spain waged between insurgent Carlists and the government Cristinos. The Carlists managed to secure reliable territorial control only over one part of Spain (upland Navarra and rural parts of the neighbouring Basque provinces). Although pockets of armed Carlism flourished elsewhere in Spain, especially in Catalonia, Aragón and Galicia, these insurgents were ineffective at coordinating actions. The Carlist court in the Basque country tried to break its strategic blockade by launching a series of expeditions into Cristino-held territory in the hope of destabilising the Madrid regime and consolidating distant insurrections. This chapter explains how and why these expeditions scored tactical victories but strategic failures. In particular it argues that Carlist raiding strategy was a failure, for its use of violence against real and imagined enemies in marginal and Cristino areas of control alienated civilian support.

For all the diverse historiography concerning the reactionary current of the 1830s known as Carlism, the military historiography has been perhaps the least innovative. Military history has been generally out of fashion in Spanish universities, owing both to the ascendancy of social and cultural approaches and to Spain's unseemly and comparatively recent burden of militarism. This oversight would seem surprising. Carlism managed to mobilise regional armed support more successfully than the Cristino regime it fought, even though that regime increasingly claimed to represent the 'people' as it liberalised its institutions in the revolutions of 1835 and 1836. Certainly, the First Carlist War (1833–40) abounds in the Spanish historiography more broadly. For a century the war was presented in partisan terms, as traditionalists and liberals produced very learned if also tendentious histories. In the forbidding atmosphere after the Spanish Civil War, nineteenth-century history riled a Francoist dictatorship bent on viewing all Spanish history since 1812 in apocalyptic terms. Historians critical of nineteenth-century liberalism gained official blessing, especially the 'Pamplona school' of Jesuit historians who defended the 'renovating' power of the old order against the 'innovations' enforced by liberals who were supposedly besotted with 'Frenchified' ideals. Some of these historians were sympathetic to the most extreme form of the old order, Carlism.[1] These historians were countered by Marxian historians, many of whom were in exile, while others got by in Spain, who condemned the old order which spawned the Carlist insurgency while also explaining the shortcomings of Spain's 'bourgeois revolution'. Over the past forty years Spanish historiography has diversified as the old ideological straitjackets have mostly been discarded and new avenues of cultural, regional and social history opened up. But both Carlism and the First Carlist War in particular have never been the most popular subjects for contemporary Spanish historians, and hardly at all for foreign historians.

The First Carlist War historiography may be divided into five categories. The first comprised nineteenth-century dynastic and classical diplomatic, biographical and military histories, led by the historian, Antonio Pirala y Criado. The second comprised the panegyrics from Franco-era traditionalists who depicted Carlism as an organic Christian good resisting the onslaught of godless and 'Frenchified' Spanish liberalism. The First Carlist War was but one protracted episode in the wider war between Christianity and the Anti-Spain that was joined in 1808 and won by the crusaders only in 1939. The most impressive work in this vein was the 30-volume history of traditionalism edited by Melchor Ferrer from the 1940s, which was meant to be the Carlist answer to the Liberal Pirala, but which in fact lacked the latter's balance and command of primary sources. The third category comprised the 'Navarra School' of Pamplona-based neo-traditionalists, led by Federico Suárez Verdeguer. These scholars sustained a far more sophisticated right-wing analysis based on modern empirical research.[2] Their contention that Spain remained royalist, or apolitical, throughout this period rendered Liberalism an artificial and arrogant innovation. As the best neo-traditionalist scholar of the First Carlist War, Alfonso Bullón de Mendoza, put it, the Cristino Liberals were waging war against their own people.[3]

The neo-traditionalists were challenged from their own ranks by a 'heresy' of neo-Carlists writing from the 1970s who reinterpreted the nineteenth-century Carlist struggles as 'objectively revolutionary', and from the non-Carlist left by liberal and Marxist historians. Both the 'heretics' and the ideologues were interested in the socio-economic drivers of counter-revolution, and much less in its military aspects. To a large degree this focus was justified by the complexity of 1830s Carlism. There were three major 'focos' of armed Carlism, most of Navarra and the upland Basque provinces, the Aragón-Valencia uplands centred on the Maestrazgo, and the smallest zone, the Catalan far west. The motives for armed counterrevolution have been shown to be complex, certainly more complex than the victorious Liberals allowed. A Barcelona newspaper in 1840 reflected on the recently extinguished civil war, attributing Basque Carlism to the defence of 'liberties' (especially the autonomous 'fueros'), Catalan Carlism to 'religious fanaticism', and Aragonese Carlism to 'banditry'.[4] In reality the motivations for Basque Carlism were at least threefold: a foralist wing driven by defence of the region's historical autonomy, a dynastic wing driven by 'Castilian' refugees from Cristino-held Spain, and an intransigent ultramontane wing with adherents from within the Carlist Basque country and from beyond. Catalan Carlism was driven in part by religiosity but also by economic decline in the interior where the insurgency would take hold, and socio-economic motives also pertained in the Aragón-Valencia insurgency. What united all insurgent regions was a popular tradition of armed insurrection, as witnessed in regional particularities in militia service which had been heavily bloodied in the national and civil wars of 1808–14, 1821–23 and 1827. This chapter examines how the regionalised Carlist war effort was ultimately incapable of defeating the larger and symmetrical forces available to the Cristino regime which also benefited from significant foreign support. Using British and Spanish archives, war memoirs, and secondary sources, including political science civil war theory, this article shows how the greater recourse to violence exercised by insurgents raiding beyond the Carlist regions alienated civilian support and hastened their strategic defeat.

The growing insurgency

The first year of hostilities opened in October 1833 witnessed the rise of the Carlist military genius, Tomás de Zumalacárregui, who used guerrilla tactics to carve out an expanding territorial control stretching from upland Navarra gradually into the neighbouring Basque provinces proper (Guipúzcoa, Vizcaya, Álava). The Cristino counter-insurgency efforts over 1834 progressed from policing measures (such as confiscating arms and horses) to raids and blockhouses.[5] But in a series of defensive victories by the summer of 1835 Zumalacárregui expanded Carlist control over virtually all of upland Navarra and the Basque provinces minus the Cristino provincial capitals. Despite Zumalacárregui's death during the failed Carlist siege of Bilbao in June 1835, the Carlist military genius had bequeathed the insurrection a regular 'Royal Army' of about 35,000 men, supported by a significant local arms industry, irregular supplies from sympathisers breaching the French border and the Anglo-Cristino naval blockade, and conscription. By summer 1835 Spain's civil war had become a major feature of European diplomacy as Madrid secured indirect military support from France, Britain and Portugal (the recruitment of auxiliaries in those countries for service in the Cristino army and a naval blockade of coastlines close to the Carlist insurrection). The Carlists got fewer military volunteers from such anti-liberal powers as Prussia, Austria and the Italian states, but they never lost hope that Chancellor Metternich or the Holy See might defy the Atlantic powers by recognising the Carlists.

Carlist military successes fuelled the revolutionary crisis affecting Cristino Spain between 1835–37. Desertion became rife as government army logistics broke down while the Carlists took as much as they pleased, and military commanders always had one eye on the revolutionary threat posed by the National Militia, the liberals' paramilitary force which controlled urban spaces and pressed demands for political reform. The war by 1836 produced a subsistence crisis in government-held areas adjoining insurgent zones of control. In April 1836 a Cortes deputy from embattled Navarra, angered by the revolutionary rhetoric of Madrid, said 'the best law is worse than useless if it distracts attention from the wants of a starving population overrun by a bloody and remorseless enemy'.[6] The subsistence crisis was compounded by poor harvests which by 1837 had grown so severe that insurgents were even seizing oxen for food.[7] Populations in conflict zones were frequently 'recon-centrated' to fortified centres, obliterating the subsistence base of one area and burdening the next. By March 1836 Carlist activities in Lérida province had driven some 4,000 families from their homes to fortified centres. Cortes deputy, Castells, complained how '300 villages have not eaten bread for three months', and that the July harvest would be exposed to Carlist depredations.[8]

Cabrera and the insurgency in Valencia-Aragón

By 1836 the worst of the subsistence crisis was affecting eastern Spain. Unlike in the case of the River Ebro and the insurgent Basque provinces, clear areas of territorial control had not been established in the east, and the war continued to have a brutal character distinguished by raids and reprisals against both combatants and civilians. The Carlist zone in Aragón-Valencia was centred on the Carlist 'capital' of Morella (Valencia). It was intimately connected to the person of Ramón Cabrera, from 1835 commander-in-chief of Carlist forces in this zone, and second in Carlist legend only to the great Tomás de Zumalacárregui in terms of military prowess. Socio-economic motivations for supporting the Carlist revolt in Aragón-Valencia included defence of the use-ownership rights of peasants who enjoyed a de facto if not de jure ownership of the lands they farmed in a 'common law' arrange-ment which was threatened by the liberal property revolution's drive towards contracts and cash transactions.[9] Added to this was an active banditry tradition enabled by a culture of horsemanship and rugged terrain, as well as the enticement offered by the wealthy Valencian huerta. Prussian Carlist volunteer, von Rahden, contrasted the sturdy religiosity of the

Maestrazgo Carlists with the 'volatility' of the Cristino coastal plain.[10] Whereas only 1.05% of the city of Valencia's population became Carlist militants, and 2.2% in the Valencia countryside, in Teruel province (Maestrazgo), the figure was 4.6%.[11]

Ramón Cabrera showed genius for maximising a limited economic base for the needs of his war effort. Like his comrades in the larger zone of Basque-Navarra Carlism he had the advantage of internal lines. The influential British ambassador, George Villiers, commented on the Carlists' use of intelligence, including spy networks and flash telegrammes, to offset their numerical disadvantage. The Cristinos, according to Villiers, 'are facing a harder task than even the French occupation forces twenty-five years earlier'.[12] The Cristino press compared Cabrera's organisational genius to that of Abd el-Kader, the emir proclaimed sultan of Algeria in 1832 who waged defensive war against the French conquest.[13] Cabrera's forces fortified villages in the Maestrazgo, and turned the invariably wealthy homes of cowed, killed or fled liberals into barracks, stores or blockhouses. Rahden remarked on the 'cemetery-like' appearance of the Maestrazgo villages, including a complete absence of windows and chimneys. Upon the poor roads connecting Carlist villages depended the entire Carlist logistical efforts. In front-line areas Carlists used hillpaths driving precarious mule-trains in dangerous single file while the valley-bound Cristinos passed below.[14]

Cabrera's efforts were often discredited by the Cristino government which continued to associate the Aragón-Valencia Carlism with banditry. Certainly the proximity of the bandit-friendly Maestrazgo to the wealthy citrus groves of the Valencian *huerta* offered huge scope for wrongdoing, not least because the *huerta* itself was worked by labourers who continued to suffer the legacy of some of the worst feudal conditions in Spain and thus had little incentive to defend landowners' property. A month after the war ended in 1840 the Barcelona press lamented the failure of Cristino militia detachments to protect the *huerta* from Carlist raids. Whereas villages loyal to Madrid faced constant depredations, villages in the Carlist Maestrazgo flourished, witnessing unprecedented flows of cash. At the centre of these ill-gotten gains was Cabrera's capital of Morella, flowing with cash and chattels to pay soldiers and protect the civilian economy.[15] Cabrera's punishment system relied heavily on pillage. Loyal villages were protected and rewarded for steadfast defence, whereas hostile villages faced attacks and depredations, compounding the logic driving violence in supporting Cabrera's cause.[16] The primitive war effort in the Maestrazgo was bolstered by another 'logical' result of violence, the enlistment of Cristino prisoners of war who, according to Rahden 'always abounded in our depots' and who needed little persuasion given than that they could lawfully be executed at any time, as the quarter offered by the 'Eliot Treaty' did not apply to the east. During 1838 a Carlist siege train was constructed by prisoners of war.[17]

Territorial control was maintained by arming farmers to protect 'roads' linking Morella to Cantavieja, and to Teruel and Daroca, confining Cristino security to the environs of the provincial capitals of Valencia and Zaragoza. Closer to Cabrera's capital of Morella permanent fortresses became a more common way of protecting territory and communications in interlocking patterns dominating vantage points and maximising the defensive fire-power of artillery. The more remote hilltop areas were secured by guerrilla patrols which often engaged with Cristino raiding parties.[18] Mutually rein-forcing insurgent fortifications and flying patrols made it impossible, accord-ing to a German Carlist volunteer, for the Cristino forces to take either Cantavieja or Morella on their own: both needed to be taken simultaneously.[19] Cristino Captain-General San Miguel's success at storming Cantavieja at the end of October 1836 occurred only due to the diversion of insurgent troops employed in the Gómez Expedition.

Even though Cabrera's system of territorial control was remarkable, anti-Carlist commentators ridiculed the 'banditry' of Cabrera's Carlists. George Villiers in 1835 commented upon Cabrera's '4,000 vagabonds armed with pikes, sticks and knives ... led by officers whose only care is to plunder and desolate the countryside'.[20] Cabrera himself came to personify the barbarism associated with Carlism, as the Cristino press dubbed him the 'Tiger of the Maestrazgo', an adage that stuck. Villiers wrote that Cabrera came from the 'dregs of the people', but also warned that the infamous execution of his mother, a high-profile victim of the 'law of hostages' operated by both sides, 'placed the Cristino executioners on his level'.[21] Much of the brutality in the eastern zone was caused by the absence of any covenant safeguarding the lives of captured prisoners. The Basque-Navarra zone of operations from April 1835 was covered by the 'Eliot Treaty', named after the emissary of the Duke of Wellington who brokered a deal in which both sides agreed to respect the lives of enemy soldiers captured in the northern zone and their rights to receive quarter and imprisonment in depots from where they would be exchanged at regularly agreed intervals. No such covenant covered Cabrera's zone. Moreover guerrillas, militiamen and foreign auxiliaries were excluded from the Treaty's provisions even in the north. Casa-Eguía, Cabrera's chief minister in the east, was unbending in his response to the Cristino Commander-in-Chief's complaint at the egregious killing of captured foreign auxiliaries: 'If they did not come, they would not meet it. The remedy is in their hands, but they wish for it, and are doubtless excited thereto by the money they receive and have enlisted themselves as adventurers and mercenaries'.[22]

Cabrera's ruthlessness was matched by audacious military strategies which exploited the insurgent advantages of interior lines and initiative. Carlist Prussian volunteer, Wilhem von Rahden, identified two typical Cabrera strategies: 1) to attack enemy forces as quickly as possible in order to wrong-foot their numerical superiority, and 2) to entice enemy forces by apparently offering an exposed flank (a tactic known in Spanish as the 'llamada').[23] Cabrera executed this strategy using tactics which seemed to hail from a bygone era of warfare. Such tactics as driving herds of cattle to leave false footprints to shake off pursuing enemy troops, or ordering caracole cavalry charges against Cristino infantry in order to cover a retreat across a river, supported the strategy of deflecting superior enemy numbers.[24] For their part, the government's National Militia adapted its regulations in order to meet the challenge of multiple Carlist threats. The legal exclusivity was entrenched early in 1837, when National Militia members were given their own prisons, exemptions from billeting soldiers, and the power to designate a fortified safe-house in each village in the event of invasion by insurgents.[25] The spatial aggression of the Carlist insurgents thus forced the nominally superior government forces onto the defensive. Carlist tactics served to win numerous engagements with usually larger Crisitno forces and to expand Carlist control over the Maestrazgo. By 1837 Cabrera's success had become prominent enough for the king to include Cabrera's forces in his calculations to win the war via a march on Madrid, known as the 'Royal Expedition'.

Limits of territorial control

As in several other civil wars involving an insurgent side, the Carlist methods of controlling civilian populations ranged from 'hearts and minds' to indiscrimate violence. As Stathis Kalyvas has explained, insurgents' violence arises in inverse proportion to their territorial control. Kalyvas identifies five types of territorial control, ranging from areas entirely controlled by the incumbent side, to areas entirely insurgency, with three progressive types in between, including, significantly for this study of raiding into 'neutral' areas, the third type of fully contested territory where neither side seemed to have the upper hand.[26] Fully contested areas, or areas mostly or entirely controlled by the Madrid government, were more likely to be subjected to indiscriminate Carlist violence. In these areas civilians and especially their paramilitary National Militia proxies were identified as a collective enemy because of their geographical location and political affiliation. By contrast, areas fully under Carlist control, like large areas surrounding rural Navarra, and to a lesser extent areas mostly under Carlist control, such as the fortified villages between Cantavieja and Morella (Aragón), were subjected to a range of political and religious forms of propaganda designed to shore up support, maximise military resources, and minimise defection. Carlist violence in these areas tended to be selective, akin to law enforcement.

The royal expedition

In May 1837, in response to revolutionary crisis and diplomatic feelers to the Queen-Regent about a compromise peace, the Carlists launched the 'Royal Expedition'. This large-scale raid towards Madrid also served the logistical problem of Carlist Spain's embattled redoubts in the Basque country and Maestrazgo. Both the Carlist Basque country and Maestrazgo were straining under the demands of the war economy and militarisation. Hence the paradox of the strategic need to revert to the raiding and insurgency styles of warfare that characterised the first 18 months of Zumalacárregui's command.

The difference during the 'deep war' of 1835–37 is that the raids would be long-range and strategic in aims. The earliest raid launched in 1835 from the Basque country into Catalonia had been predicated on meeting substantial indifference or defection from the government forces. The Carlist court had calculated that some 3,000 out of the 15,000 enemy militia garrisoning Catalonia were royalists forced to take up arms, and a further 5,000 indifferent, leaving 7,000 'fierce liberals and murderers'.[27] The 1836 Carlist Gómez Expedition had successfully traversed Spain and fascinated the world. Galicia in north-western Spain, which had been the strategic aim of this raid, had harboured localised insurgencies which the Carlist leadership could never fully exploit owing to fragmentary support from Portuguese legitimists (miguelistas) across the border and government interceptions of privately-contracted arms supplies by sea.[28] And General Espartero, Commander-in-Chief of the government Army of the North, pursued the Expedition across the north, preventing Gómez from consolidating the insurgency in Galicia. The Carlist raiders were deflected southwards, where Gómez met lighter resistance and where he managed a series of temporary occupations of urban centres, including provincial capitals and Mediterranean ports, embarrassing the Cristino regime and poisoning the revolution in its liberal politics.[29] But, strategically, the Gómez Expedition achieved very little. Even though the raiders returned to the Basque country intact, having dazzled the rest of Europe, they had achieved no strategic breakthrough. The Basque country remained blockaded, and the Cristino Army of the Centre had made gains against the denuded Carlist forces in Aragón.[30] Cabrera was ordered back from the Expedition with a token guard and an assumed name ('Llorens'). But despite a number of tactical successes in minor battles en route, his arrival was too late to save Cantavieja.[31]

In May 1837 a new raid was planned, this time with greater strategic aims as betrayed in its 'Royal' title (the Royal Expedition). Don Carlos's advisers were acutely aware of the blockade facing the Basque country. In March 1837 mutinies over pay affected the Carlist Royal Army, and the king was told that the Basque country could sustain only 15 days' more operations.[32] Other calculations were diplomatic: feelers between agents of the Cristino government and the Queen-Regent's native Naples had concluded that María Cristina was willing to strike a peace involving her exile and her daughter's future marriage to a Carlist prince. Metternich's Europe pledged no more financial support for the Carlists unless they established themselves beyond the River Ebro, and King Louis-Philippe of France hinted that he might liberate France from the pro-Cristino 'Quadruple Alliance' if Don Carlos could clear the French frontier of Cristino control. Europe's conservative powers riled at Liberal Spain's proclamation in the summer of 1836 of the revolutionary Constitution of 1812, and proved more willing to challenge Anglo-French supremacy in Spain. Thus when the Royal Expedition was launched from the Carlist capital of Estella on 15 May 1837, it looked in some ways more like a regime in waiting than an invasion. It had no artillery arm until it reached the stores at Cantavieja (Aragón), and was overburdened by Carlist bureaucrats and their dependents.

The debilitation of the Cristino army, caused by a year of liberal revolutions and mutinies, could not disabuse the Carlists of their notions. The Carlists also had a moral firepower to compensate for artillery: religion. Two years of legislative and violent anti-clericalism in Cristino Spain had plunged ordinations to the priesthood, alienated monastic properties, and disturbed the religious practice surrounding parish priests (who, for example, were barred from practising if they did not swear allegiance to the queen). The Carlist press had trailed stories of abandoned parishes in the wake of the schism with Rome, the powerlessness of the bishops, the expulsion or imprisonment of priests by Cristino soldiers and militia, and, above all, the *mendizabalista* disentailment: indeed some 400 parishes closed due to lack of ministers during the first half of 1836 alone.[33]

Despite the mobilisation of some 12,000 National Militia from Huesca and Zaragoza, and the pursuit by General Espartero's Army of the North, the first Expedition victory went to the Carlists. At the battle of Huesca on 24 May 1837, some one thousand Cristino soldiers and militia were either killed or captured as the Carlist invaders turned difficult terrain and a swollen river to their advantage. The Cristino political community evacuated Huesca ahead of the insurgents' occupation, leaving the bishop (who had protested his 'ill health') behind. Perhaps the religious strategy was working. Don Carlos remained for three days in the city before moving on to take Barbastro, which surrendered without a shot being fired. En route the hungry Carlists resorted to routine excesses against villages. Don Carlos again heard Mass in this ancient city's cathedral whilst the Cristinos tried to regroup. General Oráa, respected by the Carlists as the 'grey fox', rushed forward with 12,400 infantry, 1,400 cavalry and artillery towards the Carlist forces which were now fortifying Barbastro. The Carlists during this battle on 2 June 1837 were in roughly equal numbers to the Cristinos, though vastly inferior in artillery. But despite this disadvantage, the Carlists proved to be expert in active defence, defeating the Cristinos' advance in detail by again using the geographical contours of the river bank to their advantage. The Carlist victory was marked by a 'civil war within a civil war', as the Carlists' 'foreign legion' (some 850 defectors from the foreign auxiliaries) exchanged fire with their former comrades, and the killing zone echoed with cries in French and German.[34]

Violence beyond the insurgent community

The Carlists' onward march into Catalonia was conditioned more by environmental than military factors. As the Carlists were always a battlefield state, exposed at any moment to Cristino incursions, the fluidity of territorial control dictated options for repression and control. Where Carlists expected to remain largely in control, as one of the two of the five zone models described by Stathis Kalyvas, the Carlists were incentivised to behave as 'stationary' insurgents, moderating their military and logistical demands as much as possible in order to shore up popular collaboration and the intelligence networks this offered. The short-term costs for 'stationary' insurgency in terms of forgoing pillage were high, yet the long-term benefits promised to be substantial. By contrast, areas beyond Carlist political control which offered only temporary occupation at best, were more likely to be used for short-term benefit, namely by pillage and intimidatory killings, prioritising the short-term gains over longterm costs. In these areas, Carlists were indeed 'bandits' for they were targeting individuals who were either explicitly beyond the insurgent political community (in the case of volunteer army officers and National Militia more than the case of conscripts) or implicitly by being civilians under Cristino control.[35] The Madrid government often exercised the same implicit calculations in their strategy. When General Espartero's forces occupied the key Basque coastal town of Guetaria, he believed the inhabitants could be won to the Cristino political community, and accordingly ordered draconian executions of some of his scouts who had been found guilty of brutalising civilians and their religion.[36] But the Carlist mountain fastnesses of Catalonia and Aragón seemed resolutely outside the Cristino community, and government actions reflected this. Deforestation edicts in the Catalan far west aimed to rob the insurgents of natural cover and also equated the savagery of the landscape with that of the insurgents.[37] The Prussian volunteer for Carlism, August von Goeben, recalled how Cristino counter-insurgents behaved with more brutality the further into enemy territory they reached, solidifying victims' support for the insurgency in the medium and long term: '(Cristino) troops used terror; all inhabitants, it was said, were Carlists and had to be destroyed, plundered, violated, and their homes burnt down. Several hundred, left with nothing, later joined the Carlists'.[38] But the insurgent Carlist side, unlike the incumbent nominally in control of most territory, had to make these calculations on a daily basis.

Particular tragedies, like the loss of 300 men drowned crossing the swollen Cinca, mixed with the general war for food. Wilhelm von Rahden's fond memory of Don Carlos subsisting on a pan of fried potatoes per day obscures the calamity facing villagers in the invaders' path.[39] Meanwhile, the entourage of favourites, place-seekers, priests, wives and other dependants gave a desperate – and hungry – quality to the insurgents' political community which was only partly obscured by the ostentatious thanksgiving and Masses held in 'liberated' villages in their path.[40] One ironic feature of religious liberation was the tendency for insurgents to deface churches by fortifying them as strong-points and smelting their bells to forge artillery.[41] The illusions of a receptive political community were frequently burst when villagers tried to conceal foodstuffs from their 'liberators' and exposed themselves to outright pillage.[42] Other inhabitants staved off the worst by ostentatiously welcoming the expedition, in reality to keep the Carlists as much at arms' length as possible. A visiting Polish aristocrat reported how villages were so exhausted by marches and countermarches that they kept a secret dual regime for appearances' sake. The approach of Cristino troops would be greeted by the local constitutional authorities, whereas the approach of Carlist insurgents would be greeted by a priest released from hiding for this purpose.[43]

Proximity and clerical militancy made the Catalan far west the first destination for the expedition. But northern Aragón had first to be traversed and the king's subalterns complained that agriculture in the Huesca area was too poor to support a sudden human influx. They also complained that the region was thickly garrisoned by Cristinos whose strength threatened to turn the merest defeat into a rout. The Carlists' raiding strategy worked to their advantage. As the British ambassador and key ally of the Cristinos observed: '(The Carlists') system in Aragón is the same as in Navarra: to deceive the Queen's Generals by false information, and to harass the troops by constant marches and counter-marches and then to beat them in detail'.[44] Carlist movements proved self-sustaining, either by not outstaying their welcome passing through friendly villages or by pillaging conquered population centres outside their political community. The Cristino counterinsurgency, by contrast, was stymied by poor intelligence and logistics, and a pay crisis which drove a series of army mutinies amidst the wider political upheaval of Liberal Spain.

The Carlist expedition, for its part, managed to keep advancing on a nutritional shoestring. Either conviction of coercion brought peasant guides to the aid of the invaders, who scouted out isolated food sources, especially sheep, ahead of the vanguard.[45] In reality, the revolutionary crisis affecting the cities during 1836–37 paralysed the government counter-insurgency, as Captains-General, like the Baron Meer in Barcelona, feared deploying more than token forces in the countryside for fear of losing control of their population centres to the radicals.[46] Thus the insurgents, untroubled by political factionalism while on campaign or by responsibilities for garrisoning cities, retained the upper hand when raiding. After traversing deep into Catalonia, the insurgents succeeded in making contact with the next Carlist safe zone in the Maestrazgo.[47] Linking up with Cabrera's vanguard at Xerta (Catalonia) on 29 June, the whole Expedition managed to cross the River Ebro, frustrating the Catalan counter-insurgency. Only the large towns and coastlines seemed beyond the insurgents' reach. The former were well garrisoned and possessed urban geography which rendered superior

Carlist tactics as useless, and the latter could always count on amphibious supplies and reinforcements landed by the British and Cristino navies.[48]

But the further away from its political community that the insurgents raided, the more formidable the resistance they faced. At Chiva (Valencia) on 15 July 1837 the exhausted and badly supplied bulk of the Carlist army was defeated by a smaller Cristino force, and only a rearguard action led by Cabrera himself prevented the flight from turning into a rout.[49] The insurgent retreat northwards through the Maestrazgo inflicted the direst suffering on soldiers and civilians alike of the whole 1837 campaign, as villages were subjected to pillage twice over by both sides. The better logistics supporting Generalísimo Espartero's counterinsurgency won over villagers outside the Carlist home region, as the Cristinos offered producers warehouses and receipts for impounded harvests, better prospects than those offered by famished insurgent raiders. Only reliable insurgent political communities, such as the devastated Teruel region, continued to aid the insurgents, further help coming from a careerist dispute on the Cristino side between Generalísimo Espartero and his jealous subordinate, General Oráa. Thus when Don Carlos halted his retreat at Villar de los Navarros (Aragón) in order to give battle on 24 August 1837,[50] the Carlists secured a major defensive victory including a booty of prisoners and badly-needed guns and supplies.[51] Carlists for once had numerical superiority in this battle: 11,000 infantry and 1,000 cavalry defeated slightly more than half the number of Cristino attackers over the course of five hours.[52] The fate that befell the Cristino prisoners who were moved around the starved countryside performing forced labour was horrific in the extreme. Most would die of hunger and typhus over the next months, as starvation rations forced them to scavenge for unripe root crops and, eventually, to commit cannibalism against their demised comrades. Cabrera condemned to execution some cannibals caught in flagrante, and yet the emaciated men could not even stand to receive the bullets who were finished off after hours of cruelty. Six months later, after an outcry from both sides, and despite the Eliot Treaty's provisions not applying to the eastern zone of operations, the surviving minority of captives were exchanged.[53]

The fateful Carlist victory at Villar de los Navarros coincided with another raid launched by a smaller force launched from Navarra under the command of Juan Antonio Zaratiegui. Originally intended to operate as a feint, Zaratiegui's forces faced denuded Cristino opposition and therefore acquired a momentum of their own. Traversing parts of Castile yet unmolested by war, the raiders made rapid progress. Early in August Zaratiegui conquered Segovia on the approaches to Madrid around the same time as the Royal Expedition approached the capital from the east.[54] News in the capital of Zaratiegui's approach caused panic. The Cristino front appeared to be evaporating across Castile. Cabrera's occupation of Guadalajara on the approaches to Madrid turned the political community Carlist, as the *caudillo* greeted the inhabitants from the town-hall balcony amidst public dances and open shops whilst the small Cristino garrison cowered in a tower one hundred metres away.[55] On the 6 August the government imposed a state of siege in Madrid, decreeing a wide range of pro-Carlist activities and opinions to be punishable by councils of war.[56]

With insurgent momentum restored deep into the enemy's political community, the Cristino side regrouped in the capital and often surrendered surrounding population centres after offering token resistance.[57] Valladolid was occupied in a gentlemanly manner, conforming to the dual regime explained earlier, as Liberal members of the local government resigned and retired to Madrid whilst pro-Carlist administrators took their place at Zaratiegui's service. This accommodation spared the city from pillage and atrocities.[58] But good manners ceased once a Cristino counter-thrust obliged Zaratiegui to abandon Valladolid to Cristino counter-occupation, and to join Don Carlos' expedition. The Cristino defence continued to be paralysed: only General Zurbano, a 'mere guerrilla chief' according to Ambassador Villiers, showed consistent success launching raids into the Carlist Basque Country, playing the insurgents at their own game.[59] But Cristino units were mostly paralysed on the approaches to Madrid itself. Generalísimo Espartero thought his command too weak to dare to punish a spate of politically-charged Cristino mutinies outside Madrid. Evaristo San Miguel, Madrid's military governor, admitted that the gravity of the Carlist invasion outweighed the need to punish indiscipline.[60]

By 10 September 1837 the insurgents seemed to be about to achieve a victory beyond comprehension. They stood in front of Madrid's city walls in an eerie standoff amidst vague reports of a diplomatic solution.[61] Whilst the Cristino civil and military authorities made a show of strength, some radicals even demanding the preparation of the city for siege warfare, the Queen-Regent played her double game. Ultimately the diplomatic initiatives came to naught. In a curious series of events which began as a civil-military dispute in Madrid, Espartero's Army of the North succeeded in putting the Carlist vanguard into flight. Espartero had been drawn into a standoff with the capital's revolutionary dictatorship of General Seoane over the political leadership of the revolution, which since 18 July 1837 had seen its Constitution of 1812 moderated by a bicameral legislature and other brakes on radicalism that were part of the Constitution of 1837.[62] Espartero's mobile forces marched on the capital in a show of strength against the revolutionary elements in Seoane's garrison. Fearing the worst, Don Carlos ordered a retreat from the city walls. Espartero inadvertently was heralded as the saviour of the liberal revolution and the Cristino regime, and the Queen-Regent hurriedly buried her diplomatic scheme.[63] Don Carlos, for his part, finally lost his nerve, and withdrew from the capital, even though the king's belligerent nephew, Prince Sebastián, had wanted to fall on Espartero's flank.[64]

Redeploying to Alcalá in a bid to fall on Espartero's flank, the Carlist army then received an order to retreat further still, causing outrage in the ranks. Rumours ran that Don Carlos had ordered the retreat to a more salubrious location in order to hear Mass. Others asserted that he wanted to avoid bloodshed. Madrid was now free of external threats. Antonio Quiroga made a triumphalist victory declaration couched in terms of popular defence and remarking the Queen-Regent's constitutional patriotism by having inspected the capital's Militia in person.[65]

The Carlists now lost all momentum in retreat, suffering rearguard defeats and an emboldened counter-insurgency. Villiers reported: 'Espartero has completely dispelled the prestige which the Carlists had created in the different villages where from the vast numbers in which they presented themselves they were looked upon as invincible'.[66] The panicked Carlists lost arms and men in growing numbers. In desperation, the king invited Cabrera to replace Moreno as the Carlist Commander-in-Chief but Cabrera – who had also reached Madrid and was poised to storm the capital as part of a pincer action – refused in thinly-disguised disgust at the failure of the king's leadership. Cabrera, after all, had wanted a lightning descent on Madrid lasting only two weeks as opposed to the leisurely, 'throne-and-altar' carnival march which the army's supreme commander, Prince Sebastián, had undertaken under the king's influence.[67] Carlist foreign auxiliaries, especially the sizeable German contingent were exasperated with the absence of the strategic vision worthy of Cabrera.[68] The artillery expert, Wilhelm von Rahden, did not understand the decorative time-wasting of the Carlist political community, and was already demoralised that earlier Carlist victories had been squandered with masses and festivals of grace rather than a more purposeful thrust.[69]

Meanwhile the Carlist retreat ploughed north to the safety of the Basque provinces, Espartero hard on its heels. Although the Cristinos could not prevent the Carlist forces of Zaratiegui and the king regrouping in their retreat, Espartero was ruthless in his pace and in the way he treated civilian populations in his path (not least because an unusually high number of Castillians had defected to the Carlists), sometimes promising the death penalty against hoarders of food and drink. Routed Carlists stumbled aimlessly into villages, and those who fell out from fatigue were shot by their own side in order to discourage desertion.[70] The dark side of raiding backfired onto the unhappy Carlist soldiers, as ruthlessness hitherto imposed on enemy communities was repaid in the insurgents' retreat. But Espartero faced problems of his own. His campaign was slowed down by the desperate hunger and neglect faced by his own men, as even the shoes which the hated *asentistas* (private contractors) supplied often had wooden or even cardboard heels which fell apart soon into campaigning.[71] Despite this, Espartero still managed to work wonders for the Cristino cause. On 4 October he gained a close victory at Retuerta after committing his reserves at the most opportune moment.[72] The tired remains of the Royal Expedition thereafter escaped back across the River Ebro and into their safe zone,

Cabrera and the path to defeat

Thus the Royal Expedition failed in circumstances which harked back to Early Modern warfare, of dynastic misunderstandings, lumbering and slow logistics, and religious piety. The retreating Carlists suffered a particularly dire lack of footwear, impeding their ability to counterattack or to acquire redress given that village shoemakers now feared the Cristino counterinsurgency more than the panicked insurgents. Usually the Carlists resorted to theft and pillage, or as Villiers reported: 'every village through which the Queen's army has passed lately the population has been found barefooted'.-[73] The collapse of the insurgents' political community obscured the logistical crisis besetting the Cristino counter-insurgency. General Espartero's pursuit was hamstrung by the private contractor system – the derided *asentistas* – who faced pillage and Carlist interceptions en route to supplying field armies, and delivered insufficient food and clothing.[74]

Militarily the expedition was a strategic disaster for Carlism, and revealed the fundamental dilemma affecting the rebels, who were secure in their mountain fastnesses yet too weak to carry to carry insurgency warfare decisively into Cristino territory. Don Carlos' retreat marked a critical third and final stage in Carlist politics which would pave the way for peace. The ignominious failure to take the capital saw international support for Carlism ebb, and no sooner had the Expedition retreated across the Ebro than politics moved in a radical direction. Uranga's limited offensives demonstrated the structural reality that the Carlists were virtually unbeatable in the internal lines of their Basque and eastern fastnesses, but frail in longer-range operations like the Royal Expedition. On 29 October 1837, the so-called Manifesto of Arciniega saw radical *apostólicos* dominate the king and launch public proclamations describing the Expedition's failure as the work of the hated peace faction known as *transaccionistas*. The term *ojalatero* ('if only') became popular in Carlist discourse, deriding the shirkers within the Carlist political community and the growing peace faction at the top.[75] Moderates were purged from the cabinet and many exiled, whilst such diehard Carlists (*apostólicos*) as Teijeiro and Guergué took their place and thereafter maintained a stranglehold on Carlist politics. The diehards would antagonise the army, undermining its Commander-in-Chief, Rafael Maroto, and ultimately result in a civil war within the civil war. That Maroto would win, and thus clear the way for peace, could not have been predicted by anyone in late-1837.[76] The defeat of the Expedition thus meant victory for the Carlist hardliners in the Basque country. But few Basques were fooled by the king's proclamation at Arceniega that he had 'returned only momentarily' and – bizarrely – that the Royal Expedition was a promising dress rehearsal for a future offensive of 'national liberation' (rather than the conclusive march on Madrid spoken about only weeks earlier).[77]

The Royal Expedition set the Basque country on a path of stagnation from which it would never recover, despite two more half-hearted expeditions launched from there in 1838 (the García and Negri raids). The fact that most of the Expedition's victories had been won on Aragonese territory consolidated Cabrera's leadership in the Maestrazgo and created a functioning and centralised Carlist state in the region. This offensive stalemated the Basque country but energised the Carlist Maestrazgo.[78] The 'Supreme Royal Governing Junta of Aragón, Valencia and Murcia' (Real Junta Superior Gubernativa de Aragón, Valencia y Murcia), set up in 1837, gave Cabrera an effective power of veto on all matters: whereas he could vote on any matters he saw fit, the Junta was forbidden from interfering in any way with military matters. Moreover, the propaganda value of local Carlist victories (especially at Huesca, Barbastro and Xerta) consolidated a phenomenon underway since the Liberals' confiscation of monastic properties in progress since 1836, namely, the growing enlistment of the conscript class into Carlist rather than Cristino ranks.[79] In many ways, the defeat of the Royal Expedition turned Carlism's gravity eastwards, away from the stalemate in the north.

During 1838 Cabrera's territorial control reached its zenith, and he 'conducted a villainous war to the death', escalating reprisals against prisoners and refusing mediation to bind the Eliot Treaty's humanitarian provisions to his zone.[80] One count taken in 1837 saw Cabrera's army totalling 11,418 infantry 1,282 cavalry, 337 artillery-men and 22 guns. He had shown foresight in turning the recaptured Cantavieja into an arsenal. This force was enough for Cabrera to take Morella on 25 January 1838, which now became his *de facto* capital.[81] But the demands for blood and treasure turned Cabrera into a military dictator who overruled the civilian junta and even subjected the Church to arbitrary rule. Churches and monasteries fell victim to scorched earth tactics just like all other buildings, bells were smelted into weapons, and for all the religiosity of Carlism, Ramón Cabrera had little patience or compassion for militarily 'useless' clerics. When Don Carlos protested at Cabrera's execution of a priest found guilty of theft, the Commander-in-Chief replied 'Your Majesty is being misled. The man I shot was no priest but a thief('). Cabrera demoted the pro-Carlist Bishop of Mondoñedo when he complained to Don Carlos of Cabrera's expulsion of 'useless' friars from Morella once they refused to bear arms in defence of the besieged capital.[82]

But Cabrera's secure position inside his political community in rural Aragón could not withstand the overwhelming weight of Cristino pressure bolstered by its overwhelming media and demographic resources. In June 1838 Carlists sent out feelers to Livorno, Italy, seeking foreign support for a new Carlist effort in Catalonia. But these feelers were intercepted by Cristino authorities and British vessels.[83] In December 1838 worse was to come. A private Carlist arms dealer seeking 15,000 firearms in Britain was frustrated when British authorities informed the Cristinos and put the blockading Royal Navy on watch for suspicious cargoes approaching the east coast of Spain.[84]

Defeat became inevitable once the Basque provinces agreed to conditional surrender in the August 1839 'Embrace of Vergara'. Generalísimo Espartero's overwhelming army descended on Cabrera's eastern zone over the winter of 1839–40. Cabrera's government descended into a virtual reign of terror, and Cabrera himself fell gravely ill in February 1840. Upon his recovery by May most of the east had fallen to government troops. By June 1840 Cabrera had interned himself in France and the First Carlist War was over. Thus the Carlist effort never squared the circle of defensive superiority versus offensive vulnerability. Raiding as a form of warfare is ultimately limited in its strategic and tactical effectiveness, as its employment is in itself usually a symptom of the military inferiority of the raiders.[85] By the same token, the irregular warfare that dominated the Carlist effort in the northern zone during 1833–35, and in the east from 1833–37, followed a pattern of tactical success which in turn guaranteed strategic failure. Both Zumalacárregui and, to a lesser extent, Cabrera, welcomed the transition from irregular to regular organisation of Carlist forces. The Carlists, after all, were supposed to be a 'regime-in-waiting' worthy of the support of Absolutist Europe, not the bandits and brigands of Cristino propaganda. But for all their advantages of interior lines, defensive topography, morale and greater military effectiveness, the Carlists could never outweigh the demographic, economic and military superiority of a Cristino regime in receipt of decisive foreign support.

Notes

1. E.g. Bullón de Mendoza, *Primera Guerra carlista*.
2. Lawrence, *Nineteenth-century Spain*, 64.
3. Bullón de Mendoza, *Primera Guerra carlista*, 685.
4. Cruz, "Un retrat del general carlista Ramon Cabrera," 102.
5. A.H.N., Estado, 8755: En territorio navarro, docs. 36 and 111: 14 April 1834 and 4 August 1834 letters from Viceroy of Navarra to comandante militar de armas de Puente de Reina.
6. T.N.A., FO/72/458, Doc. 92: 17 April 1836 letter from Villiers to Lord Palmerston.
7. Miraflores, *Memorias del reinado de Isabel II*.
8. T.N.A., FO/72/458, Doc. 92: 17 April letter from Villiers to Lord Palmerston.
9. Marichal, *Spain (1834–1844)*, 118–123. Marichal suggests that the hostility of the peasantry in the Maestrazgo towards change was the main cause of popular Carlism here. Such contemporary liberals as Evaristo San Miguel, on the other hand, thought that "agricultural improvements" in this bleak,

sparsely-populated, zone, could win over the population to liberalism (Rújula, *Historia de la guerra última en Aragón y Valencia*, LXXXVI-LXXXIX).

10. Rahden, *Cabrera*, 118.
11. Caridad Salvador, "Los carlistas de Valencia," 181–182.
12. T.N.A., FO 72/459, Doc. 149: 18 June 1836 letter from Villiers to Lord Palmerston.
13. Cruz, "Un retrat del general carlista Ramon Cabrera," 100.
14. Rahden, *Cabrera*, 77–90.
15. *Diario de Barcelona*, 12 August 1840.
16. Cruz, "Un retrat del general carlista Ramon Cabrera," 101.
17. Rahden, *Cabrera*, 81.
18. Ibid., 110–117.
19. Ibid., 66–68.
20. Lawrence, *Spanish Civil Wars*, 66.
21. T.N.A., FO72/458, Doc. 51, 7 March 1836 letter from Villiers to Lord Palmerston.
22. T.N.A., FO/72/459, Doc. 107: 2 May 1836 letter from Villiers to Lord Palmerston.
23. Rújula, *Cabrera*, 143–44.
24. Ibid., 39.
25. Bullón de Mendoza, *Primera guerra carlista*, 204–206.
26. For a conceptual explanation of the Kalyvas thesis, see Kalyvas, *Logic of Violence in Civil Wars*; and Kalyvas, "Micro-Level Studies of Violence in Civil War," 658–668.
27. Bullón de Mendoza, *Primera guerra carlista*, 213.
28. Comesaña Paz, "Armas inglesas para don Carlos," 731–758.
29. Bullón de Mendoza, *Expedición de Gómez*.
30. Rahden, *Cabrera*, 46.
31. Pirala, *Guerra Civil*, III, 284–289.
32. Lawrence, *Spain's First Carlist War*, 167.
33. *Gaceta Oficial*, 9 August 1836.
34. Lawrence, *Nineteenth-century Spain*, 81; and Lichnowsky, *Erinnerungen*, 137.
35. For a conceptual analysis of the short-term versus long-term 'investment' made by insurgents, see Stewart and Liou, "Do Good Borders Make Good Rebels?" 284–301.
36. *Eco del Comercio*, 2 January 1836.
37. Lawrence, *Spain's First Carlist War*, 125.
38. Goeben, *Vier Jahre in Spanien*, 338.
39. Rahden, *Cabrera*, 220.
40. Pirala, *Guerra civil*, IV, 108–115.
41. Rahden, *Cabrera*, 49–50.
42. Lichnowsky, *Erinnerungen*, 238–245.
43. Dembowski, *Dos años en España durante la guerra civil*, 40.
44. T.N.A., FO 72/483: 1 September 1837 letter from Villiers to Palmerston.
45. T.N.A., FO 72/483, No. 269: 9 September 1837 letter from Villiers to Palmerston.
46. Pirala, *Guerra civil*, IV, 115–122.
47. Ibid., 127–132.
48. Ibid., 154–160.
49. Pirala, *Guerra civil*, IV, 144–154; and Oyarzun, *Historia del carlismo*, 81.
50. The battle of Villar de los Navarros is also known as the Battle of Herrera.
51. Pirala, *Guerra civil*, IV, 154–160.
52. See note 44 above.
53. Lawrence, *Nineteenth-century Spain*, 83.
54. Burgo, *Historia de la primera guerra carlista*, 247; and Pirala, *Guerra civil*, IV, 177–183.

55. Rahden, *Cabrera*, 60.
56. Bullón de Mendoza, *Primera guerra carlista*, 191.
57. Pirala, *Guerra civil*, IV, 193–199.
58. Oyarzun, *Historia del carlismo*, 79, 82 83; and Pirala, *Guerra civil*, IV, 206–207.
59. See note 45 above.
60. Pirala, *Guerra civil*, IV, 215–218.
61. Janke, *Mendizábal*, 250–252.
62. This charter provided one Cortes representative per 50,000 inhabitants, enfranchised all men paying at least 200 reales in annual taxes or receiving an annual private income of at least 1,500 reales, which amounted to 1 inhabitant in 48 enjoying full citizenship (whereas under the 1834 Royal Statute this figure had been 1 in 213), (Palacio Atard, *La España del siglo XIX*, 200–202).
63. Espadas Burgos, *Baldomero Espartero*, 61; and Pirala, *Guerra civil*, IV, 431–439.
64. Aróstegui, Canal, and Calleja, *Guerras carlistas*, 61; Clemente, *Guerras carlistas*, 111; and Pirala, *Guerra civil*, IV, 227–228.
65. Pirala, *Guerra civil*, IV, 230–231.
66. T.N.A., FO 72/483, No. 272: 23 September 1837 letter from Villiers to Palmerston.
67. Holt, *Carlist Wars*, 172.
68. Lichnowsky, *Erinnerungen*, 134.
69. Burgo, *Historia de la primera guerra carlista*, 212–213.
70. T.N.A., FO 72/483, No. 275: 26 September 1837 letter from Villiers to Palmerston.
71. Pirala, *Guerra civil*, IV, 232–237.
72. Ibid., 237–239.
73. T.N.A., FO 72/483, No. 290: 7 October 1837 letter from Villiers to Palmerston.
74. Pirala, *Guerra civil*, IV, 237–239; for a more positive assessment of the private contractor system in the Early Modern era, see Parrott, *The Business of War*.
75. Pirala, *Guerra Civil*, II, 260–265.
76. Pirala, *Guerra civil*, IV, 246–255; Julio Aróstegui Sánchez, 'La aparición del carlismo y los antecedentes de la guerra' in *Historia de España: La era isabelina y el sexenio democrático (1834–1874)*, 121–122; and Canal, *El carlismo*, 98.
77. Oyarzun, *Historia del carlismo*, 105–106.
78. Rújula, *Historia de la guerra*, LXXV-LXXVI.
79. Remírez de Esparza, *Carlismo aragonés*, 46–59; and Burgo, *Historia de la primera guerra carlista*, 185–186.
80. A.H.N., Emigrados, 8119, 29 December 1838 letter from Spanish government to British minster in Madrid.
81. Pirala, *Guerra civil*, IV, 408–41; and Oyarzun, *Historia del carlismo*, 157.
82. Cruz, "Un retrat del general carlista Ramon Cabrera," 103–104.
83. T.N.A., FO 72/516: 6 June 1838 letter from Villiers to Palmerston.
84. T.N.A., FO 72/500: 6 December 1838 letter from Villiers to Palmerston.
85. Inbar, *International Affairs*, 1428–1431.

Disclosure statement

No potential conflict of interest was reported by the author.

Bibliography

Primary sources

A.H.N. (Archivo Histórico Nacional, Madrid).

Dembowski, Carlos. *Dos Años En España Durante La Guerra Civil, 1838–40*. Madrid: Editorial Crítica, 2008.

Diario de Barcelona.

Gaceta Oficial.

Goeben, August Karl von. *Vier Jahre in Spanien. Die Carlisten, Ihre Erhebung, Ihr Kampf Und Ihr Untergang*. Hanover: Hahn, 1841.

Lichnowsky, Felix. *Erinnerungen Aus Den Jahren 1837, 1838 Und 1839*. 2 vols. Frankfurt-am-Main: Sauerländer, 1841.

Rahden, Wilhelm Baron von. *Cabrera. Erinnerungen Aus Dem Spanischen Bürgerkriege*. Frankfurt-am-Main: Friedrich Wilmans, 1840.

T.N.A. (The National Archives, Kew).

Secondary sources

Aróstegui, Julio, Jordi Canal, and Eduardo G. Calleja. *Las Guerras Carlistas: Hechos, Hombres E Ideas*. Madrid: La Esfera de los Libros, 2003.

Bullón de Mendoza, Alfonso. *La Primera Guerra Carlista*. Madrid: Actas, 1992.

Bullón de Mendoza. *Expedición de Gómez*. Madrid, 1984.

Burgo, Jaime del. *Para La Historia De La Primera Guerra Carlista*. Pamplona: Diputaciòn Foral de Navarra. Instituciòn Prìncipe de Viana, 1981.

Canal, Jordi. *El Carlismo*. Madrid: Alianza Editoria, 2004.

Caridad Salvador, Antonio. "Los Carlistas De Valencia. La Reacción En Una Ciudad Liberal (1833–40)." *BROCAR*, 36 (2012): 161–183. doi:10.18172/brocar.1568.

Clemente, Josep Carles. *Las Guerras Carlistas*. Barcelona: Península, 1982.

Comesaña Paz, Alfredo. "Armas Inglesas Para Don Carlos: El Incidente De La *Express Packet*." *Hispania* LXXVIII, no. 260 (septiembre-diciembre, 2018): 731–758.

Cruz, Nuria Sauch. "Un Retrat Del General Carlista Ramon Cabrera." n edited by Daniel Montaña and Josep Rafart, 93–108. *El Carlisme Ahir I Avui*. I Simposi d'Història del Carlisme, 11 de maigde, 2013.

Espadas Burgos, Manuel. *Baldomero Espartero: Un Candidato Al Trono De España*. Madrid: Diputación Provincial de Ciudad Real, 1986.

Holt, Edgar. *The Carlist Wars in Spain*. London: Putnam, 1967.

Inbar, Efraim. "What after Counter-Insurgency? Raiding in Zones of Turmoil", *International Affairs* 92, no. 6 (2016): 1428–1431.

Janke, Peter, *Mendizábal Y La Instauración De La Monarquía Constitucional En España, 1790–1853*. Madrid: Siglo XXI de España Editores, 1974.

Kalyvas, Stathis. *The Logic of Violence in Civil Wars*. Cambridge: Cambridge University Press, 2006.

Kalyvas, Stathis. "Micro-Level Studies of Violence in Civil War: Refining and Extending the Control-Collaboration Model." *Terrorism and Political Violence* 24, no. 4 (Sep./Oct. 2012): 658–668. doi:10.1080/09546553.2012.701986.

Lawrence, Mark. *The Spanish Civil Wars: A Comparative History of the First Carlist War and the Conflict of the 1930s*. London: Bloomsbury, 2017.

Lawrence, Mark, *Spain's First Carlist War, 1833-1840*. Basingstoke: Palgrave Macmillan, 2014.

Lawrence, Mark. *Nineteenth-century Spain: A New History*. London: Routledge, 2019.

Marichal, Carlos. *Spain (1834–1844): A New Society*. London: Tamesis, 1977.

Miraflores, Manuel Pando Fernández Penedo de. *Memorias Del Reinado De Isabel II*. Madrid: Ediciones Atlas, 1964.

Montaña, Daniel and Josep Rafart, eds. *El Carlisme Ahir I Avui*. Avià: I Simposi d'Història del Carlisme, 11 de maig de 2013.

Oyarzun, Ramón. *Historia Del Carlismo*. Madrid: Editorial Pueyo, 1965.

Palacio Atard, Vicente. *La España Del Siglo XIX*. Madrid: Espasa-Calpe, 1978.

Parrott, David. *The Business of War: Military Enterprise and Military Revolution in Early Modern Europe*. Cambridge: Cambridge University Press, 2012.

Pirala, Antonio. *Historia De La Guerra Civil Y De Los Partidos Liberal Y Carlista*, 6 vols. Madrid: Turner/Historia, 1984.

Remírez de Esparza, Francisco Asín. *Aproximación Al Carlismo Aragonés Durante La Guerra De Los Siete Años*. Zaragoza: Librería General, 1983.

Rújula, Pedro, ed. *Historia De La Guerra Última En Aragón Y Valencia (escrita Por F. Cabello, F. Santa Cruz Y R. M. Temprado)*. Zaragoza: Institución 'Fernando el Católico', 2006.

Stewart, Megan A. and Yu-Ming Liou. "Do Good Borders Make Good Rebels? Territorial Control and Civilian Casualties." *Journal of Politics* 79, no. 1 (Jan. 2017): 284–301. doi:10.1086/688699.

Zamora, Jover and José María, ed. *Historia De España: La Era Isabelina Y El Sexenio Democrático (1834–1874)*. Tomo XXXIV, fundada por Ramón Menéndez Pidal. Madrid: Espasa-Calpe, 1981.

Holmes' front: constructing a new face of battle for America's Civil War

Susan-Mary Grant

ABSTRACT

In his seminal study of the changing nature of warfare between Agincourt and the Somme, military historian John Keegan proposed that future historians might consider combatants' emotions in their assessments of the impact and nature of conflict. Recent years have witnessed the emergence of the history of emotions as an analytical approach, but rarely, if ever, is this directed toward the study of military history, far less the history of insurgencies and counter-insurgencies. This paper examines America's civil war (1861–1865) as a case study of the ways in which an emotional history approach might illuminate not the physical experiences of but rather the immediate and longer-term reactions to counter-insurgency conflict through a focus on one specific individual, the future Supreme Court Justice Oliver Wendell Holmes, Jr. It proposes that Holmes, whilst not a man of the ranks, nevertheless can reveal the wider ramifications of civil war and its emotional impact, both individually and culturally. As a relatively limited internecine war, one not fought by professional armies but by volunteer forces, America's civil war highlights the ways in which the soldier's response points us toward the kind of emotional revolution that has, to date, mainly been located within the European nations.

Writing several decades ago now, historian Marvin Cain argued that America's mid-nineteenth-century civil war needed its own 'Face of Battle.' He was, of course, invoking military historian John Keegan's ground-breaking 1976 work, *The Face of Battle: A Study of Agincourt, Waterloo and the Somme*. In this, Keegan sought to shift the focus of military history away from the grand narrative and toward the experiences of the typical soldier. In particular, and this was where Cain's perspective on 'motives and men' came in, Keegan was interested in the reactions of those who found themselves at 'the sharp end' of war. Keegan's was a work that prompted a long-needed reappraisal of the 'face of battle,' but it was also a product of its time and might, perhaps more accurately, have been entitled the mind of battle, since that, in the end, was what it was really about. In this respect it was very much a bridging book, a study that both broke with the military narratives of the past and foreshadowed the contemporary fascination with both the long-term physical and psychological costs of conflict.[1]

Keegan's selection of Agincourt, Waterloo and the Somme was not accidental. In part it reflected the cultural import of these battles in British history; and in part, by metaphorically standing in one geographical location, Keegan gained a valuable vantage point from which to assess the changing nature of European warfare over some four centuries via these three flashpoint conflicts. Although influenced by Keegan's approach, this paper, located as it is in the relatively limited temporal and physical space that contained America's civil conflict, has of necessity adopted a rather different approach. It examines the experiences and changing attitudes of one of the Civil War's more famous soldier-writers, future Supreme Court Justice Oliver Wendell Holmes, Jr., over the course of three, for him, crucial military encounters: Balls Bluff in 1861; Antietam in 1862; and, finally, the Wilderness (Virginia Overland) Campaign of 1864.

Through Holmes' reactions to the Civil War, this paper seeks to trace the changing face of battle as that was experienced and expressed by a single individual using the interrogative tools developed by scholars, Barbara Rosenwein among others, studying the history of emotions. This is not a meaningless academic exercise, not least because Keegan himself, the preeminent military historian, highlighted the need for at least 'some exploration of ... combatants' emotions,' and noted that so long as this were not simply an excuse for 'the indulgence of our own' it was 'essential to the truthful writing of military history.'[2]

It must be noted at the outset that there are some obvious limitations to the approach taken here. First, Holmes was not a man of the ranks; he was, by the end, a lieutenant colonel. His experience of warfare was relatively restricted. He fought for the Union, and solely in the Eastern Theatre of the conflict. Indeed, his battlefield experience did not go far beyond Virginia; only Antietam took him, briefly, into Maryland. So his personal experiences in and reactions to the war derived from a limited, albeit active, military geography. The second issue to be aware of is that in America's civil war, when writing about commanders and troops, one is frequently not writing about men with any kind of professional military training or military-informed assumptions about command and its complications. The regular army of the United States was part of the Civil War narrative, certainly, but in terms of size was swamped by the volunteer forces raised by both sides to defend and attack, respectively, the Union.

Finally, the nature, and the (re)naming of the internecine conflict that the United States experienced in the mid-nineteenth century makes it far from clear in respect of whether it was a war or an insurgency, a rebellion within a single nation or a conflict between two nations. As described by politicians and its participants alike, it began as a civil war, moved, from a Union perspective, to a rebellion, and then returned to a civil war, the better to achieve sectional reconciliation but also to downplay the role that slavery had played in the war's causes and emancipation in its outcome. Yet to this day the Confederate bid for separate statehood is still described as an insurgency, and federal opposition to this as counter-insurgency, despite the fact that, with the exception of some guerrilla attacks and tactics deployed on the border, notably in Western Virginia, Kansas and Missouri, both sides sought to fight a conventional war; neither deployed anything approximating asymmetrical tactics on the battlefield. But there was a problem as regards the validity of both sides' claims to statehood. For the Union, there was a certain Constitutional vagueness surrounding the right of secession that only victory in warfare resolved; in brief, the Union War was effectively state-making as a counter-insurgency tactic. For the Confederacy, the war was 'not an *intra*national but an *inter*national one,' yet the more established nations of Europe failed to see it that way. And this confusion, or rather these alternative possible interpretations of the war informed Holmes' reactions to it, and offer us an indicator of some of the complexities of the emotional fallout from internecine insurgencies, from wars that devi-ate, in crucial ways, from clearly conventional conflicts.[3]

No more was the Civil War fought through a series of discrete battles that, in Keegan's words, obeyed 'the dramatic unities of time, place and action.' A glance at one of the many composite maps of that conflict might suggest that it was precisely that, but these virtual recordings of the war can be misleading. And historians are increasingly turning their attention to these maps' geographical silences, their visual absences, often in search of what Keegan described, but dismissed, as 'the sort of sporadic, small-scale fighting which is the small change of soldiering,' recognising as they do that too fixed a focus on an individual battle as a stand-alone set piece action risks losing sight both of the war and the men who fought it. In the case of America's civil war, too, it is sometimes the case that it is only in hindsight that an armed encounter comes close to conforming to Keegan's definition of a battle; as was the case with Gettysburg in July of 1863. Generally regarded as the central, crucial battle of the war, the initial encounter between Union and Confederate troops was accidental and its outcome restricted to a specific time and place only because the Union's Major General, George Meade, failed to pursue Confederate commander Robert E. Lee back into Virginia. Whatever else Gettysburg was, it was not a 'deliberate' encounter, planned for by the commanders of each side who anticipated pitting 'their wits against each other to make their plans suc-ceed' on that spot, at that time.[4]

Balancing some of the difficulties is the fact that, first, although at the time of writing *The Face of Battle*, Keegan regarded the First World War as the first conflict in which we can clearly 'hear the voice of the common man' and America's civil war as offering only 'infant murmurs' In this regard, in fact this is very much not the case. With a literacy rate of some 90 percent in the Union, and some 80 percent in the Confederacy, and with some 3 million serving, it was not surprising that millions of letters moved to and from the front during the American Civil War; too many for General William T. Sherman's liking, since some of these in all innocence, but also in considerable detail, described location, military manoeuvres, and where the armies were scheduled to deploy next. The American Civil War is a virtual treasure trove for anyone looking to gain access to the sentiments of the 'common man.' And the study of such letters has become a mini-industry of its own in the years since Keegan dismissed the Civil War's epistolary output as negligible, to such an extent, indeed, that 'this literature has reached a point of diminishing returns,' according to one leading Civil War historian.[5]

And second, the post-war narratives and memoirs that emerged from America's civil war were relatively balanced in terms of according a voice to commanders and their armies. *Century Magazine*'s four-volume *Battles and Leaders* publication that appeared in the late 1880s had begun, as the name suggests, as a series of articles in the magazine written mainly by the officer class; mainly, but not entirely. As the opening volume indicated, the series included contributions from, and relied on the support of surgeons as well as soldiers, sometimes non-combatant women as well as fighting men, and 'veterans, from the highest rank to the lowest.' Whether this relatively democratic dissemination of Civil War voices can be ascribed to a uniquely American republican reading appetite, what Keegan rather admiringly described as 'the spirit of American life,' or whether it owed more to America's post-war's burgeoning publication industry combined with an awareness that the nation's Civil War generation was passing away may be a debatable point.[6]

By examining Holmes' contribution to such literature, both the letters he wrote and the diary he kept during the war along with, or perhaps especially contrasted against the ways in which he invoked the conflict in his later years within the context of military history inflected by the interrogative tools introduced by the history of emotions, we can begin to trace the outline of the face of battle from one soldier's perspective. Holmes' war, and the shift in his response to it, also helps us to trace the changing face of the Civil War over the four years in which it was fought. In that sense, Holmes here is a conduit as well as the main character; his war a foreshadowing of the modern world's reaction to the small wars and insurgencies that characterise contemporary conflict, and the personal costs of these on both combatants and civilians when the face of battle is no longer simply that of the soldier.

Ball's Bluff

For historians looking for the face of battle, the soldier arrives fully-fledged on the battlefield, and his reactions, assuming we have some record of them, can usually only be assessed within that military moment. There is no before and, often, no after. In the case of Oliver Wendell Holmes, we have some evidence of the road that took him to war, and we should take advantage of that to trace the emotional landscape that he traversed before launching ourselves, as Holmes' regiment did, onto the slopes of Ball's Bluff on the shores of the Potomac in October of 1861. We should do what the officers singularly failed to do that morning and take our bearings; not of the landscape of that first battle, but of the internal landscape of Holmes himself, to assess by what route he arrived at this, his first and almost destined to be his last battle in the Civil War. And the first question to ask is: prior to joining the Union army, what was Holmes' emotional response to the idea of conflict?

Looking back at the Civil War from a distance of over half a century, Holmes observed to his close friend and frequent correspondent, the British political theorist Harold Laski, that despite the carnage of the First World War, human beings continued to romanticise conflict. Not 'only all society but most romance rests on the death of men,' he wrote, 'and where the most men have died there is the most interest.' Holmes was undoubtedly speaking from experience. Over fifty years before, his decision to volunteer to fight for the Union was informed by just such a romantic outlook, one inspired by a youthful enthusiasm for fantasy battle. Holmes was hardly alone in this respect; Victorian versions of medieval military narratives, especially as imagined in the work of Scottish author Sir Walter Scott and English Poet Laureate Alfred, Lord Tennyson were universally popular. In their fiction, the face of battle was that of the chivalric knight, an invention that Scott himself described as no more than 'a beautiful and fantastic piece of frostwork, which has dissolved in the beams of the sun.' Not that this particular fantasy ever fully dissolved for Holmes. In his seventieth year he still indulged himself in 'periodic wallows in Scott,' explaining to a correspondent that 'the old order in which the sword and the gentlemen were beliefs is near enough to me to make this their last voice enchanting in spite of the common sense of commerce.'[7]

It is important to stress that the mythical worlds in which Holmes and others absorbed themselves as adolescents, and in Holmes' case for much of his life, were not literary imaginaries, emotional spaces, written for children, nor even for the educated elite of antebellum American society. It is certainly the case that, as far as the kinds of emotional communities that Rosenwein identified are concerned, Holmes was born into and raised in an unusually tight-knit example not just of an emotional but also an economic and cultural community. Bostonians of Holmes' class, as one of his biographers noted, 'lived in the same neighborhoods, went to the same schools, read the same books, shared the same ignorances as well as knowledge, belonged to the same clubs, dined at the same houses, donated to the same charities, and married their cousins.' Holmes' reading matter, significant, according to philosopher Martha Nussbaum, as a window into emotions that constituted 'parts, highly complex and messy parts, of ... reasoning itself,' was not, however, confined to this Bostonian Brahmin class. A wealth of evidence from soldiers' letters suggests that many sought inspiration for what was, for most, an entirely new experience in 'stories of the ancient world, of the Spartan soldiers and their women rather than in any contemporary military works as such. People knew how they were supposed to act in times of war and crisis,' Civil War historian Edward Ayers has noted; they 'followed the scripts and took great pleasure in speaking lines they already knew.'[8]

And lest we conclude that the classics were, even then, somewhat rarified reading matter for the masses, evidence from Union draft board examinations later in the war, by which point federal officials, somewhat dismissively, believed they were drawing on 'the dregs of the population of the cities,' revealed individuals who saw themselves as a modern-day Achilles and whose emotional expectations of battle were primed not by histories of America's eighteenth-century revolution, but by the literary landscapes of different lands and times. Nor should we forget the fascination that Napoleon held for the antebellum generation; not simply in respect of his tactical influence on military leaders, on which much has been written, but as regards his, and his army's reputation's more widespread resonance within the public imagination. In common with the French Revolutionary troops examined by Alan Forrest, America's civil war armies comprised educated men, lawyers as well as labourers, teachers as well as tradesmen, accountants as well as artisans. And, like Forrest's French soldiers, this was not a 'silent army'; these were men educated to be citizens, with expectations and opinions that informed their reactions to the war.[9]

Undoubtedly there were class differences within Union ranks, divisions between what Lorien Foote sums up as 'the gentlemen and the roughs' that 'reflected the socioeconomic diversity of the northern population.' Yet both, the gentlemen and the roughs alike, were expected, and expected themselves, to conform to an ideal of the citizen soldier of the republic rooted in the Cincinnatian ideal, the 'symbolic link between patriotism and the plough' established during the American Revolution, epitomized by George Washington himself, and aspired to in the concept of the largely volunteer forces that fought the Civil War. This was, in its broadest sense, and cutting across class lines, an example of one emotional community of the later antebellum era, in which men, mainly, shared the 'same norms of emotional expression and value' in relation to conflict.[10]

Nuancing that community, and complicating Holmes' story, as indeed that of many Union soldiers at the start of the Civil War, was the role that abolition played in his self-identification with the warrior-heroes of legend. In terms of peer-pressure and influence, both Holmes' close friendship with Harvard colleague, Quaker, and active abolitionist Norwood Penrose 'Pen' Hallowell and his familial links with leading abolitionist spokesman and activist Wendell Phillips were undoubtedly significant. But arguably more dominant in his decision to fight for the Union was his sense, as a youth, of what achievement in life might mean, and the importance of military success in this respect. Recalling a time long before the Civil War, long before Harvard, when he had seen a parade (most probably commemorating the Mexican-American conflict of 1846–48) wending its way through the streets of Boston, Holmes remembered being 'most impressed by the part played by a carload of veterans. I got the notion, which has persisted,' he recalled, 'that the glory of life was to be carried in a civic procession, in a barge, as a survivor – I did not inquire too curiously of what.'[11]

Holmes' progress along the path toward that imagined future civic procession was slow to get started. He secured, in part through his father's influence, a commission in the Twentieth Massachusetts, the so-called 'Harvard Regiment,' a nod to, and by some other regiments a damning indictment of the 'blue-blooded' background of most of its officers. Its commander, Colonel William Raymond Lee, was a West Point graduate, giving the regiment the benefit of having a military professional in charge when many other regiments had politically-appointed, and wholly inexperienced leadership. It also had the benefit of experienced regimental surgeons, one of whom, Henry Bryant, had previously served with the French Army in Algeria. Countering these benefits, the regiment was operating at about two-thirds strength when it left Boston for the front.[12]

At first this was of little significance. Holmes' first experience of military life was far from the front, at which point his emotional state can best be described as one of anticipation and, in his own words, impatience. During basic training at Fort Independence in Boston Harbour he wrote of 'enjoying the life much,' and of living in 'expectation of a fight very soon.' At the same time, this enthusiasm for new experiences was constrained by Holmes' expectation that his regiment would not fight; that they would serve as reserves. War was not real to him at that point. There was no fear, because there was, as he saw it, no danger. His first experience of conflict was a static one; largely what he had anticipated. It was war at a distance, quite literally. 'It seems so queer to see an encampment & twig men through a glass & think they are our enemies,' Holmes wrote to his mother. He reported seeing 'one man in a straw hat sitting unconcernedly on his tail apparently a guard on duty for the secesshers (*sic*). Men and horses,' too, he noted, 'are seen from time to time from the tops of the trees,' but as 'firing across the river is forbidden,' all that Holmes could do was 'sit & look & listen to their drums.' He was also able to hear 'some of our pickets talking' to their Confederate counterparts, but that was as close to the 'reality' of life as a soldier as he got in the late spring and summer of 1861.[13]

Holmes' reaction at this point, however, indicates one of the emotional complexities of irregular warfare, of conflict not between different nations but between different parts of the same nation. Indeed, for Holmes, unlike for some others, there was no sense of fighting for the nation at all, no concept of a Union worth defending. He simply never mentions it. For him, it was idealized conflict abstracted from any cause; conflict against countrymen, and for that reason somehow less real at its outset than it would later become. Like the respective armies' pickets that Holmes overheard, still behaving as brothers rather than belligerents, the early stages of the Civil War seemed hardly a war at all. The assumption on the Union side had been that a few, brief battles would persuade the secessionists of the error of their ways and secure their return to the federal fold. Holmes probably shared that assumption. But in this he was wrong.

By the time Holmes wrote to his mother again, or at least the next extant letter in sequence that we have from him, he was in a hospital bed in Camp Benton, 'wounded but pretty comfortable,' two days after the Battle of Ball's Bluff. What is most notable about this letter is that Holmes almost instantly – in his first sentence – sought to assure her that he both 'felt and acted very cool and did my duty.' He was keen to convey to her that he had been something of a hero, 'out in front' of the regiment, where he was initially hit by a spent shell. Advised to 'Go to the Rear,' Holmes refused and instead 'rushed to the front' where, as he reported, 'I waved my sword and asked if none would follow me,' at which point, perhaps unsurprisingly, he was shot again; twice. What his mother made of the information that followed, namely that he had considered taking a bottle of laudanum that his father had given him, so that he might die without pain, is not recorded. He decided not to do so, as he felt he would survive. He repeated his 'conviction I did my duty handsomely,' and pronounced himself 'happy' about this. But what followed, with no indication of emotion at this point, was a brief list of all those who had not survived, or had survived seriously wounded: one of the Captains, John C. Putnam, lost an arm; one of the Sergeants, John Merchant, a man whom Holmes had personally recruited, was dead.[14]

Ball's Bluff swiftly gained a notoriety of sorts, drawing comparisons with the Charge of the Light Brigade, in part because both arose over a misunderstanding about orders, in part because of the futility and relatively high human cost of the Union attack on October 21st, 1861, the near certainty of its defeat that day on the banks of the Potomac that separated

Maryland from Virginia. But it was not for this reason that Holmes later recorded his experiences during that battle in some detail in a diary entry. Perhaps inevitably, given that he had been wounded near the start of the battle, Holmes' memories of it were closer to medical than military, but still grounded in the kinds of imaginative martial imagery that he had absorbed earlier in his life. Unlike several of his comrades who later composed detailed accounts of the battle itself, Holmes was entirely focused on his own experiences and on the 'intensity of the mind's action and its increased suggestiveness, after one has received a wound.' This prompted him to recall a children's book about the English Civil War, Frederick Marryatt's *Children of the New Forest*, in which a central character 'died with terrible haemorrhages & great agony,' and he drew parallels with his own situation. But his emotional response to his wounds recorded in his diary largely simply echoed the letter to his mother. Believing he was dying, his response was to consider that he would die 'as a soldier ... shot in the breast doing my duty.' He reported himself not afraid, but 'proud.'[15]

Antietam

Holmes' emotional community during the Battle of Ball's Bluff could best be described as his immediate regiment, imbued as this was with the kind of romantic nationalist ideals that Kanisorn Wongsrichanalai has examined in respect of college-educated New England officers. There was no wider emotional community in the form of the nation that he identified with; not then, and not later. But there was already a clear sense in which he regarded himself as set apart from that regimental community by the simple act of survival. On the single occasion in his diary that he does not focus on himself or, rather, that the sentiment he reports, a 'sickening,' is prompted by the pain of others, is when he sees John Putnam's amputated arm lying 'in a pool of blood.' The 'spectacle wasn't familiar then,' he recalls. One of the regimental surgeons, Bryant, despite his previous experience in Algeria, was almost equally shocked, as he had been expecting to have only a few casualties to contend with. Instead he was met by large numbers of wounded 'crying and shrieking' and a floor 'covered with blood.' One 'man had three balls through his head, one taking off his nose and one of his eyes' while 'another man was lying near him with brain projecting from a wound in the side of his head.' Lieutenant William Lowell 'Willie' Putnam 'was lying near the fireplace with his intestines projecting from a wound in his abdomen.' He did not survive.[16]

Holmes' pride in his own conduct cannot but have been tempered by such sights, by the unexpected death toll from an engagement that was neither planned nor, when underway, well executed, fought by volunteer troops who had no experience of battle before. It was a battle that, taken together with the Union defeat at Bull Run in July, prompted a change in the way that the federal government approached the war. In terms of casualties, these were disproportionately on the Union side despite both sides being evenly matched in terms of numbers with around 1,700 troops each: some 85 percent of Union forces fell, or about nine hundred men, in contrast to Confederate losses of around one hundred and fifty. Given these circumstances, it is no surprise that Holmes had a list of the dead to convey to his mother; more detailed descriptions of the wounded to confide to his diary. But it is his response to this extreme slaughter that interests us. In his letter, there is no obvious emotional reaction. In his diary, clearly written some period after the event and likely, given the internal evidence, a few years after it, he records, but only briefly, the emotional impact that the results of Ball's Bluff had on him.

There may be a missing link here, but it is another indicator that insurgencies and counter-insurgency actions within a single nation can be productive of a different effect on the troops engaged than a conflict fought on foreign soil more typically produces; an effect deriving from the proximity of home-front and battlefield, and the ease and frequency with which many troops moved between the two. Given the nature of America's civil war, when many troops, certainly socially well-positioned troops such as Holmes, could be treated at home if wounded, returning to their regiments only when recovered, some of the emotional response to combat became lost in the domestic setting and never committed to paper. More crucially, the emotional link between home-front and battlefield became significant in respect of how the individual soldier reacted to the pressures of conflict. America's civil war troops hardly had the opportunity to become absorbed in a military world, even if they did become seasoned in battle, since they inhabited it in a complete sense neither psychologically nor physically.

At the same time, it is the soldier who, while not perhaps seasoned as such has nevertheless had some of his youthful enthusiasm for conflict modified by battle's reality that has formed the focus of much of the research into America's civil war over the last few decades. Cain's 1982 'assessment of motives and men' drew a line, as it was intended to do, under those themes that had dominated American Civil War historiography up to that point. Whilst not entirely clearing the decks, it did look toward a future in which the 'face-of-battle approach' might reveal with more clarity 'the motives and purposes of the soldiers who resolved the issues of 1861–65.' Yet Cain concluded with a caveat, and its subject was, in part, Oliver Wendell Holmes. Hindsight, Cain warned, citing leading Civil War historian Allan Nevins, both inflected and infected the Union soldier's understanding of his motives during the Civil War, and he singled out Holmes as someone who shifted from 'wartime pragmatist' to 'middle-aged romantic' on the subject of the Civil War. This is, however, almost exactly the opposite of

Holmes' actual reaction to the Civil War: what Cain reads as romanticism in Holmes' later life is a much more pragmatic, outward-focussed deployment of his veteran status. In this he was hardly alone.[17]

Cain's difficulty with Holmes' perspective is hardly surprising. One of the problems in reading American Civil War soldiers is the fact that, in almost all of the literature devoted to them, a significant problem constrains our view: the tendency, near-universal, for America's civil conflict to be discussed entirely in isolation from all other wars, and those elements that now most interest historians – soldier motivation, the effects of disease, the environmental impact of the war – located solely within the United States in the mid-nineteenth century. Yet Holmes' response to Ball's Bluff, and the more general motivation that apparently informed his behaviour, differed little, if indeed at all, from the reactions of troops at Waterloo, indeed from many troops in many conflicts. As Keegan discusses the matter, it 'was the receipt of wounds, not the infliction of death, which demonstrated an officer's courage' at Waterloo, and this 'demonstration was reinforced by his refusal to leave his post even when wounded.' Officers in the early nineteenth century, Keegan observes, 'were most concerned about the figure they cut in their brother officers' eyes. Honour was paramount, and it was by establishing one's honourableness with one's fellows that leadership was exerted indirectly over the common soldiers.'[18]

The subject of honour among Civil War soldiers has interested historians for quite some time; at first in the broader context of masculinity, and more recently in the specific context of class relations within the Union Army. And in that latter respect, Holmes' behaviour at Ball's Bluff fits tidily into the profiles of officers and troops highlighted by both Wongsrichanalai and Foote in their respective discussions about manhood and honour in the Union Army. Foote, in particular, stresses that 'Northern men, to a greater extent than their southern counterparts, did not conform to a singular understanding of manhood or to a uniform ideal of what constituted manly behaviour.' At the same time, she emphasises that there was a public-facing aspect to masculinity, and the 'reputation for an upright character was an essential mark of a gentleman.' And Keegan, too, stresses that reputation is rarely clear cut in its attainment. He draws a distinction between 'classically Heroic' expressions of honour at Waterloo and the 'Homeric' ideal that relied both on single combat and public recognition of victory. Civil War soldiers such as Holmes, however, effectively straddled these subtleties of distinction between a classical age that had inspired their youth and the already publicity-hungry nineteenth century. Just as the 'facts of death' at Waterloo became 'invested, by those who recount what they witnessed, or even perpetrated, with a tinge of Romantic regret,' so Civil War soldiers faced the ages through public accounts of their actions, their honour, and their purpose. For Holmes, in this respect as in others, it was the Battle of Antietam that proved the turning point.[19]

It is, however, almost impossible to see Holmes at Antietam, the bloodiest single day of America's civil war, blurred as he has become by what might be described as the emotional prism of parental pride; except that it was a little more complicated than that. In the aftermath of Ball's Bluff the leading New England magazine, *Harper's Weekly*, ran an article headlined 'New England Never Runs,' in which Holmes was held up as evidence of 'the steadfast bravery of the Massachusetts boys,' each one of whom was a 'hero' and exhibited 'the quality of which invincible armies are made.' *Harper's* identified Holmes as exemplary in this respect because the wound he had sustained at Ball's Bluff was in his chest ... no, not in the back. In the breast is Massachusetts wounded, if she is struck. Forward she falls, if she fall dead." Having witnessed so many of their comrades do precisely that at Balls' Bluff, it is doubtful that survivors of that engagement appreciated the somewhat mawkish tone of *Harper's* article. In later years, Holmes did recall the publicity *Harper's* had accorded him, but in relation to Antietam on which occasion he had been 'hit in the back,' a fact that, he pointed out, had not been 'so good for the newspapers.'[20]

By the time of Antietam Holmes was a more seasoned soldier, but his letter home on the eve of that battle in September, 1862 differed little from that prior to Ball's Bluff. He did note that his health had never really recovered since the Battle of Fair Oaks in the summer, but as before Ball's Bluff he reported being 'in reserve' although this time he was anticipating being in a battle. Holmes' emotional reliance on his parents and on home was evident: 'Write just as regularly (as you have *not* lately),' he chided them, 'whether I answer or not – I want letters.' His next missive, again as at Ball's Bluff, was to tell his parents that he had been wounded. He strove to reassure them: 'Usual luck – ball entered at the rear passing straight through the central seam of coat & waistcoat collar coming out towa (sic) the front on the left hand side,' he observed, 'yet it don't seem to have smashed my spine or I suppose I should be dead or paralysed or something.' His bravado continued a few days later when he assured them that he was 'smoking pipes partaking of the flesh pots of Egypt swelling round as if nothing had happened to me.' He was heading home, he told them, and 'I may remark that I neither wish to meet any affectionate parent half way nor any shiny demonstrations when I reach the desired haven.'[21]

Holmes was too late. His father, having received word by telegraph of his son's situation, had already set out for the front. But in the end this was no private, familial affair. A few months later, Holmes, Snr. published a lengthy description of his trip in *The Atlantic Monthly*, an account that was as much travel narrative as it was an account of the search for his son. It opened in a high emotional register: 'In the dead of the night which closed upon the bloody field of Antietam,' Dr Holmes began, 'my household was startled from its slumbers by the loud summons of a telegraphic messenger. The air had been heavy all day with rumors of battle, and thousands and tens of thousands had walked the streets with throbbing hearts.' And it soon lost sight of Holmes, Jr., as his father pursued his peripatetic path over the landscape of the war, never allowing his readers to forget that his journey was undertaken 'with a full and heavy heart' and with blood that that was 'chilled' by 'what were perhaps needless and unwise fears.' He travelled in the company of others who had received similar telegrams after Antietam, one of whom, William Dwight, received word before they had got very far at all that his son, Wilder, had died. Holmes shared with his readers his awareness that 'empty words of consolation' were useless in such a situation, that grief of this nature was a private matter; 'borne' by men, 'felt' by women.[22]

Holmes, Snr.'s reaction, both to the receipt of the telegram and to what he witnessed *en route* to Antietam, would not have differed much if at all from civilian reactions to other conflicts, albeit that the speed of information available to him had not been available to those whose relatives fought at Agincourt, or Waterloo, or in the French Revolutionary armies. His descriptions of battle and its aftermath, a 'red vortex' followed by the 'long, diverging rays' of the wounded straggling along the roads after it, 'flushed with fever or pale with exhaustion or haggard with suffering' strove for visual effect. By contrast, his description of how he searched with a lantern among the wounded with 'a kind of thrill' as he 'looked upon the features it illuminated' strove to convey personal emotion, the hope combined with fear that he was experiencing. When he does finally find his son, however, emotion is put aside in favour of a contrived, overtly masculinised, in a nineteenth-century context, pretence at its absence: 'How are you, Boy? How are you, Dad?'[23]

Dr Holmes' intention in writing such an account was, as he saw it, a way of personalising a conflict that was in danger of becoming too abstract for the northern population. Photographer Matthew Brady had exhibited his grim collection, 'The Dead of Antietam,' in New York, but the living, the wounded were another matter. Holmes was aware that the 'great caravan of maimed pilgrims' was so large that it made 'a joint-stock of their suffering; it was next to impossible to individualize it and so bring it home as one can do with a single broken limb or arching wound.' He was, in short, seeking to show the northern readership of *The Atlantic Monthly* the face of battle on a human, emotional scale. At the same time, he sought to elevate it beyond that with its concluding Biblical invocation of the Prodigal 'son and brother [who] was dead and is alive again … was lost and is found.' From individual soldier to spiritual symbol, Antietam was the turning point for Holmes' public persona. It is also a reminder that the face of battle in America's civil war was especially difficult to separate from the society that had sent the soldier into battle in the first place in a war such as this was, fought mainly by volunteer troops on American soil, where the emotional timbre of the battlefield achieved such sympathetic resonance across the home-front.[24]

The Wilderness

Viewing Antietam from a distance of nearly three decades, the author Edward Bellamy echoed Dr Holmes' sense of democratic community embroiled in civil conflict. The Civil War soldier, he recalled, 'had staked for his country not only his own life but the earthly happiness of others also, having been fully empowered by them to do so ... In offering up their lives to their country,' he intoned, 'they had laid with them upon the altar these other lives which were bound up with theirs, and the same fire of sacrifice had consumed them both.' It is doubtful if Holmes, however, understood his relationship with his family in this way. They had provided emotional sustenance in the early stages of the war, but by its final years he no longer regarded the conflict from the same perspective as his parents. This was a role reversal of sorts. Uncertain about the war at the outset, Dr Holmes became increasingly focussed on its moral dimension. How 'idle it is to look for any other cause than slavery as having any material agency in dividing the nation,' he suggested the year following Antietam. Dr Holmes' was capable of seeing the war as one whose moral purpose had changed: from a war to save the Union, it had, by 1863, become a war to end slavery. His son, by contrast, became increasingly embittered in the aftermath of Antietam; more cynical in his thinking, the abolitionist sentiments that may, at least in part, have taken him into battle in the first place had dissipated in the face of conflict's physical and psychological costs.[25]

Antietam therefore became a turning point for Holmes as for the Union. As the latter moved more towards the eradication of slavery following the issuance of the Preliminary Emancipation Proclamation in September, to be followed by the Emancipation Proclamation itself at the start of 1863, Holmes moved in the opposite direction, along with his regiment that, by this point, was overtly hostile to abolition. The predominant emotional tone in Holmes' correspondence from this period is one of frustration, almost anger: anger at the 'thick-fingered clowns we call the people,' particularly as 'represented at political centres' where they were, in his view, 'vulgar, selfish & base'; anger at the African Americans whose cause had become that of the Union; had, indeed, always been part of the Union if not always overtly. In separating himself from that cause, Holmes was, in a sense, separating himself from the new-found enthusiasm of the northern home-front for emancipation, from its revived faith in its armies and the cause they were fighting for. Holmes had none. He described the Union Army as simply 'tired with its hard, & its terrible experience & still more with its mismanagement.' He also no longer regarded the war as a rebellion, but an attempt, a futile one in his view, to subjugate 'a great civilized nation.' By the end of 1862, the emotional tone of Holmes' correspondence was almost wholly negative: dominated by words like 'anxiety,' and 'miserable,' and phrases such as 'helpless hopelessness' and 'disappointed expectations.'[26]

In part Holmes' emotional downturn can be ascribed to his own ill-health, but in part to the death of comrades whose demise he was reporting more frequently despite his attempts to claim that 'it's odd how indifferent one gets to the sight of death – perhaps,' he added, 'because one gets aristocratic and don't value much a common life.' His mood, perhaps inevitably, was not entirely consistent; there were times when he expressed the desire to 'be ready and cheerful,' and admitted to being 'a totty bit melancholy just now' but assumed it was 'only a passing cloud.' The general emotional trajectory, however, remained downwards. There seems, too, to have been a greater willingness to admit feeling despondent to female family members, and especially to his mother, than to Dr Holmes. Removed from combat for the best part of a year due to an injury sustained at Marye's Heights (the second Battle of Fredericksburg) in May of 1863, by the time Holmes returned to active duty his regiment was poised to take part in the Wilderness Campaign at the start of May, 1864, part of General Ulysses S. Grant's Overland Campaign against the Army of Northern Virginia. At this point he described himself as 'well and in excellent spirits.' There was some suggestion from his diary – never diagnosed – that Holmes may have been suffering a mild case of 'soldier's heart' (Da Costa's syndrome) at this time, but his general tone was, for most of May, neutral. He restricted himself to descriptions of the various engagements, and self-reflection was, unusually for him, either kept to a minimum or absent altogether. This did not last.[27]

Ultimately, Holmes may have simply become overwhelmed by the appalling human cost of Grant's relentless Overland Campaign, a method of warfare generally taken to more closely foreshadow that of the First World War than previous Civil War battles. Union casualties in the Wilderness in the first two weeks of May, 1864 were just under eighteen thousand; Holmes reported that his own Corps had lost some ten and a half thousand men. He recorded spending his time 'burying dead,' who were 'piled in the trenches 5 or 6 deep – wounded often writhing under superincumbent dead – The trees,' he observed, 'were in slivers from the constant peppering of bullets.' His tone, by this point, was as of a man thinking (and writing), metaphorically, through gritted teeth: he was having a 'terribly tiresome time,' and the marches were 'fatiguing.' He became more direct in letters home. 'Before you get this you will know how immense the butcher's bill has been,' he wrote to his parents on 16 May 1864, and 'I have not been & am not likely to be in the mood for writing details.' These 'nearly two weeks have contained all of the fatigue & horror that war can furnish … nearly every Regimental off – I knew or cared for is dead or wounded … I have made up my mind,' he told them, 'to stay on the staff if possible till the end of the campaign & then if I am alive, I shall resign.'[28]

By the conclusion of his military career, Holmes had become estranged from family and regiment alike. 'The duties & thoughts of the field,' he told his parents, 'are of such a nature that one cannot at the same time keep home, parents and such thoughts as they suggest in his mind at the same time as a reality.' The 'intense yearning' for home that 'immediately precedes a campaign,' he observed, vanished in the horror of the fighting. He still relied on his parents' letters, but was to discover that they did not support his decision to leave the army. He read their opposition as a slight on his 'soldierly honor.' But he was not, he told them, 'the same man ... and I *will not acknowledge the same claims upon me under those circumstances* that existed formerly.' He assured them that he was not 'demoralized,' but that he had 'started this thing a boy' and was 'now a man.' Consequently, as he saw it, his 'duty' had 'changed.' He tried to explain to them that he could 'do a disagreeable thing or face a great danger' when he understood it to be his duty, but the 'duty of fighting' had now 'ceased' for him. The war had caused him 'much suffering of mind and body,' and he felt that he could no longer 'endure the labors & hardship of the line.' By the 19th of July, 1864, Holmes' war was over, and he was home.[29]

Holmes' return from the battlefield did not mean that a switch had been flipped in respect of his emotional responses to the Civil War. Indeed, for Holmes' scholars, the Civil War is usually deemed to have coloured his emotional response to the rest of his very long life. This is largely in part because Holmes referred back to it frequently, providing a wealth of evidence of the war's long-term impact on one combatant's post-war emotional landscape. But in this case, the most obvious evidence may be misleading. For a generation of Americans, the Civil War became the core experience of their lives, a generation that Holmes described as 'set apart by its experience' and which he later urged 'to keep the soldier's faith against the doubts of civil life.' Scholars who consider both of these statements, delivered in the course of two very famous Memorial Day addresses in 1884 and 1895 respectively, have, perhaps inevitably, focused on what they reveal about Holmes' perspective, the long-term impact emotional that the Civil War had on him. If the spotlight is turned away from Holmes, however, to light up not the speaker but his audience, the war's wider generational impact becomes clearer; and Holmes' role as a reflection of an evolving post-war emotional community, rather than a single individual romanticizing his own past, becomes clearer.[30]

The face of battle, in short, needs now to be placed in context; understood in relationship to the wider public culture and to the degree of emotional investment that Americans, combatant and non-combatant alike, had in the Civil War. One early piece of evidence in support of the significance of this lay in the public-facing, literary representations of Holmes the soldier. What *Harper's Weekly* and Dr Holmes had initiated during the war, authors such as Henry James continued after it. James' short story, 'A Most Extraordinary Case,' published in *The Atlantic Monthly* in 1868, detailed the experiences of a fictional Civil War veteran that, in its details, clearly was based on Holmes, a man described as 'a singularly nervous, over-scrupulous person,' who for three years 'had been stretched without intermission on the rack of duty.' The subsequent 'sense of lost time' that James' character experienced, remained 'his perpetual bugbear.' In this respect, this quasi-fictional character's sense of lost time resonated not just with the relatively restricted community of non-fictional Civil War veterans but with the wider society.[31]

When Holmes delivered his 1884 address, the audience did comprise veterans: members of the John Sedgwick Grand Army of the Republic (GAR) Post No. 4 at Keene, New Hampshire. It took place against the rise of the GAR, the largest and the most politically influential Union veterans' organization that, by 1890, consisted in some 400,000 members organized in seven thousand posts. Together with the success of Century Magazine's 'Battles and Leaders' series of the 1880s, the power of the GAR both reflected and reinforced a more widespread interest in the Civil War some two decades after that conflict had ceased, the rise of the veteran voice in contemporary culture. In such a social and political climate, it was hardly surprising that so professionally successful a veteran as Holmes was by that time might be invited to address the GAR; and it was no surprise what he chose to say on that occasion, or that the tone of his address was overtly emotional, almost sentimental. He evoked specific emotions directly: 'love and grief' and ascribed these to 'heroic youth.' 'Through our great good fortune,' he came near to preaching, but to an audience of the converted, 'in our youth our hearts were touched with fire. It was given to us to learn at the outset that life is a profound and passionate thing.' It was a cliché, but no less powerful for that, when he assured them that warfare is productive of 'the closest tie which is possible between men – a tie which suffering has made indissoluble for better, or worse.'[32]

In 1895, by contrast, Holmes was addressing a rather different audience; the graduating class at Harvard. By then, his public persona, besides its professional, judicial dimensions, was very much that of Civil War veteran; an echo of a, by then, distant conflict, that had a powerful emotional resonance for a generation that had never known battle. And Holmes played up to this. He was hardly the only Civil War veteran to do so, but he did it spectacularly well. The 'soldier's faith' that he invoked for the youth about to embark on their lives was overtly emotional in tone even as it purported to be quite the opposite in its attacks on the 'literature of sympathy' that critiqued battle and sacrifice. Holmes' proposed to an audience of young men that they inhabited a fantastical realm, 'a place hung about by dark mists, out of which come the pale shine of dragon's scales, and the cry of fighting men, and the sound of swords.' This was more than a retreat into the past. It bears comparison with John Keegan's description of the contemporary English response to Agincourt as an emotive reaction informed by, and essentially located in art and literature. It was warfare 'vividly visualized,' an echo of 'a fading national myth,' except that, in Holmes' case, the myth was a sentimentalized, emotive evocation of another time and place that had never even existed in the United States; that had never, in fact, ever existed at all.[33]

Conclusion

Holmes' Civil War letters and his diary, albeit one that he edited, or more accurately redacted some of the latter after the war, offer us at least a glimpse into one man's emotional response to battle. Further, they provide a consistency of insight from the perspective of an individual soldier who was unusually prone to a greater degree of self-reflection than many. In short, Holmes not only wrote about his experiences, he wrote down how he felt about those experiences. Speaking to Nussbaum's point about the need for 'texts that contain a narrative dimension,' Holmes's letters, shorter than some, are more to the point than most. But the letters and diaries are documents; they are, to use Julie Livingston's concept, already flattened, their descriptions of wounding 'the human body ... rendered as text,' no more than a pale imitation of what Holmes was actually feeling, or what he was actually prepared either to tell his parents or to leave to posterity. They also point to one perspective, and no more than that; they give us a clue into a single face of battle, not a generalised insight into conflict. From them we can glean some evidence as to Holmes' expectations, to an extent his motivations, his behaviour in combat, his fears and frustrations, but what we cannot do is what *Harper's Weekly*, his father and William James did try to do at the time; we cannot make him into the universal Union soldier.[34]

If we look up from a narrow focus on Holmes, however, the value of his letters and diaries, and the value of approaching these with an eye to the history of emotions, becomes far clearer. Then we can begin to trace, in the mid-nineteenth century United States, the beginnings of the kind of 'emotional revolution' that, to date, has been largely located solely in the European nations, and rarely in relation to warfare. Consisting, as William Reddy defines it, in a 'cult of sensibility,' this has, as he notes, been broadly applied: to 'Methodism, antislavery agitation, the rise of the novel, the French Revolution ... and the birth of Romanticism.' Holmes' emotional response to the American Civil War suggests that it may be time to consider the emergence of this kind of sensibility in relation to conflict; not in the individual trauma of its aftermath, but in its wider applicability to the societies that experienced conflict; the 'gradual and thorough alteration in emotional common sense' that conflict produces. This approach at least hints at the creation of a new, northern emotional community structured around the Civil War, similar to, if not quite as aggressive as that in the states of the former Confederacy, where the very public, racially exclusive emotional community that constituted the Lost Cause has exercised scholars' attention ever since the war ceased, and continues to do so.[35]

From the moment he joined his regiment as a volunteer soldier, Holmes was engaged in what Reddy sums up as 'a kind of emotional self-shaping' in the context of the martial environment and what he understood that to require in behavioural terms. After the war, as he gradually evolved into the symbolic Union veteran, his 'self-shaping' returned to its origins, absent the uncertainty, the fear, the anger and the frustration that had typified his military career. This paralleled the process by which both North and South, seeking to incorporate what had been a brutal internecine conflict into a common history, sentimentalised the Civil War in ways that downplayed the realities of race and emancipation, treason and loyalty to construct a coherent, martial narrative around which both sides could cohere. Largely viewed by historians as an entirely negative process because of its denial of the racial realities of the post-war South and the denial of the racial rights that so many Union troops had, in fact died for, the alternative approach in a civil conflict remains difficult to envisage if neither side is prepared to engage in a realistic appraisal of, and an attempt to resolve the original causes of the conflict. In the war's aftermath, the United States failed to address, or even acknowledge what one doctor, writing at the start of the Second World War, described as the 'disintegrating emotions' that would only be productive of 'a peace more bitter and vindictive' in the years to come. That, in many respects, was America's fate after 1865; even if, with its focus on the war's veterans, that fact was obfuscated for decades.[36]

In this regard, the face of battle in America's civil war became, in the war's aftermath, more akin to a military mask behind which former enemies could disguise the war's causes and its consequences and satisfy a nation increasingly hungry for accounts of martial fervour as an acceptable face of combat. In presenting himself to the public as a Civil War veteran, even when he had been elected onto the Supreme Court, Holmes also reveals the extent to which he, in common with many other veterans, had become locked in a martial persona. For Holmes, there was doubtless a degree of pragmatism in promoting his past rather than his profession as his dominant public image, in aligning himself with what he termed 'the armies of the dead,' but this tendency also spoke to his having internalised, to some extent at least, his temporary identity as a soldier. In his later years, he continued to describe himself as 'an old warrior who cannot expect to bear arms much longer,' at a time when he had already outlived most of the old warriors of his generation. In his case, as doubtless in many others, the face of battle was the only one that he felt comfortable enough to show the world.[37]

Notes

1. Cain, "A 'Face of Battle' Needed"; and Keegan, *Face of Battle*, 36.
2. Rosenwein, *Emotional Communities*; and Keegan, *Face of Battle*, 17.
3. Foster, "What's Not in a Name," 416–7, 421–2; Rafuse, "Confederate Insurgency"; Storey, "I'd Rather Go to Hell"; and Armitage, "Secession and Civil War," 43–4.
4. Keegan, *Face of Battle*, 1, 31; and Parish, *American Civil War*, 285.

5. Keegan, *Face of Battle*, 17; McPherson, *What They Fought For*, 4; and Gallagher, *The Union War*, 61.

6. Editors, *Battles and Leaders*, Vol. 1, x; and Keegan, *Face of Battle*, 55–6.

7. Holmes to Laski, 28 July 1927, in Howe (ed.), *Holmes-Laski Letters*, II, 966; Scott, "Essay on Chivalry," II, 261; and Holmes to Baroness Moncheur, quoted in Wilson, *Patriotic Gore*, 747.

8. Baker, *Justice from Beacon Hill*, 22–3; Nussbaum, *Upheavals of Thought*, 3; and Ayers, *Presence of Mine Enemies*, 150–1.

9. Baxter, *Statistics, Medical and Anthropological*, lxxv; Crosby, "Surgeon's Report," 187; Bonura, *Under the Shadow of Napoleon*; Lee, "Le Culte de Napoléon," 148–49; Forrest, *Napoleon's Men*, x; and Berkovich, *Motivation in War*, 3–5.

10. Foote, *Gentlemen and Roughs*, 4; Linenthal, *Changing Images of the Warrior Hero*, 51–2; Kammen, *Season of Youth*, 100; Snyder, *Citizen-Soldiers*, 86; and Rosenwein, *Emotional Communities*, 2.

11. Holmes, "Reflections on the Past and Future," 163.

12. Novick, *Honorable Justice*, 38; and McPherson, *Battle Cry of Freedom*, 329.

13. Holmes to mother, 1 May, 8 September 2023 September 1861, in Howe, *Touched With Fire*, 3–4, 5, 11–12.

14. Holmes to mother, 23 October 1861, in Howe, *Touched With Fire*, 13, 18.

15. Holmes Diary No 2., in Howe, *Touched With Fire*, 23, 27.

16. Wongsrichanalai, *Northern Character*, 2–3, *Passim*; Holmes Diary No 2., in Howe, *Touched With Fire*, 25; Bryant quoted in Miller, *Harvard's Civil War*, 80–1; and Putnam, *William Lowell Putnam*, 9.

17. Cain, "'A Face of Battle' Needed," 26–7.

18. Keegan, *Face of Battle*, 165.

19. Foote, *Gentlemen and Roughs*, 3, 86; Wongsrichanalai, *Northern Character*, 159; Keegan, *Face of Battle*, 167; and Bederman, *Manliness and Civilization*, 18–22.

20. *Harper's Weekly*, "New England Never Runs," 9 November 1861; Holmes to Pollock, 28 June 1930 in Howe (ed.), *Holmes-Pollock Letters*, II, 269–70.

21. Holmes to parents17, 18 and 22 September 1862, in Howe, *Touched With Fire*, 63–7.

22. Holmes, Snr., "My Hunt," 738–9, 741.

23. Ibid., 743–4, 746, 769.

24. *New York Times*, 20 October 1862; Holmes, Snr., "My Hunt," 743–44.

25. Bellamy, "An Echo of Antietam," 381; and Morse, *Life and Letters*, I, 43–4.

26. Holmes to Amelia (sister), 16 November 1862; to mother, 12 December, in Howe, *Touched With Fire*, 71, 73, 74–5.

27. Holmes to mother, 14 December 1862; to father, 29 March 1863; to parents, 3 May 1864; Diary 7 May 1864, in Howe, *Touched With Fire*, 78, 91, 103, 108.

28. Holmes Diary, 12 May 1864; to parents, 16 May 1864, in Howe, *Touched With Fire*, 117–9, 121–2.

29. Holmes to parents, 16 May; 30 May; 7 June 1864, in Howe, *Touched With Fire*, 123, 135, 142–3.

30. Holmes, "Memorial Day Address, May 30 1884"; and Holmes, "Memorial Day Address, 30 May 1895," 80–7, 87–95.

31. James, "A Most Extraordinary Case," 466; Novick, *The Young Master*, 98–9; Edel, *The Untried Years*, 232; and Halperin, "Henry James's Civil War," 28.

32. Holmes, "Memorial Day Address, 1884,", 15.

33. Holmes, "Memorial Day Address, 1895," 73–4, 78, 80–1; Keegan, *Face of Battle*, 61; and Kagan, *Eye of Command*, 89.

34. Nussbaum, *Upheavals of Thought*, 1–2; and Livingston in Eustace et.al. "AHR Conversation," 1488.
35. Reddy, *Navigation of Feeling*, x, 143.
36. Reddy in Eustace et.al. "AHR Conversation," 1497; Reddy, *Navigation of Feeling*, x; Watson, "Disintegrating Emotions," 663; and Janney, *Remembering the Civil War, Passim*.
37. Holmes, "Memorial Day Address, 1884", 13; and Holmes, *Collected Legal Papers*, ix.

Disclosure statement

No potential conflict of interest was reported by the author.

Bibliography

Armitage, David. "Secession and Civil War." In *Secession as an International Phenomenon: From America's Civil War to Contemporary Separatist Movements*, edited by Don. H. Doyle, 37–55. Athens: The University of Georgia Press, 2010.

Ayers, Edward L. *In the Presence of Mine Enemies: The Civil War in the Heart of America, 1859–1863*. New York: W. W. Norton, 2004.

Baker, Liva. *The Justice from Beacon Hill: The Life and Times of Oliver Wendell Holmes*. New York: Harper Collins, 1991.

Barton, Michael. "Journalistic Gore: Disaster Reporting and Emotional Discourse in the *New York Times*, 1852–1956." In *An Emotional History of the United States*, edited by Peter N. Stearns and Jan Lewis, 155–172. New York: New York University Press, 1998.

Battles and Leaders of the Civil War. Vol. 4. New York: The Century Company, 1887.

Baxter, Jedediah Hyde, ed. *Statistics, Medical and Anthropological of the Provost-Marshal-General's Bureau*. 2 Vols. Vol. 1. Washington, DC: Government Printing Office, 1875.

Bederman, Gail. *Manliness and Civilization: A Cultural History of Gender and Race in the United States, 1880–1917*. Chicago: University of Chicago Press, 1995.

Bellamy, Edward. "An Echo of Antietam." *Century Magazine* 38, no. 3 (1889): 374–381.

Berkovich, Ilya. *Motivation in War: The Experience of Common Soldiers in Old-Regime Europe*. Cambridge: Cambridge University Press, 2017.

Bonura, Michael A. *Under the Shadow of Napoleon: French Influence on the American Way of Warfare from the War of 1812 to the Outbreak of WWII*. New York: New York University Press, 2012.

Cain, Marvin R. "A 'face of Battle' Needed: An Assessment of Motives and Men in Civil War Historiography." *Civil War History* 28, no. 1 (1982): 5–27. doi:10.1353/cwh.1982.0059.

Crosby, Dixi. "Surgeon's Report: New Hampshire, Third District." In *Statistics, Medical and Anthropological of the Provost-Marshal-General's Bureau. 2 vols*, edited by J. H. Baxter, 187. Vol. 1. Washington, DC: Government Printing Office, 1875.

Edel, Leon. *Henry James: The Untried Years, 1843–1870*. Philadelphia: Lippincott, 1953.

Eustace, Nicole, Eugenia Lean, Julie Livingston, Jan Plamper, William M. Reddy, and Barbara H. Rosenwein. "AHR Conversation: The Historical Study of Emotions." *The American Historical Review* 117, no. 5 (2012): 1486–1531. doi:10.1093/ahr/117.5.1487.

Foote, Lorien. *The Gentlemen and the Roughs: Violence, Honor, and Manhood in the Union Army*. New York: New York University Press, 2010.

Forrest, Alan. *Napoleon's Men: The Soldiers of the Revolution and Empire*. London: Bloomsbury, 2006.

Foster, Gaines M. "What's Not in a Name: The Naming of the American Civil War." *The Journal of the Civil War Era* 8, no. 3 (2018): 416–454. doi:10.1353/cwe.2018.0049.

Gallagher, Gary W. *The Union War*. Cambridge, MA: Harvard University Press, 2011.

Halperin, John. "Henry James's Civil War." *The Henry James Review* 17, no. 1 (1996): 22–29. doi:10.1353/hjr.1996.0005.

Holmes, Oliver Wendell, Sr. "My Hunt After 'The Captain'." *The Atlantic Monthly* 10, no. 62 (December, 1862): 738–764.

Holmes, Oliver Wendell. *Collected Legal Papers*. New York: Harcourt, Brace and Howe, 1920.

Holmes, Oliver Wendell, Jr. "Reflections on the past and Future: Remarks at a Dinner of the Alpha Delta Phi Club, Cambridge, September 27, 1912." In *The Occasional Speeches of Justice Oliver Wendell Holmes*, edited by Mark DeWolfe Howe, 163–167. Cambridge, MA: The Belknap Press of Harvard University Press, 1962.

Holmes, Oliver Wendell, Jr. "Memorial Day Address, May 30 1884." In *The Essential Holmes: Selections from the Letters, Speeches, Judicial Opinions, and Other Writings of Oliver Wendell Holmes, Jr.* 1992. Reprint, edited by Richard A. Posner, 80–87. Chicago and London: The University of Chicago Press, 1996.

Holmes, Oliver Wendell, Jr. "Memorial Day Address, May 30 1895." In *The Essential Holmes: Selections from the Letters, Speeches, Judicial Opinions, and Other Writings of Oliver Wendell Holmes, Jr.* 1992. Reprint, edited by Richard A. Posner, 87–95. Chicago and London: The University of Chicago Press, 1996.

Howe, Mark DeWolfe, ed. *Touched with Fire: Civil War Letters and Diary of Oliver Wendell Holmes, Jr., 1861–1864*. Cambridge, Mass.: Harvard University Press, 1946.

Howe, Mark DeWolfe, ed. *Holmes-Laski Letters: The Correspondence of Mr Justice Holmes and Harold J. Laski*. Vol. 2. Cambridge, MA: Harvard University Press, 1953.

Howe, Mark DeWolfe, ed. *Holmes-Pollock Letters: The Correspondence of Mr. Justice Holmes and Sir Frederick Pollock, 1874–1932*. Vol. Two Volumes in One. 2nd ed. Cambridge, MA: The Belknap Press of Harvard University Press, 1961.

James, Henry. "A Most Extraordinary Case." *The Atlantic Monthly* 21, no. 126 (April, 1868): 461–485.

Janney, Caroline E. *Remembering the Civil War: Reunion and the Limits of Reconciliation*. Chapel Hill: The University of North Carolina Press, 2013.

Kagan, Kimberly. *The Eye of Command*. Ann Arbor: The University of Michigan Press, 2006.

Kammen, Michael. *A Season of Youth: The American Revolution and the Historical Imagination*. New York: Alfred A. Knopf, 1978.

Keegan, John. *The Face of Battle: A Study of Agincourt, Waterloo and the Somme*. 1976. Reprint, London: Pimlico, 1991.

Lee, Kennett. "Le Culte de Napoléon Aux États-Unis Jusqu'à La Guerre De Sécession." *Revue de l'Institut Napoleon* 122 (1972): 145–156.

Linenthal, Edward T. *Changing Images of the Warrior Hero in America: A History of Popular Symbolism*. New York and Toronto: The Edwin Mellon Press, 1982.

McPherson, James M. *Battle Cry of Freedom: The Civil War Era*. New York and Oxford: Oxford University Press, 1988.

McPherson, James M. *What They Fought For, 1861–1865*. Baton Rouge: Louisiana State University Press, 1994.

Miller, Richard F. *Harvard's Civil War: A History of the Twentieth Massachusetts Volunteer Infantry*. Hanover and London: University Press of New England, 2005.

Morse, John T. *Life and Letters of Oliver Wendell Holmes*. Vol. 2. London: Sampson Low, Marston and Company, 1896.

Novick, Sheldon M. *Honorable Justice: The Life of Oliver Wendell Holmes*. New York: Dell Publishing, 1989.

Novick, Sheldon M. *Henry James: The Young Master*. 1996. Reprint, New York: Random House, 2007.

Nussbaum, Martha C. *Upheavals of Thought: The Intelligence of Emotions*. Cambridge: Cambridge University Press, 2001.

Parish, Peter. *The American Civil War*. New York: Holms and Meier, 1975.

Putnam, Mary, and Trail Spence Lowell. *William Lowell Putnam*. 1862. Reprint, Cambridge, Mass.: Riverside Press, 1864.

Rafuse, Ethan S. "Why the Confederate Insurgency Failed: Another Take on the Essential Question." *North and South: the Official Magazine of the Civil War Society* 7, no. 1 (January, 2004): 24–33.

Reddy, William M. *The Navigation of Feeling: A Framework for the History of Emotions*. Cambridge: Cambridge University Press, 2001.

Rosenwein, Barbara H. *Emotional Communities in the Early Middle Ages*. Ithaca: Cornell University Press, 2006.

Scott, Sir Walter. "Essay on Chivalry." In *Miscellaneous Prose Works*, edited by Scott. Vol. 28. Edinburgh: Robert Cadell, 1834–36. II

Snyder, R. Claire. *Citizen-Soldiers and Manly Warriors: Military Service and Gender in the Civic Republican Tradition*. Lanham, MA and Oxford: Rowman & Littlefield, 1999.

Storey, Margaret M. "'i'd Rather Go to Hell': White Unionists, Slaves, and Federal Counter-Insurgency in Civil War Alabama." *North and South: the Official Magazine of the Civil War Society* 7, no. 6 (November, 2004): 70–82.

Watson, David. "Disintegrating Emotions in War." *The British Medical Journal* 2, no. 4107 (1939): 663. doi:10.1136/bmj.2.4096.77.

Wilson, Edmund. *Patriotic Gore: Studies in the Literature of the American Civil War*. 1962. Reprint, London: The Hogarth Press, 1987.

Wongsrichanalai, Kanisorn. *Northern Character: College-Educated New Englanders, Honor, Nationalism, and Leadership in the Civil War Era*. New York: Fordham University Press, 2016.

Memory, magic and militias: Cora Indian participation in Mexico's wars, from the reforma to the revolution (1854-1920)

Nathaniel Morris ⓘ

ABSTRACT

Mexico's Cora Indians have played an outsized role in national history, thanks to their skilful use of guerrilla tactics and success in forging strategic alliances with outside forces in defence of their cultural, territorial and political autonomy. Cora participation in elite struggles between Liberals and Conservatives (1850–73), and subsequently in the Mexican Revolution (1910–20), helped to shape the way that both conflicts played out in Western Mexico. Such participation also allowed Cora communities to keep hold of traditional landholdings in the face of political and economic reform, while sowing the seeds for the foundation of the Mexican state of Nayarit.

The Cora Indians – or, in their own tongue, *Naayari* – are an Uto-Aztecan people native to the mountains of north-west Mexico who today number around 30,000 individuals. They earned a reputation as fierce fighters during the Colonial period, and held out against the Conquistadores until 1722, when their priest-kingdom was amongst the last of Mexico's native polities to fall to Spanish forces. For a small people – who in the late nineteenth century numbered between three and five thousand individuals – the Coras have played an outsized role in Mexican history both at regional and national levels, thanks to their skill as guerrilla fighters, their relative unity as a people, and their determined efforts to preserve their cultural, territorial and political autonomy, often via strategic alliances with outside forces.

To this end the Coras took part in the Mixtón war of 1532–42 between the Spaniards and an alliance of rebellious western Mexican tribes; held up the conquest of their homeland and nearby regions until the seventeenth century; and played an important regional role as insurgent fighters during the independence struggles of the early 1800s. Above all, however, this article will focus on the hitherto little-studied issue of Cora participation in the conflicts that wracked Mexico from the Liberal 'Reform' era of the 1850s, to the end of the Mexican Revolution in 1920: a tumultuous period that saw wars between Conservative and Liberal factions of the national elite; between the nascent Mexican nation-state and the forces of Maximilian's French empire; between a consolidated Liberal dictatorship and revolutionary insurgents; and, in the western Mexican state of Jalisco, between local elites based in the small trading city of Tepic (which lay close to the Cora homeland), and those of the state capital, Guadalajara, far away to the south.

Frequently portrayed by nineteenth-century politicians, journalists and scholars alike as apolitical 'savages' isolated in their mountain strongholds,[1] Cora fighters came to make up the backbone of the agrarian army of bandit-turned agrarian revolutionary Manuel Lozada. For reasons distinct from those of Lozada's mestizo followers (whom historians have tended to uncritically lump together with the Coras to form a homogeneous, 'agrarista' mass),[2] the support of independent-minded Cora communities for Lozada helped him carve an autonomous peasant republic out of northern Jalisco between 1854 and 1873, while concurrently helping to shape the course of the French Intervention in western Mexico from 1861–1867. This sowed the seeds for the separation from Jalisco of their homeland and the surrounding areas, which became a federally-administered military district. After Lozada's final defeat and death in 1873, some Coras switched sides and helped to prop up the local rule of the Liberal central government; while others took part in local uprisings that allowed their communities to keep hold of traditional landholdings in the face of Liberal reforms and the colonisation attempts of Spanish-speaking *mestizo* Mexicans.[3] Finally, at the outbreak of the Revolution in 1910, a few Cora individuals joined the insurgent forces fighting to overthrow Porfirio Díaz. Their involvement – which, with the exception of this author's own work, has been little examined by historians[4] – paved the way for their more systematic participation in the civil wars between rival revolutionary factions that consumed the country between 1914 and 1920. At the national level, this participation helped the Carrancistas to defeat their Villista enemies and cement their control over Mexico, and also won the Cora homeland and nearby regions the status of a new Mexican state, named 'Nayarit' after the Coras' own name for themselves.

Pagan priest-kingdoms and Cora costumbre

The Cora homeland – or 'Sierra Cora' – forms part of Mexico's Sierra Madre Occidental mountain range. Spanning a total area of some 5,000 km^2, it is located almost completely within Nayarit, making up nearly eighteen percent of the state's total area, including some or all of the municipalities of El Nayar, La Yesca, Acaponeta, Huajicori, Ruíz and Rosamorada. The volcanic plateaus that

dominate the region are overlooked by peaks that rise up to 3,340 m above sea
level, and are cut through by rivers whose canyons drop down to around 400 m
above sea level. Areas above 1,000 m are temperate, while at lower altitudes the
prevailing climate is sub-tropical and often extremely hot. The climate also
fluctuates between a wet season between June and September, and a harsh
dry season that lasts from October until May.

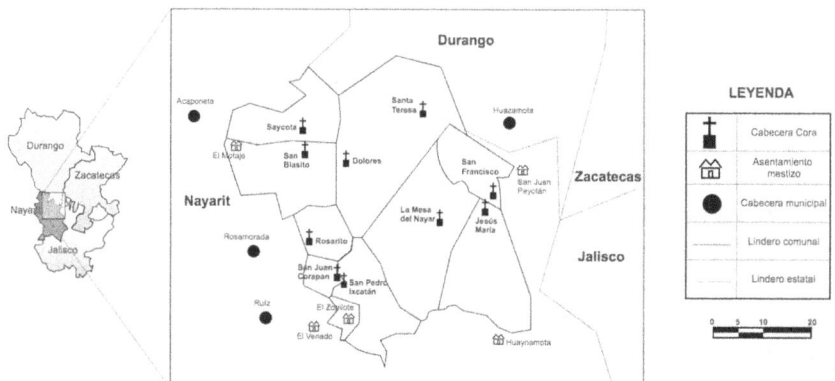

The Coras themselves are a people forged in war. During the early sixteenth
century, in the context of the upheaval caused by the arrival of the Spaniards in
what is now western Mexico, the Coras' ancestors, previously divided between
numerous smaller polities, united under the rule of a single politico-religious
leader known as the 'Tonatí,' or 'Rey Nayar.' The inhabitants of this priest-
kingdom included not only recalcitrant or apostate Indians from across western
Mexico, but also escaped African slaves and renegade mulattos and Spaniards.[5]
Although neither Catholic missionaries nor Spanish soldiers were permitted to
enter the Tonatí's kingdom, which remained independent of Spanish rule until
1722, his subjects traded in salt and livestock with the inhabitants of nearby
coastal regions, and some worked seasonally in the mines of Durango and
Zacatecas.[6]

As a result of such contacts, the Tonatí's people became adept at blacksmi-
thing and raising (and raiding) cattle, horses and mules, and also began to adopt
elements of the Catholic faith into the framework of their own customary
practices (which they refer to, in Spanish, as *costumbre*).[7] However, the Cora
priest-kingdom was ultimately held together by resolutely non-Christian costumbre, including ritual dances called *mitotes* practiced at the level of both descent-
groups and entire communities; faith in the power of ancestors and pre-Hispanic
gods; cyclical conceptions of history; and a belief that shifting subsistence agriculture, hunting, and gathering were sacred activities in which all 'real people' –
that is, the Coras and their indigenous neighbours, the Huichols, Tepehuanos and

Mexicaneros – were obliged to participate. Throughout the sixteenth and seventeenth centuries, mass mitote ceremonies – which at this point featured human sacrifice – were overseen by the Tonatí at La Mesa, which became the focal point of a 'centralized political and religious tradition'[8] uniting a population organised at a lower level into clans or descent-groups defined – as they continue to be today – by blood-ties and participation in family-level mitotes (which coincide with the stages of both the agricultural year and the human life cycle).[9] Although the military capacity of this unified population was key to their resistance to Spanish attempts to conquer them, Colonial-era missionaries and soldiers preferred to use geography to explain the continued independence of the Tonatí's kingdom,[10] describing the region as 'furious and horrible ... It is not even possible to ride horseback in this country as the abruptness of the terrain is hard on the horses and the steepness of the slopes frightens the horsemen.'[11]

As mentioned above, the Cora priest-kingdom was the last Indian kingdom in Mexico to fall to Spanish forces, two centuries after the conquest of the Aztec empire.[12] However, even after Jesuit missionaries congregated the conquered Coras into missions at Jesús María, La Mesa, Santa Teresa, San Juan Peyotán, San Juan Corapan, Rosarito, Huaynamota, and San Pedro Ixcatán, they continued to practice pan-communal rituals that facilitated the political and military cohesion of the Cora population as a whole.[13] Within each of the new mission settlements – some of which doubled as Spanish military garrisons – the cults of statues of Catholic saints – known as *santitos* – were also absorbed into Cora costumbre.[14] Even after the Coras returned to their dispersed ranches following the expulsion of the Jesuits from Mexico in 1767, they continued to use the missions and their churches 'as ceremonial and assembly centres.'[15] By the late eighteenth century, the Coras thus possessed 'a half-digested Christianity but a well organised and functioning civil-religious hierarchy,'[16] which held together their disparate descent groups and diverse, autonomous communities in the absence of the centralised rule of the Tonatí. This allowed them to continue to mobilise effectively against external threats, often 'taking advantage of the disputes between Spanish authorities to establish interethnic and also inter-class alliances,' as demonstrated by wholesale Cora partipation in wider regional rebellions in 1723, 1724, 1758, and 1801.[17]

The Mexican war of independence and the rebellion of Manuel Lozada

Less than a hundred years after the conquest of the Tonatí's kingdom, the outbreak of the War of Mexican Independence in 1810 provided the Coras with yet another opportunity to use their skills as guerrilla fighters to claim autonomy for themselves and their communities. The majority of the Cora population supported regional pro-independence forces, and although a Franciscan royalist did his best to organise the defence of the Sierra, he was soon forced to flee the region in the face of the Cora insurgents' superior numbers.[18] More than a decade of local fighting followed this early defeat, during which both the royalist and pro-independence sides employed guerrilla tactics during six-month campaigning seasons. In line with Cora customs, these started not in October, with the harvest, but rather in November, after ceremonies that rendered the corn crop edible had been celebrated. The local campaigning season meanwhile ended at the beginning of the rains, in May, when the Sierra became impossible for Royalist cavalry to negotiate. For the Coras, rituals also defined the timings of military offensives, and shaped preparations for raids, or measures to defend communities from raiders. Even the directions in which civilians fled from their attackers were informed by older routes of pilgrimage, which linked the communities of the Sierra Cora to the sacred caves and forest groves that now became the hideouts of Cora families.

Royalist forces, in order to survive in the harsh terrain of the Sierra, and, they hoped, cut off supplies to their enemies, focused their energies on 'confiscating all of the grain which the Coras had stored ... and capturing their cattle.'[19] But Cora fighters saw off such threats by 'making a fearful noise and launching a rain of rocks and arrows down upon the soldiers, after which they vanished.'[20] The use of rocks and boulders as weapons of war is, of course, as ancient as war itself, and has been used to great avail by guerrilla forces across the world; perhaps most famously in the Battle of the Roncevaux Pass in 778, when, as commemorated in the medieval 'Chanson de Roland,' Basque guerrillas obliterated the rearguard of Charlemagne's Imperial army as it passed through their territories. Such tactics remained extremely effective in the mountainous Cora homeland – especially against outsiders who were unaccustomed both to the local terrain and the rigours of guerrilla warfare – and Cora fighters continued to use them with success throughout the rebellion of Manuel Lozada, and into the twentieth century.

In 1821, thanks in large part to the support of indigenous groups like the Coras, the insurgents finally won the day against the Spaniards and established the new, independent nation of Mexico in place of the Spanish colony of 'New Spain.' But the defeat of royalist forces did not mean that the Coras would return to their previous status as supposedly 'conquered' Indians. Instead, Cora participation in local uprisings and their canny political manoeuvring in the context of national-level conflicts between Liberals and Conservatives for control of the young nation allowed them to prevent government officials and Catholic priests alike from asserting control over their communities.[21]

In particular, Cora support for the revolutionary movement of mestizo peon-turned-bandit chief and agrarian reformer Manuel Lozada – who ruled the area that is today the state of Nayarit as an independent peasant republic between 1853 and 1873 – allowed the Coras to continue to maintain their political and religious autonomy into the 1870s. Thus they were able to avoid the most deleterious effects of the mid-century Liberal Reform Laws, which, seeking to create a nation of prosperous smallholders from a country still dominated by corporations, mandated the division and distribution of communal and Church-owned lands, opening up traditionally Indian territories across Mexico for purchase by speculators and large landowners.

Manuel Lozada rose to power on the back of his opposition to these reforms. This opposition which garnered him widespread popular support throughout the Seventh Canton of Jalisco. From humble beginnings as a bandit, in 1856 Lozada gained political prominence through an alliance with a Conservative faction of the commercial elite of Tepic, the Seventh Canton's capital city. In exchange for helping these Conservatives in their fight against both the Liberals of Tepic and those of Jalisco's state capital, Guadalajara, they kept Lozada well supplied with cash, arms and powder, and 'capitalised on the discontent of the Cora communities ... which they channelled into a struggle for the defence of the conservative national project that began to take shape during those years.'[22]

However, Lozada and his local followers – made up of mestizos and Cora and Huichol Indians – proved more independent-minded than their new Conservative allies had envisaged. Lozada followed perfectly the trajectory of the social bandit as outlined by Hobsbawm – 'outlaws whom the state regards as criminals, but who are considered by their people as heroes, as champions, avengers, fighters for justice, perhaps even leaders of liberation,' and who, in societies resisting the 'encroachments and historical advances of central governments ... may even be helped and supported by the local lords.'[23] As well as attacking Liberal militias and federal troops, Lozada helped his followers take back territory lost due to the Liberal Reform, and to defend other lands threatened by these same laws; in the Sierra Cora, Lozada's support enabled a 'multitude of indigenous rebels' to attack mestizo colonists who had begun to seize Cora lands around the community of Jesús María. They killed some of them, forced the rest to flee the region 'in the greatest misery,' and took the local priest prisoner, on the grounds that he had encouraged mestizo settlers and tried to curb Cora religious autonomy.[24] In response to these and numerous other perceived 'outrages,' Federal troops and Jalisco's state militia intervened in the late 1850s on behalf of landowners attacked by Lozada's supporters, whom they pursued deep into the mountains of the Sierra Cora. The sudden appearance of government soldiers in their homeland only increased Cora support for Lozada. In 1858, the comisario municipal de San Juan Peyotán reported that all of the Cora communities, without exception, 'have declared in favour of "the wicked Lozada."'[25]

Cora guerrillas proved an invaluable addition to Lozada's forces. Their hit-and-run tactics terrified government forces struggling to penetrate the 'rocky and heavily forested terrain' of the Sierra.[26] Just as they had done a few decades before, during the struggle for Mexican independence, Cora fighters responded to the Federals' bullets and futile artillery barrages with primitive but effective weapons such as boulders: 'The cyclopean rock, the enormous mass of granite, is detached, using levers, from its socket ... and flies, pushed by unseen hands, knocking down horses and not stopping until it reaches the foot of the mountain, clotted with blood and splattered brains.'[27] They also used fire against their enemies to great effect, taking advantage of the aridity of the Sierra during the local campaigning season, which coincided with the driest months of the year. The Mexican novelist Mariano Azuela, in a short story based on his reading of reports on Lozada's rebellion, describes in terrifying detail a night-time ambush of government forces, during which Cora guerrillas set alight the scrub surrounding their enemies' camp. As the flames lit up the darkness of the mountains, bullets flying amidst the panicked men and burning horses, one survivor reminisced that 'no eramos gente, señor, sino demonios en el infierno.'[28]

Frustrated by these attacks, as well as by 'the apathy and lack of cooperation of the inhabitants of the Canton,'[29] Lozada's enemies demanded that Jalisco's state congress authorise their complete destruction of Lozada's main strongholds – including several of the Cora communities of the Sierra – and the dispersal of all their inhabitants, 'who in their totality are composed of bandits who have desolated the Canton, sowing death, dishonour, misery and terror and every class of the post punishable iniquity throughout.'[30] The congress refused, however, and in autumn 1857 Jalisco's Liberal leaders agreed a truce with Lozada, according to which his supporters – no longer classed as 'bandits' but as 'indigenous rebels,' the change of tone reflecting the change of government strategy – were to give up their arms, in exchange for individual pardons and Guadalajara's pledge to 'carry out the measuring, surveying, and demarcation of the lands of the haciendas of this Canton, in virtue of the question of territory being the cause of this uprising.'[31]

Jalisco's Liberals refused to offer Lozada himself an amnesty, however, and instead condemned him for causing 'a scandal infringing the law and offending the most basic moral principles,' and for committing 'crimes of all kinds at multiple junctures and without any political project.'[32] Lozada therefore remained under arms, and soon threw himself and his supporters into the new conflict between Liberal and Conservative elites – the 'War of Reform' – that broke out a few weeks later. Lozada and Carlos Rivas – the scion of an elite Conservative family in Tepic and Lozada's second-in-command – 'organised in the Sierra, the valleys and the lowlands the contingents that would make up the Conservative forces in the Seventh Canton.'[33] Just like contemporary Yeke warlords of Central Africa described elsewhere in this volume – who in the 1870s took advantage of an alliance with Portuguese traders to found an independent state in Congo's Katanga region[34] – Lozada's supposedly 'backward' followers owed much of their political and military success to their homeland's incorporation into global trade networks, which had enabled a clique of regional families, 'organised around the commercial interests of the Barrón y Forbes trading house of Tepic,' to amass sufficient wealth to fund Lozada's insurgency.[35] But the support of so many Coras for Lozada enabled Liberal journalists and politicians to condemn his cause as a 'caste war' that aimed at the 'destruction of the white race,' building on Colonial ideas of the mountains of Nayarit as a remote and impenetrable bastion of savagery.[36]

However, while dismissed as an outburst of 'barbarism' by Mexico's Liberal elites, regarded by the nation's Conservative leaders and their wealthy partisans in Tepic as a local extension of their own movement, and represented more recently by many historians as the near spontaneous uprising of a disaffected and homogeneous 'indigenous peasantry,'[37] the rebellion of Lozada and his supporters – grouped together as 'the United Pueblos of Nayarit' – was in fact a far more complex affair. In aiming to restore to the Seventh Canton's formerly influential peasant communities the lands and political and religious autonomy they had lost to Liberal reforms, it was more than just another outburst of rural discontent: as scholars such as Meyer have shown, it was in fact a movement whose declared political goals outlined an alternate model of state-formation for elites seeking to create a new Mexican nation from the ground up.[38] Rather than seeking to destroy the country's indigenous and peasant communities, and their traditions of local rule and direct democracy, Lozada's political model envisaged these corporate entities as the building blocks from which the new nation could be fashioned. In line with this model, Cora support for the movement was harnessed under one of their own: Dionisimo Gerónimo, a Cora warlord from La Mesa, the old capital of the Cora priest-kingdom. Gerónimo acted as the chief politico-military leader of a league of autonomous Cora communities, internal control of which was shared by elderly ritual specialists, middle-aged members of directly democratic cargo-systems governments, and communal 'military commands' headed by 'Indians named by Lozada.'[39] All of these assorted political and military authorities were furthermore subject to the overall, centralised control of a single 'governor of the Indian territory' elected once a year, on 15 January,[40] a system legitimised by the development and formalisation of complex internal, and also inter-communal, ritual practices.

Although Cora participation in the 'United Pueblos' movement 'implied subjecting themselves militarily to Lozada's General Command in [the lowland village of] San Luis, and politically to its political captaincy in Tepic,' the recent work of Regina Lira demonstrates that the Coras' own internal politico-military organisation in many ways echoed arrangements in the Sierra in the late eighteenth century, when each Cora community had fielded a militia that answered to a central military command in Jesús María. During the Lozada era, each of the Cora communities continued to contribute its contingent to a combined Cora militia force, maintaining 'a list of men old enough and in good enough physical condition to fight, ten percent of whom remained always at arms, according to a system of rotation that guaranteed a minimal but permanent mobilisation, as well as the entire contingent's familiarity with military practices.' Each community also took care of feeding its militia, and ensured they were well-armed with 'a lance, sword, machete, dagger, pistol and rifle.' The only difference was that now, the Cora militias were organised around two military commands: one, as before, in Jesús María, with a second established in Dionisio Gerónimo's stronghold of La Mesa, each of which was overseen by a council 'composed of a military commander who functioned as president, and two subordinate captains who served as secretaries.'[41]

Between late 1857 and late 1860, Lozada and the United Pueblos battled for control of the Seventh Cantón against Liberal guerrillas led by Lozada's nemesis, Ramón Corona. Although at local level the latter found it impossible to overcome 'the enormous resistance ... offered by the mountainous terrain he trod and the warlike race he tried to subjugate,'[42] by the end of 1860, the war had been won at national level by the Liberals, who took power in Mexico City and installed Benito Juárez as President of the Mexican Republic. In the Seventh Canton, which Conservative leaders had earlier declared independent of Jalisco in exchange for Lozada's support, Lozada proclaimed his submission to the Liberals and looked to sign an armistice with Juárez and his new government. He had demonstrated little real ideological attachment to the Conservative cause beyond a shared antipathy to the Liberals, to whom he was nevertheless now willing to claim fealty; his real interest lay in maintaining his influence in the Seventh Canton, now in a limbo between independence and re-incorporation.

However, Lozada refused to give up his weapons, while the new, Liberal Governor of Jalisco, Pedro Ogazón, refused to acknowledge any directives issued in the Seventh Canton before the Liberal victory. Negotiations between the two sides quickly broke down, and in February 1861 Ogazón declared that anyone who did not immediately and unconditionally submit to his government would be 'considered as bandits' and 'irrevocably shot at the moment of their apprehension,' while ordering the forces he sent to the region, composed of five thousand men,[43] to completely destroy Lozada's strongholds and distribute their property amongst their neighbours.[44]

There followed another year of harsh warfare in the Sierra Cora. The anti-Lozada tradition consistently ignores the savagery of the Liberal campaign, as outlined by Ogazón's above decree, or bandit-turned-Liberal commander Antonio Rojas' plan to invade the Cora homeland and 'not leave before exterminating this accursed race of bandits ... and throw from their homes these Indians and push them into the neighbouring states.'[45] Even these measures could not end the war in the Seventh Canton decisively for the Liberals, '"unable to land the decisive blow because the Indians roam from hill to hill."'[46] But it was only after the arrival of French, British and Spanish troops near Veracruz early in 1862 that they agreed to negotiate with Lozada's forces.

The ostensible aim of the foreign expeditionary force was to collect on unpaid debts that Mexico had racked up during the War of Reform; but before long the French decided to mount a full-scale invasion of the troubled republic, in order to establish in its place a profitable monarchical protectorate – the so-called 'Second Mexican Empire' – under an Austrian prince, Maximillian of Hapsburg. In the face of this new threat to Liberal power, General Ogazón met with Lozada, who, in clear acknowledgement of his cause's coherent local political and socio-economic objectives, agreed to demobilise his forces in exchange for an amnesty for his supporters, and the formation of a new, neutral Canton-wide government, which would ensure 'the defence of the Indians in territorial conflicts with the neighbouring landowners.'[47]

Despite this treaty, however, Liberal attempts to return the Seventh Canton to Guadalajara's control continued. Lozada's arch-enemy, Ramón Corona, moved large numbers of troops in and out of the region, while positions in the new Cantonal authority were given to Lozada's old Liberal opponents, who enacted a plan to colonise the Sierra Cora with armed American settlers.[48] These would be given land and other economic incentives in exchange for their 'service against the Indians when thus required by the nation's government.'[49] In the midst of the French intervention at national level, but faced with the potential of a rather more threatening foreign invasion on his own doorstep – that of American settlers committed to destroying the Coras – Lozada was forced to accept that the Liberals had reneged on their treaty commitments, and in August 1863 instead recognised as Mexico's legitimate government the Emperor Maximilian and his French backers, who were now ensconced in the interior of the country.

During the five years of war between Maximilian, French expeditionary troops and Conservative forces on one side, and Benito Juárez and his Liberal loyalists on the other, the Sierra Cora served as a 'zone of refuge, rearguard actions and guerrilla warfare.' Various Cora communities were raided by Liberal militias under the command of Ramon Corona.[50] Such attacks prompted the ever more direct participation of Cora guerrillas within 'the military forces of the Seventh Canton ... as part of the Imperial Auxiliary Troops of Maximilian ... receiving five-year salaries and temporary discharges during sowing and harvest times.'[51] But testament to his supporters' concerns for local, rather than national, issues, Lozada paid little attention to French requests for active armed support, and instead concentrated on defeating Corona, taking Tepic, and reasserting order in the Seventh Canton. He remained the master of his own kingdom, pursuing his own agenda, which his ostensible superiors within the French expeditionary forces had no choice but to accept (a policy that, as described by Mario Draper elsewhere in this issue, influenced the attitudes of both French and Belgian forces in their African wars of conquest in the ensuing decades).[52] Thus General Felix Douay, the senior French officer in the west of Mexico, pragmatically refused to interfere in Lozada's running of the Seventh Canton (which had been officially severed from Jalisco by the new government), admitting that 'General Lozada is not sufficiently subservient to my actions for me to be sure that my orders will be carried out.'[53]

Lozada engaged politically with the Imperialists just as he did militarily – when it suited him. Thus it was only in 1865 that he sent his most trusted aide, Carlos Rivas, to Mexico City; his objective was to establish the Seventh Canton permanently as the independent and sovereign state of Nayarit, and to approach Maximilian about the issues that had dominated Lozada's struggles since 1857 – land and indigenous autonomy. An agreement was made: Nayarit was declared a new Department, separate from Jalisco, and the process of communal territorial demarcation was set to begin immediately, the task falling to a committee under the authority of the Ministry of Public Works.[54] Such contacts between Lozada and Maximilian likely influenced the latter's interest in and sympathy for indigenous issues, which the next year resulted in his passing a wealth of other national-level legislation designed to protect Indian communal lands, and other laws that 'conceded relative liberty to *peons*, and lands to the indigenous communities that lacked them.'[55]

However, despite the Imperial government's support, Lozada was too politically adroit to continue in his adherence to Maximilian's cause after Napoleon III abruptly withdrew French military support in late 1866, and it became clear that the course of the civil war had turned and the Liberals were likely to triumph. Lozada slowly disassociated himself from the Imperialists, declaring, on 1 December 1866, the *'Act of Neutrality of the pueblos of Nayarit.'* After suffering a series of defeats throughout early 1867, Maximilian was captured by Liberal forces in May and, the next month, was executed along with two of his top generals, Miramón and Mejía. However, Lozada, having freed himself and his followers from any commitments to either of the warring national parties, continued to control a swathe of western Mexico, and began the last phase of his career, as the independent *cacique* ('political and/or military boss') of Nayarit.

Several years of peace in Nayarit followed, during which time Lozada stepped up the redistribution of lands, and his Cora supporters set about further institutionalising their political, cultural and territorial autonomy through the creation of new civil-military authorities within their communities, whose legitimacy was bolstered through the invention of new ritual practices. With the death of President Juárez in July 1872, however, Lozada's position again became precarious. Internal differences within his ever-heterogeneous political movement – which included feuds between rival pueblos as between rival chiefs – manifested themselves with the rebellion of Andres Rosales and Praxedis Núñez, two of Lozada's lieutenants, who took with them a thousand men and the town of Atonalisco. After Lozada defeated them militarily, they fled to the protection of Corona, and would prove instrumental in their former leader's eventual downfall. At the same time, with Lozada's protector in Mexico City gone, Liberal elites in Guadalajara and Tepic seized on the opportunity to change the Supreme

Government's attitude towards affairs in the Seventh Canton, publishing a glut of damning and openly racist local dispatches in the national press. 'Here there has triumphed at last the *huarache* [a pejorative term for "Indian," after their leather sandals],' reported the Liberal newspaper *Juan Panadero* from Tepic; 'the entire local government is composed of ignorant unknowns ... and the hacendados are still being robbed of their property. Various families are fleeing.'[56]

As a result, Lerdo declared that Lozada's strongholds must submit themselves to the constitutional order, with Lozada's government replaced with the authority of a *jefe político* named by the Supreme Government, and all questions of land boundaries passed to official tribunals, 'without that practiced by [Lozada's] commissions taking place.'[57] He also ordered the movement of state militias from Zacatecas and Sinaloa to Guadalajara, where they joined forces sent from Colima and Jalisco in anticipation of an attack on Lozada's strongholds. In response, Lozada and his compatriots made the decision to go to war with Lerdo's government at the traditional new-year meeting of all of Nayarit's Lozadista communities, with a first strike launched from the sierra viewed as preferable to waiting to be attacked. On 17 January Lozada proclaimed his 'Plan Libertador' ('Programme of Liberation') to the nation, calling for, amongst other things, the abolition of the land taxes, the governance of the nation by sovereign municipal governments chosen via free elections, and the formation of a council made up of three representatives from each state, again named via free and direct elections, who would decide 'what form of popular representation to give the nation, whether with the character of republic, empire or kingdom, as long as that chosen allows for the nation's true aggrandizement and a lasting peace.'[58]

As Lozada's forces took Tepic and then marched towards Guadalajara, the 'Plan Libertador' was essentially ignored by the regional and national press, which preferred to condemn Lozada's movement on the grounds of the supposed savagery of the Coras and other indigenous peoples of the Sierra who supported him. Thus *Juan Panadero* printed a plea to the government of Jalisco, 'that you take the necessary precautions, so that the appalling tendencies of the Indians are not increased,' without any mention of any political plan whatsoever.[59] And J.M. Vigil, in a front page editorial in *el Siglo XIX*, wrote of 'these savage tribes of Álica amongst whom there does not exist even the shadow of individual rights, nor the most remote façade of respect for the interests of society,' and characterised the conflict as the 'crusade of civilisation against the most disgraceful barbarity,' without a single reference to the openly declared goals of Lozada and his supporters, Cora and mestizo alike.[60]

For around a century afterwards, such Liberal propaganda defined interpretations of Lozada and his movement; for history is written by the victors, and on 28 January 1873, Lozada was defeated by Corona at La Mojonera, at the gates of Guadalajara. His primarily guerrilla forces were poorly trained and ill-equipped for a conventional battle against a fortified defensive force, and were routed with heavy losses. His movement shattered with the weight of the blow and Lozada was forced back into the Sierra Cora, where his dwindling forces faced off against 'ten thousand soldiers led in three sections which penetrated the Sierra de Nayarit along different paths.'[61] As the months went by, he was abandoned by one after another of his chiefs, including, finally, Dionisio Gerónimo and his Cora fighters. After several more months of pursuit, Lozada was eventually captured, taken back to Tepic, and executed. President Lerdo appointed a *Jefe Politico* (political chief) in his place, although in much of what had been the Seventh Canton – now officially a Federally-administered 'Military District' named for its capital, Tepic – Lozada's former lieutenants remained the paramount local authorities.

Lozada lost more than just a battle at La Mojonera; he lost a war of words that had raged since 1857. To many today, Lozada remains a traitor who sold himself to the French; he is never a marginalised campesino rising up in the name of the exploited, both indigenous and mestizo, against unfair Liberal reforms. Perhaps the charge of treachery stands; but if Lozada was a traitor to Mexico, he was never a traitor to his real homeland – for he stayed true to the *pueblos* of Nayarit, and above all to the Coras. What use had they for a newly created nation that had already declared them obsolete? What was siding with the French to a people threatened by armed American settlers hired to carry out a genocide in the Sierra?

Thus in the Sierra Cora today, Lozada is still venerated by local people, associated with Jesus Christ and believed to have been betrayed and killed by 'Jews' not in Jerusalem, but in Mexico City; while many of the Coras' Huichol neighbours also associate Manuel Lozada with 'the promethean deer god Kayumarie.'[62] Furthermore, although after the fall of the Second Mexican Empire most of the indigenous groups that had supported Maximilian and the French – such as Sonora's Opata people – were crushed militarily and lost nearly all of their remaining lands and political autonomy,[63] Cora support for Lozada, followed by their defection to the government side at the last minute, allowed their communities to retain their *de facto* political and territorial autonomy into the late nineteenth century. Not a single Cora appears to have been involved in litigation with a hacienda during the Lozada era, suggesting that, unlike most of the mestizo communities that supported Lozada, his Cora followers had successfully managed 'the recovery of their territories from the hands of their neighbours since the events of 1858.'[64] Meanwhile, after Lozada's death, Cora leader Dionisio Gerónimo remained the Sierra Cora's chief politico-military authority, charged by the federal government with keeping the local peace and ensuring that his compatriots abstain from active rebellion against the Liberal order, in exchange for federal non-intervention in local issues.

Part 3: the Porfiriato

The support of the Cora communities for Manuel Lozada had not only allowed them to expel mestizo settlers from their territories, but also made the application of early Liberal 'reform laws' in the Sierra del Nayar impossible. New pressures were unleashed on the region after Porfirio Díaz overthrew President Lerdo in November 1876 and established a new Liberal regime, which developed into a durable military dictatorship that lasted until it was overthrown during the Mexican Revolution in 1911. However, despite the backing of Díaz's regime for the division and privatisation of communal lands in the Territory of Tepic, which was one of the ten entities most affected by the disentailment of communal lands during the Porfiriato,[65] continued Cora recalcitrance, together with the difficulties of transport and communication presented by the rough terrain of the region, obstructed the surveying and dividing of Cora communal lands.[66] Thus the Cora communities escaped largely unscathed, with the exception of San Juan Peyotán, which was fully colonised by mestizos.[67]

Similarly, although 'auxiliary courts' were set up in each community to represent the Porfirian state, these were either ignored by the Cora population, or alternatively their judges 'obeyed in all respects the community's [Cora] governor,'[68] having been integrated into existing Cora authorities 'through strange ceremonies.'[69] Throughout the Porfiriato, then, each Cora community retained its control of large tracts of land, as well as a high level of cultural and political autonomy from the state. The Coras remained fiercely defensive of this fact: the Norwegian explorer Carl Lumholtz, who visited the Sierra in the late nineteenth century, noted that, in general, the Coras had 'very strong objections to unions with [mestizos],'[70] while contemporary ethnologist Konrad Preuss reported that the Cora "do not wish to know anything of the outside world," describing the Coras as 'haughty subjects, who handle weapons very well and whom one does not wish to bump into in the mountains.'[71] And although the Porfirian state by now regarded the Sierra Cora as 'pacified territory,'[72] and Lumholtz and various other foreign visitors, in line with the positivism and social Darwinism that dominated contemporary views of indigenous societies,[73] predicted that the Coras would soon become culturally and racially extinct,[74] the Czech doctor and anthropologist Ales Hrdlicka was ultimately correct when he noted that the Coras had 'not given up the thought of armed resistance' to the threats that outsiders posed their lands, cultures and ways of life.[75]

Part 4: the revolution

In October 1910, after fraudulent elections in which Porfirio Díaz won a landslide victory, dissident political leader Francisco I. Madero issued his 'Plan de San Luis Potosí,' urging the nation to rise up against the dictatorship on 20 November. Across the country, peasants, workers and members of the provincial middle classes answered Madero's call, setting in motion the Mexican Revolution. The uprising soon reached the mountains separating Durango and Zacatecas from the then-Federal Territory of Tepic. Although in response an expeditionary force of cavalry and rurales set out from Tepic for the Sierra Cora, 'with the object of protecting those distant pueblos and avoid the invasion that the rebels of

Durango and Zacatecas might carry out,'[76] it failed to achieve its objectives. Rebel forces took the Sierra and marched on Tepic itself, which fell to Maderista forces in May 1911, without a shot being fired.[77] A few days later Madero and Pascual Orozco defeated the Federal army at Ciudad Juárez, and forced President Díaz from power.

However, the power-hungry Pascual Orozco soon revolted against Madero, and was seconded in Tepic by an ex-Federal Lieutenant, Miguel Guerrero, who raised a force that raided along the western edge of the Sierra Cora.[78] Guerrero then joined forces with Camilo Rentería in Huaynamota, in the Sierra Cora itself,[79] and together the rebels attacked Tepic on 29 April. Although they were repelled with losses – to the extent that the Federal commander claimed that Rentería himself had died in the battle[80] – the city's elite were alarmed by the sudden appearance of a rebel army at their gates, and the Cámara Nacional de Comercio de Tepic warned President Madero that:

'this city does not possess a sufficient garrison of federal soldiers commanded by a skilful military chief ... [given] the threat posed by the numerous armed subversives and the Indians of the Sierra ... It is well known that the intention of ex-Lieutenant Guerrero is to take control of Tepic and gather together here the elements necessary to then attack the city of Guadalajara, and we invoke the unfortunate memory of the chieftain Lozada, to remind you of the damage that a resolute man can cause the nation once he has taken Tepic.'[81]

References to Indian rebellion had long been used by the elites of both Tepic and Guadalajara to tarnish political opponents like Manuel Lozada,[82] and to elicit Federal intervention in local disputes by invoking the spectre of a Yucatan-style 'caste war,'[83] the Cámara's appeal probably also reflected the real presence of Coras from the area around Huaynamota amongst Rentería's troops. But just as the memory of Lozada remained a byword for the dangers of Indian rebellion amongst Tepic's elite, so too it seems likely that those Coras and Huichols who decided to join Rentería – himself a descendent of one of Lozada's lieutenants – were encouraged to do so by memories of their former successes under Lozada.[84]

The fears of Tepic's elite were soon confirmed, for news of Renteria's death proved premature, and it transpired a month later that the rebel leader was in fact alive and well and back in Huaynamota at the head of two hundred fighters.[85] Rentería and Guerrero's forces skirmished with Federal troops throughout the summer, and by July the number of 'bandits' active in the Territory had reached more than eight hundred men.[86] The use of the term 'bandit' by the authorities to disparage revolutionary forces was common throughout Mexico in this period, but it is important to note that many early revolutionary leaders – including such major figures as Emiliano Zapata – had much in common with Hobsbawm's figure of the 'social bandit', and often struggled to form cohesive social movements from the even more 'bandit-like' local rebels who formed the backbone of their forces.[87] In the Sierra Cora, the depredations of the latter turned local people against Rentería, and, angered by Rentería's attempts to levy 'forced loans' on them, attacked his men and forced them to flee the region in disarray.[88]

In February 1913, over the course of the ten days known as the *Decena Trágica*, President Madero was deposed and then assassinated by his foremost military chief, General Victoriano Huerta. Revolutionary military commanders and reformist statesmen including Venustiano Carranza, Francisco 'Pancho' Villa, Emiliano Zapata and Álvaro Obregón turned their arms on Huerta and eventually defeated his Federal army in July 1914. However, tensions then grew between Carranza and Obregón on one side, and Villa and Zapata on the other. A new civil war soon broke out between the rival revolutionary factions. After Villa was heavily defeated in the Bajío in spring 1915, some remnants of his dispersed army took refuge in the Sierra Cora, pursued by Carrancista forces. This development served to further increase contacts – both friendly and hostile – between the Coras and armed outsiders, and paved the way for the local establishment of Carrancista-armed self-defence militias, which, according to one rather disapproving observer, transformed 'savages with bows and arrows' into 'savages with carbines and mausers.'[89] In turn, this led to the rise of a new generation of Cora leaders who can be defined, according to Alan Knight's cacical grading system, as bottom-level, local (or even sub-local) 'caciquillos,' some of whom survived the revolution to emerge, in the 1920s, as higher-level 'municipal' or even 'regional' caciques.[90]

Some Cora Defensas were formed after the Carrancistas occupied a community, met with the traditional authorities, charged them with selecting a suitable leader or picked one out from amongst them, and made them responsible for organising a militia. In other cases, regional Carrancista commanders lent official recognition to pre-existing Cora paramilitaries, on condition they pledged their allegiance to Carranza. The Carrancistas would then hand over whatever arms they could spare (generally bolt-action *cerrojo* rifles), although there were often fewer weapons than needed, and many Defensa members initially armed themselves with bows and arrows, slings, machetes, or the ever-versatile *coa* digging stick. As in other peripheral regions of Mexico, the size of the Cora Defensas varied from five to more than forty members,[91] and often grew or shrank according to the availability of weapons, the severity of external threats, and the value of the prizes to be won through armed action.[92]

Those who ended up heading the Defensas, whether elected by their communities or hand-picked by the Carrancistas, tended to be young men able to mediate between their communities and the outside world. Given the long histories of contact between the Coras and the regional mestizo population, and the way in which the shamanic practices of the former often required individuals to take on multiple contradictory identities, there were always a few Indians in each community able to act simultaneously as 'Indian *and* mestizo'.[93] In a few cases, these men were themselves mestizos, albeit with close ties to the local Cora majority. If they had some military training, their command of a Defensa was all the more authoritative; so much the better if they were wealthy (in this period synonymous with cattle ownership), and therefore had the means to supply some of a Defensa's material needs, and more to lose from Villista raids.

The Defensas of the Sierra Cora played an important part in Carrancista efforts to combat remnant Villistas and generalised banditry in the Territory of Tepic. After Carranza elevated the Territory to the category of a 'Free and Sovereign State' in January 1917 – named Nayarit after the Cora endonym, *Naayari*[94] – they consolidated their power through an alliance with the new state's governor, General Jesús M. Ferreira. By early 1919, when General Francisco de Santiago seized power in Nayarit,[95] *de facto* control over the entire Sierra Cora lay in the hands of two Carrancista-aligned Defensa commanders: Mariano Mejía and Eutimio Domínguez.

Domínguez was a full-blooded Cora born in 1882.[96] His father was also a native of San Juan and a relatively wealthy rancher, while his mother, according to some accounts, was originally from San Francisco – reflecting the strong links of trade, kinship and ritual which still linked together the different Cora communities. Domínguez seems to have been an early participant in the revolution in Nayarit, eventually returning home with a wounded knee (which would never fully heal, forcing him to spend most of the rest of his life on horseback),[97] and the rank of Colonel in Carranza's Constitutionalist army.[98] There he established a Defensa to defend both his community, and his father's cattle, from the raids of the 'bandidos' then active in the area.[99]

After his father's death in one of these raids, Domínguez inherited what was left of his herds.[100] This enabled him to sponsor ceremonies that increased his status within the descent-groups groups that he belonged to, while his status as a ranking military officer, and his command of a well-armed Defensa, limited the opposition of rival families or factions to his increasing influence in San Juan and neighbouring communities. He not only appeared to his Cora compatriots as 'one of us', but also as a natural leader possessed of supernatural powers. He was known to spend much of his time fasting and praying to his community's santitos, 'so that nothing would happen to his men.' Thanks to Domínguez's asceticism, the five devoted bodyguards who always surrounded him in battle were reputed to have been practically bullet-proof, their white cotton shirts and trousers 'all shot full of holes, but not even a scratch on their bodies.'[101]

However, Domínguez also spoke good Spanish, had learned to read and write while in the army, and his intellect would later impress SEP officials, who rather patronisingly described him as a 'Cora Indian of relative culture.'[102] Like other indigenous caciques in this period, from Oaxaca to Michoacán, he used his literacy, charisma and military prowess to successfully negotiate the violent and alien world of revolutionary mestizo politics, while also playing up his ethnicity in order to present himself as a 'noble Indian'.[103] Thus Domínguez acted as a bridge between Coras and outsiders, allowing him to consolidate his influence over the rest of the Cora Baja – whose communal Defensas Rurales were all under his control by 1920 – while simultaneously developing his clout with the state government, which would later come to depend on him to harness the collective military potential of the Coras during the first Cristero rebellion.

Domínguez's friend and then-ally Mariano Mejía was an even more powerful figure. Unlike the Cora Domínguez, Mejía was born 1879 in increasingly 'mestizoised' Huaynamota,[104] and oral sources agree that his mother was a local 'Indian',[105] and his father a mestizo rancher. Although unable to speak the Cora language,[106] and despite traditional Cora distrust towards mestizos, Mejía nonetheless managed to become one of the many 'white and mestizo ethnic brokers' to raise a powerful and predominantly Indian force during the Revolution,[107] and by 1920 he represented Carranza's government throughout the entirety of the Cora Alta and beyond into neighbouring parts of Durango and Jalisco.[108]

Mejía had been orphaned while still a young boy, after 'revolutionaries' – most likely ex-Lozadista rebels – killed his father, and his mother fled the community.[109] Mejía's uncle, a priest, adopted the boy and took care of his education,[110] which allowed him to become, like Domínguez, one of the few literate natives of the Sierra Cora at this time. Mejía arrived in Jesús María some years before the outbreak of the Revolution, after his uncle was sent there to serve as a parish priest. Mejía was able to use the inheritance left him by his late father to buy cattle and set himself up as a rancher.[111] His herds increased rapidly, and he became a rich – and thus influential – local figure, hiring Coras and mestizos from across the region to work on his ranch,[112] and thus establishing patron-client ties with his neighbours.[113] He also married a Cora woman and established Catholic-style ritual kinship ties with other Coras through the institution of compadrazgo.

As a rich and literate local mestizo who also commanded the respect of many Coras, Mejía had huge potential to become a 'revolutionary' cacique in the serrano tradition,[114] while his extensive herds would have been natural targets for the Villista bands roaming the Sierra. It is not surprising, then, that at some point during the revolutionary upheaval,[115] Mejía obtained arms from the Carrancistas for himself and eleven men – probably selected, as usual, from amongst his peons, in-laws and compadres[116] – to set up a local militia.[117] Only after he had actively demonstrated his loyalty to the Carrancistas, however, was Mejía able to extend his authority beyond his base in the Jesús María region. According to local interviewees, at some point during the revolutionary civil war a rebel leader called José Gallegos, most likely a Villista from Zacatecas, took refuge in the Sierra Cora.[118] However, the Constitutionalist authorities in Tepic soon heard of his presence, and ordered Mejía to assassinate him. Mejía accordingly invited Gallegos to join his Defensa in a raid on 'bandits' supposedly hiding out in the forests around Santa Teresa, and then betrayed and murdered him as they ate lunch in the mountains.

The versions of this story told by Mejia's grandsons, Juventino Mejía Rivera and Enendino Escobedo Mejía, attribute supernatural powers to Gallegos, who was rendered invincible by a huge tattoo of a devil across his back. It was only after Mejía and his men 'cut off the skin with a knife, all that had the devil on it, that Gallegos then finally began to die.'[119] Although it is not surprising that Mejía's descendents should exaggerate the powers of Gallegos – and thus of the famous forebear who managed to dispatch him – the other details of their stories, such as the cruelty of Gallegos' murder, are typical of many local tales of the revolution and the Cristiada.[120] Meanwhile, the mythical elements of the story – and the way in which having vanquished Mejía is said to have taken on the dead Villista's powers[121] – fit within a well-documented Latin American tradition of attributing supernatural qualities to influential leaders,[122] which in turn served to bolster their charismatic power and further legitimise their authority.[123] In Mexico, Emiliano Zapata is the most famous example of how such processes of mythification and mystification function,[124] but much the same could also be said of Eutimio Domínguez, and, more importantly, of Manuel Lozada, to whom many older Coras and Huichols still attribute supernatural powers.[125]

Even discounting the similarities between the supernatural tales told about Mejía and Lozada, it seems clear enough that the general precedent set by the latter must have helped boost local acceptance of Mejía's influence some fifty years later. Just as Lozada, aided by his Cora lieutenant Dionisio Gerónimo, had overcome traditional Cora antipathy towards mestizos to rise to power over the entire Sierra Cora, so too Mejía was helped by a Cora compadre, León Contreras, to rise from commanding eleven men in Arroyo de Santiago, to controlling the Defensas of every Cora and mestizo community in the Cora Alta. His influence even extended beyond state lines as far as Huazamota and San Lucas in Durango,[126] and San Andrés in Jalisco[127]; which, significantly, were also the only communities outside of the Territory of Tepic to sign Lozada's 1870 political manifesto,[128] further reflecting the parallels between the two men's careers and spheres of influence. Nugent has noted the 'importance of popular forms of historical memory and their strategic deployment in political struggle' during the revolution,[129] and it would seem that Mejía successfully tapped into Cora memories of Lozada's 'Popular Conservative' rebellions so as to unite them against the bandits and Villistas who now threatened their communities, just as Mejía's contemporaries elsewhere in Mexico mobilised their own indigenous clients through similar references to their forebears' participation on the Liberal side during the Wars of Reform.[130]

The ability of Mejía and Domínguez to establish themselves as paramount authorities in the Sierra Cora no doubt owed much to the prior
existence of a centralised politico-military command amongst the Coras,
both during the rule of Lozada, and in the era of the Rey Nayar. In turn,
by calling together all of the Defensas of the Sierra Cora to undertake larger-
scale military operations, Mejía and Domínguez boosted their effectiveness
as military commanders, and thus the strength of their claims to legitimate
military *and* political authority over the region and its inhabitants.

Using the same guerrilla tactics that had made them such effective
fighters during Lozada's rebellion and many other, previous conflicts –
and, in addition to the occasional *cerrojo* rifle, weapons such as bows and
arrows, machetes, boulders, and the careful use of fire to burn their enemies
out of their strongholds or sow panic amongst them during night-time
ambushes[131] – they were not only able to defend their communities against
raiders, but also played an important role in Francisco de Santiago's pacification campaign in Nayarit in 1919 (to the extent that Mejía was on first
name terms with the general). Their alliance with the Revolution's 'winning
side' won the cacical regimes of Mejía and Domínguez the state government's full recognition.[132] And their unifying influence not only helped to
strengthen the Coras' collective ability to defend themselves, but also
impeded the outbreak of internecine conflicts that might otherwise have
fractured the bonds between their communities. This brought to the region
a level of political stability that local people initially welcomed after the
chaos of the early years of the revolution.

However, although the emergence of the Cora Defensas, and the rise of their
commanders to political importance, had important conceptual and historical
antecedents, local war leaders of previous eras had tended to recede into the
political background once the immediate military threat had passed. Many of the
Cora Defensa-commanders-turned-caciques of the revolutionary period, however, were reluctant to give up their newfound power and growing wealth
even after the threat of Villista raids began to fade. From 1919, as their original
raison d'être ebbed, the Sierra Cora's Defensa commanders sought to preserve
their status by entering into increasingly close alliances with the emergent
revolutionary state. In response, popular resistance to their authority grew,
engendering factional conflicts throughout the region's communities, and locally
validating Knight's general judgement that 'the only good cacique is – if not
a dead cacique – at least a short-lived cacique.'[133]

Part 5: epilogue

Although their struggles are less well known than those of the Yaqui,
between 1850 and the early 1920s the Coras successfully used guerrilla
operations and tactical alliances with outside forces to defend their lands

from rapacious hacendados, their political independence from expansionist statesmen, and their culture from the pressures of mainstream, mestizo society. Cora fighters, whether aligned to the pro-independence cause, or to Manuel Lozada, or to the Carrancistas, shared similar objectives and were united by strong, ritually-defined, political-military ties between their communities, but were also able to move quickly and effectively through the mountainous and heavily forested terrain of their homeland in small groups, giving them an important advantage over their less mobile enemies, who were usually less knowledgeable about the local landscape and less adept at guerrilla fighting.

At all three of these important junctures in Mexican history, collective memories of the priest-kingdom that existed in the Cora homeland until the early eighteenth century, filtered through a worldview that stressed the power of supernatural forces in every aspect of life, helped Cora communities to mobilise militias that would defend them against their enemies. Such tactics allowed the Coras to maintain a high level of autonomy compared to most of Mexico's other indigenous groups well into the twentieth century, and also had consequences for the nation as a whole: for instance, Cora participation in Manuel Lozada's rebellion led directly to the creation of state of Nayarit; helped the French to hold onto swathes of western Mexico even as they faced defeat in the country's interior; influenced Emperor Maximilian's indigenous-rights and agrarian policies; and cemented Maximilian's Liberal successors' beliefs in the inherent danger that Indian autonomy posed the nation-state. Later Cora backing for the Carrancistas during the Revolution helped the latter establish their control of Nayarit, defeat their Villista rivals at national level, and, it had become clear by the mid-1930s, sowed the seeds for the eventual decline of the Cora communities as politically autonomous entities, even as it won their lands some level of protection from rapacious major landowners and small-time mestizo colonists alike.

The fact is that the growing power of militia commanders in the Sierra Cora, in the context of the concerted state-building efforts of successive revolutionary regimes between 1920 and 1940, served to undermine both the internal stability of the Cora communities and the political and ritual ties that had long existed between them. Most Cora military leaders turned out to be willing to sacrifice some of their autonomy in exchange for territorial protection and personal political and economic advantage; a trade-off that was fiercely resisted by conservative Cora factions, giving rise to internecine conflict throughout the Sierra in the late 1920s, during the Cristero rebellion. However, the defeat of the conservatives by their better-armed, better-supplied 'cosmopolitan' enemies ensured that the Cora homeland would finally become an indissoluble part of the Mexican nation-state.

Thus in the face of changing national-level conditions, the same Cora militarism that had once propped up their political autonomy, ended up undermining it. Comparisons can thus be made between the Cora paramilitary organisations of the early twentieth century, and many of the indigenous militias that emerged elsewhere in Latin America from the 1950s; for example, the Peruvian *rondas campesinas* ('peasant patrols'). These played an important part in defeating the country's Maoist 'Shining Path' guerrilla movement, but also pursued communal, factional and personal agendas, and as their leaders evolved into powerful warlords, giving rise to factional tensions and social conflicts in the areas in which they were active.[134]

However, although the Sierra Cora of today is far from the priest-kingdom of the Tonatí, or even the autonomous league of indigenous republics that flourished under the regional rule of Manuel Lozada, neither is the Cora homeland as integrated into Mexico as the mid-nineteenth-century's Liberals, or their Carrancista successors, would have wanted. Thanks to historical Cora resistance and continued recalcitrance, the presence of the state in the region remains weak, and its representatives are primarily coercive forces – like the odd army battalion hunting for opium plantations – or part of agencies offering cash transfers, donations of food or blankets, and offers of political support to the Cora communities on condition that they send their children to schools where they can be transformed into 'Mexicans' through education.

Meanwhile Cora traditional governments and agrarian authorities remain influential within their communities, which continue to hold their lands collectively – anathema both to Liberal conceptions of 'civilised' land tenure and 'modern' political organisation. Meanwhile, although most Coras have now abandoned the idea of protecting their autonomy via active warfare against the state, they continue to employ Scottian 'weapons of weak' – including 'foot-dragging, noncompliance, evasiveness, and obfuscation,' and the selective use of their language as a code[135] – to defy the demands and expectations of outsiders such as government officials. Indeed, within the context of the current Mexican 'Drug War,' some of the Cora youth are winning back their forebears' reputation as 'dangerous' and 'wild' Indians, through their active participation, whether voluntary and involuntary, in the international narcotics trade, both in terms of the production of opium, and, increasingly, the trafficking of this and other drugs north towards the US. Thus the pursuit of autonomy through the negotiation of alliances with potentially threatening outside forces – whether this entails taking up arms against the state, or abandoning defeated caciques and fighting *for* the state – has at the very least allowed the Coras to defy the predictions of nineteenth century statesmen and explores alike, that 'these primitive people will soon disappear by fusion with the great nation to whom they belong.'[136]

Notes

1. González, *Ensayo estadístico y geográfico de Territorio de Tepic*, 560; Cambré, *La Guerra*, 496; see also the numerous articles attacking Lozada and his supporters in *Le Trait d' Union* and *Juan Panadero* from the mid-1850s onwards.
2. The best examples of this tendency can be seen in the work of Rendón: cf. *Rebelión Agraria de Manuel Lozada* and *Manuel Lozada y las comunidades indígenas*. A notable and comprehensive exception to this approach is to be found in the work of Regina Lira, who has recently carried out in-depth research into Cora and Huichol participation in Lozada's movement, showing that their support cannot be separated from the ethno-cultural, ritual-political and historical idiosyncrasies of each people, and, indeed, of each of the numerous communities between which they are divided. Cf. Lira, "De Buenos mexicanos."
3. In the Cora homeland and nearby regions, mestizos are primarily defined as all those local people who do not speak an indigenous language or take part in indigenous rituals. Other criteria by which Mexican mestizo identity can or should be judged vary from region to region, and are often controversial; for in-depth studies of Mexican 'mestizaje' see Lomnitz-Adler, *Exits From the Labyrinth*; Bonfil, *México profundo*; Friedlander, *Being Indian in Hueyapan*.
4. cf. Morris, 'Creating the World Anew,' and '"¿Forjando Patria?" Las políticas del Estado revolucionario y el ocaso de los vínculos intercomunales coras en la Sierra del Nayar'.
5. Coyle, *From Flowers to Ash*, 75.
6. Neurath, *Las fiestas*, 21.
7. Hinton, "Pre-Conquest Acculturation," 166.
8. Coyle, *From Flowers to Ash*, 82.
9. Jáuregui, *Los coras*, 12.
10. Coyle, "The Customs," 516.
11. Ortega, in ibid.
12. Gerhard, *La frontera Norte*, 142–5.
13. ibid., p.147; and Gómez, "Huicot," 138.
14. Hinton, "Indian Acculturation," 22.
15. Coyle, *From Flowers*, 85.
16. See note 14 above.
17. Lira, "De buenos mexicanos," 7.
18. Jáuregui, Magriñá, "Estudio etnohistórico," 64–6.
19. Santoscoy, *Colección de documentos*, 62.
20. ibid., 63.
21. Coyle, *From Flowers to Ash*, 88.
22. Lira, "De buenos mexicanos," 11.
23. Hobsbawm, *Bandits*, 20.
24. Lira, "De buenos mexicanos," 21.
25. ibid.
26. Meyer, *Manuel Lozada*, 239.
27. Ibid 220, 239 and 331.
28. ibid., 223 quoting Azuela, "El Hombre Masa," 409–10.
29. AHJ, Seguridad Pública, Rocha to Parrodí, 2 October 1857.
30. Ibid.
31. AHJ, Seg. Púb., Convenios de Paso de Caimán, 15 November 1857.
32. Ibid.

33. Lira, "De buenos mexicanos," 21.
34. cf. Giacomo Macola, 'Guerrilla Warfare in Katanga: The Sanga Rebellion of the 1890s and Its Suppression', in this issue.
35. Lira,"De buenos mexicanos," 21.
36. Quevedo y Zuvieta, *México*, 140; see also Van Oosterhout, "Popular Conservatism," 221-2.
37. cf. A. Aldana Rendón, *Rebelión Agraria de Manuel Lozada*; Rendón, *Manuel Lozada y las comunidades indígenas*; Rugerio, 'La Revuelta Agraria de Manuel Lozada y la Separación de Tepic."
38. cf. *Esperando a Lozada*; Meyer, *La Tierra de Manuel Lozada Vol. IV: Colección de Documentos para la Historia de Nayarit*; Meyer, *Manuel Lozada*.
39. 'Periódico Oficial 1878,' in Meyer, *De cantón de Tepic*, 158.
40. Alicia Hernández Chávez, 'Lozada no muere,' in ibid., 211.
41. Lira, "De buenos mexicanos," 23, citing Meyer, *Manuel Lozada*, 240-241; 251-254.
42. Quevedo y Zuvieta, *México*, 65.
43. Lira, "De Buenos Mexicanos," 22.
44. ABPE, *Colección de leyes y decretos de Jalisco*, 1860–61, 20.
45. Ibid., 497.
46. Lira, "De Buenos Mexicanos," 22, citing Peña Navarro, *Estudio histórico*, 163–165.
47. Pérez González, *Ensayo*, 416.
48. AGN, Gobernación, 1620, Lozada nullifies '*Tratos de Pochotitán*,' 13 June 1862.
49. AHJ, Fomento leg. 1864, Ogazón authorises Millen to recruit colonists, 13 October 1861; *Periódico oficial del Gobno. de Sinaloa*, 15 April 1862.
50. Lira, "De buenos mexicanos," 22-3.
51. Ibid., 23.
52. Mario Draper, "The Force Publique's Campaigns in the Congo-Arab War, 1892–1894," in this issue.
53. AGN, Archivo de Mariscal Bazaine, Vol. III, fs. 180–186 .
54. AGN, Gobernación leg. 1418, Imperial agreement with Lozada, 11 July 1865.
55. Ohmstede et al (eds.), *La Presencia del indígena en la prensa capitalina del siglo XIX*, 19.
56. *Juan Panadero*, 29 September 1872.
57. Ibid.
58. Lozada, *Plan Libertador*, 17 January 1873.
59. *Juan Panadero*, in *el Siglo XIX*, Jan 25 1873.
60. J.M. Vigil, editorial in *el Siglo XIX*, Jan 28 1873.
61. Lira, "De buenos mexicanos," 25.
62. Liffman, *Huichol Territory*, 65.
63. cf. Yetman, *The Opatas*, 243-4.
64. Lira, "De buenos mexicanos," 24-5.
65. Meyer, *De cantón*, 162-3.
66. ibid., 139-49.
67. Preuss, *Fiesta, Literatura y Magia*, 84.
68. de la Cerda, "Los Coras," 111.
69. AHSEP-84–85/C/38877, E/36, Navarro to DEFN, 26 October 1927; cf. Coyle, *From Flowers*, 135-8.
70. Lumholtz, *Unknown Mexico*, 491.
71. Preuss, *Fiesta, Literatura y Magia*, 172.
72. Pérez, *Ensayo estadístico*, 10.

73. Dawson, "From Models," 291.
74. Lumholtz, *Unknown Mexico*,i, xvi; cf. Preuss, *Fiesta, Literatura y Magia*, p.91.
75. Hrdlicka, *Observations*, 35.
76. AHSDN-C/C/105/E/XI/481.5/187, Mariano Ruíz to Sec. de Guerra, 27 March 1911.
77. Meyer, *De cantón*, 191.
78. AHSDN-C/C/105/XI/481.5/188, Unsigned military report, 28 March 1912.
79. Flores, Flores Sánchez, *Memorias políticas*, 29.
80. AHSDN-C/C/105/XI/481.5/188, Unsigned military report, 29 April 1912.
81. AHSDN-C/C/105/XI/481.5/188, Reps of Cámara Nacional de Comercio, Tepic, to Madero, 11 May 1912.
82. Pérez, *Ensayo estadístico*, 560; and López, *Fuego*, 150.
83. cf. Rugeley, *Yucatán's Maya Peasantry*, 33–60; cf. Nugent, *Spent Cartridges*, 95–6.
84. Memories of Lozada remained strong in both Cora and Huichols communities well into the twentieth century, cf. Jaúregui, Meyer (eds.), *El Tigre de Alica*; Lumholtz, *Unknown Mexico*, 1, 491–2. Indeed, in the 1900s some of his former Cora fighters were still alive and well see Preuss, *Fiesta, Literatura y Magia*, 172), while given that many Huichol Cristeros in the 1920s viewed themselves as successors of their Lozadista forebears, the same likely goes for Huichol rebels in the 1910s, cf. Meyer, *Manuel Lozada: El Tigre*, 229.
85. AHSDN-C/C/105/XI/481.5/188, Juan Castillo to Sec. Guerra, 22 May 1912.
86. AHSDN-C/C/105/XI/481.5/188, Unsigned military report, 8 July 1912.
87. Joseph, "Caciquismo and the Revolution," 200–1; cf. Hobsbawm, *Bandits*.
88. AHSDN-C/C/124/XI/481.5/250, J. Arzamendi to Sec. Gob., 23 July 1912.
89. AHAG-Totatiche/C/3/E/13, Magallanes to Arz., 6 July 1920.
90. Knight, "Caciquismo," 27–9.
91. In the Oaxacan highlands, communal Defensas averaged five or six members, but several regional caciques commanded from eighteen to thirty-one fighters (Smith, *Politics and Pistoleros*, 67). Similar ranges are given for the Defensas of the mestizo regions of Zacatecas and Jalisco that bordered the Gran Nayar (I. Landa, in Caldera, de la Torre, *Pueblos*, 49; P. Landa, in ibid., 55–56).
92. For example, when Mariano Mejía, usually the commander of an eleven-man Defensa in Jesús María, faced a serious military threat, he could assemble a combined force of around one hundred fighters drawn from communities across the Cora Alta; cf. interviews with Enendino Escobedo Mejía; Juventino Mejía Rivera; Cándido Contreras Rosales.
93. Neurath, "Contrasting Ontologies," 15 (emphasis author's own).
94. Meyer, *Del canton*, 192.
95. ibid., 193.
96. AGN-Censo/1930/Nayarit/San Juan Corapan .
97. Filiberto Sánchez, "Aurelio Chavez".
98. AHSEP-84–85/C/38876/E/17, Orozco to DEFN, 9 February 1928.
99. Interviews with Sandalio Sánchez; Filiberto Sánchez; Agustín Lamas.
100. Interviews with Erasmo González and Filiberto Sánchez;.
101. Interviews with Filiberto Sánchez, Sandalio Sánchez, and Celestino Lamas .
102. AHSEP-45/C/36301/E/29, Navarrete to DECI, 4 May 1925.
103. Purépecha leader Primo Tapia, and Juchiteco strongman Heliodoro Charis, were also notably successful in using their Indian identity to their advantage in their dealings with both local support bases and external, mestizo authorities; cf. Friedrich, *Agrarian Revolt*, 71–2; Smith, *Pistoleros and Popular Movements*, 141–2.

104. AGN-Censo/1930/Nayarit/Arroyo de Santiago.

105. Oral sources are unclear as to her actual ethnicity.

106. Interviews with Enendino Escobedo; Juventino Mejía; Cándido Contreras .

107. Many Mayos, Yaquis and Juchitecos went into battle behind mestizos during the revolution. Fallaw and Rugeley suggest that this 'suggests that Mexican armies reproduced inequality, [and] remained very much a bastion of mestizo culture' ('Redrafting History,' 16–17).

108. AHSEP-45/C/36301/E/29, San Lucas de Jalpa, to Pres., 12 March 1925; AHJ-G-9-920-921/C/52/E/11171, Clemente Villa to Gob.Jal., 17 February 1921.

109. Enendino Escobedo; Juventino Mejía; Cándido Contreras .

110. Juventino Mejía.

111. Enendino Escobedo.

112. Juventino Mejía; cf. Coyle, *From Flowers*, 185–6.

113. Scott, "Patron-Client Politics," 92.

114. Knight, *The Mexican Revolution*, i, 115–27; cf. Brewster, *Militarism, Ethnicity, and Politics*; and Rugeley, *Yucatan's Maya Peasantry*, 101.

115. Perhaps in late 1916, when the Defensas Sociales of Chihuahua were also first established by Carrancista General Enríquez (Almada, *La revolución*,ii, 326); cf. Nugent, *Spent Cartridges*, 83.

116. Torres notes that these same relationships defined factional politics amongst the Nahuas of highland Puebla; Cf. 'Nahuat Factionalism,' 467–8.

117. Enendino Escobedo; Cándido Contreras .

118. Juventino Mejía; Enendino Escobedo; Cándido Contreras.

119. Ibid.

120. cf. Caldera and de la Torre, *Pueblos*; de la Torre, *1926*.

121. Juventino Mejía Rivera.

122. eg. Reichel-Dolmatoff, *The People of Aritama*, 15–6.

123. Weber, *On Charisma*, 19; see also Lomnitz-Adler, *Exits from the Labyrinth, p.*350.

124. Brunk, *Emiliano Zapata*, 238–9; cf. Brewster, *Militarism, Ethnicity, and Politics*, 57–61.

125. Jauregui and Meyer (eds.), *El Tigre de Alica*; cf. Neurath, *Las fiestas,*229; and Alvarado, *Atar la vida*, 299–300.

126. AHED-FR/C/5/E/54, Mariano Mejía to D. Arrieta, 24 July 1919; AHSEP-45/C/36301/E/29, Residents of San Lucas de Jalpa to Gob. Dgo., 12 March 1925.

127. AHJ-G-9–920-921/C/52/E/11171, Clemente Villa to Gob. Jal., 17 February 1921.

128. Aldana, *La rebelión agraria*, 181–209.

129. Nugent, Alonso, "Multiple Selective Traditions," 244.

130. cf. Joseph, "Rethinking Mexican Revolutionary Mobilization," 157–65; Mallón, "Reflections," 98–100. Given that Lozada's 'Conservativism' and Juárez's 'Liberalism' are both remembered locally as struggles for land and autonomy, peasants on both sides of the War of Reform were obviously more concerned with these ubiquitous goals than with the political labels that elite actors attached to their causes. 'Popular Liberalism' and 'Popular Conservatism' may therefore have had more in common than is sometimes assumed. After all, as Hobsbawm has pointed out, 'a social revolution is no less revolutionary because it takes place in the name of what the outside world considers "reaction" against what it considers "progress"' (*Bandits*, 21).

131. Bruno Gómez Estrada and Nemesio Rodríguez Rodríguez, Santa Teresa, Nayarit 03/11/2013.

132. AHED-FR/C/5/E/54, D. Arrieta to Mejía, 9 August 1919.

133. Knight, "Cardenismo," 97.
134. cf. Starn, *Nightwatch*.
135. Scott, *Weapons*, xvi, 29–35.
136. Lumholtz, *Unknown Mexico*,i, xvi; cf. Preuss, *Fiesta, Literatura y Magia*, 91.

Disclosure statement

No potential conflict of interest was reported by the author.

Funding

This work was supported by the Leverhulme Trust: [Grant Number ECF-2018-466].

ORCID

Nathaniel Morris ⓘ http://orcid.org/0000-0003-0098-4604

Bibliography

Primary

Archives

Archivo General de la Nación (AGN): Gobernación, Fomento, Asuntos Económicos, Archivo de Mariscal Bazaine
Archivo Histórico de la Secretaria de Educación Pública (AHSEP). Departamento de Educación y Cultura Indígena, Departamento de Escuelas Rurales y Incorporacion Cultural Indigena, Dirección. de Educación Federal Nayarit
Archivo Histórico de la Secretaria de la Defensa Nacional (AHSDN)
Archivo Histórico de Jalisco (AHJ): Secretaría de Gobernación-Guerra, Divisiones Políticas y Territoriales, Seguridad Pública, Indios, Correspondencia
Archivo Biblioteca Pública Estatal de Jalisco (ABPEJ)

Published Primary Sources

Colección de leyes y decretos de Jalisco, 1860-61 (ABPEJ, reproduced en A. Aldana Rendón, *Rebelión Agraria de Manuel Lozada* (México, 1983))

La Chispa (ABPEJ, Hemeroteca, reproduced in J. Meyer, *La tierra de Manuel Lozada: Colección de Documentos para la Historia de Nayarit, Vol. IV* (Guadalajara, 1989))

El Estado de Jalisco (ABPEJ, Hemeroteca)

Informes militares sobre la situación en el Séptimo Cantón (AHDN, reproduced in J. Meyer, *Esperando a Lozada* (México, 1983))

Informes militares sobre la situación en el Séptimo Cantón (AHJ, Seguridad Pública, reproduced in J. Meyer, *La tierra de Manuel Lozada: Colección de Documentos para la Historia de Nayarit, vol. IV* (Guadalajara, 1989))

Juan Panadero (ABPEJ, Hemeroteca)

El Mundo (AGN, Hemeroteca)

El País (ABPEJ, Hemeroteca)

Plan Libertador proclamado en la sierra de Álica por los pueblos de Nayarit (17 Jan. 1873) (AHJ, Gobernación, reproduced in A. Aldana Rendón, *Manuel Lozada y las comunidades indígenas* (Mexico, 1983))

La Prensa (ABPEJ, Hemeroteca)

El Siglo XIX (AGN, Hemeroteca)

La Trait d'Union (AGN, Hemeroteca)

Books, Articles and Theses

Almada, F, R, ed. *La rebelión agraria de Manuel Lozada*. Vol. 2. Chihuahua: Talleres Gráficos de la Nación, 1964.

Alvarado Solis, N. P. *Atar la vida, trozar la muerte: El sistema ritual de los mexicaneros de Durango*. Morelia: Universidad Michoacana-San Nicolás de Hidalgo, 2004.

Battalla, G. Bonfil. *México profundo*. México: SEP/CIESAS, 1987.

Blanco Rugerio, M. "La Revuelta Agraria de Manuel Lozada y la Separación de Tepic." In *Nayarit del Septimo Cantón al Estado Libre y Soberano*, edited by P. L. González and J. R. M. Cervantes, 92–107. Guadalajara: Universidad de Guadalajara, 1990.

Brewster, K. *Militarism, Ethnicity, and Politics in the Sierra Norte de Puebla, 1917-1930*. Tucson: The University of Arizona Press, 2003.

Brunk, S. *Emiliano Zapata: Revolution and Betrayal in Mexico*. Albuquerque: University of New Mexico Press, 1995.

Caldera, L., and C de la Torre, eds.. *Pueblos del Viento Norte*. Guadalajara: Secretaría de Cultura de Jalisco, 1994.

Cambré, M. *La Guerra de Tres Anos: Apuntes para la historia de la Reforma*. Guadalajara: Universidad de Guadalajara, 1986 [1904].

Cerda Silva, R. de la. "Los Coras." *Revista Mexicana de Sociología* 5, no. 1 (1943): 89–117.

Coyle, P.E. "The Customs of Our Ancestors: Cora Religious Conversion and Millennialism, 2000-1722." *Ethnohistory* 45, no. 3, Summer (1998): 509–542. doi:10.2307/483322.

Coyle, P.E. "'To Join the Waters': Indexing Metonymies of Territoriality in Cora Ritual." *Journal of the Southwest* 42, no. 1 Spring, (2000): 119–128.

Coyle, P. E. *From Flowers to Ash: Náyari History, Politics, and Violence*. Tucson: University of Arizona Press, 2001.

Dawson, A.S. "From Models for the Nation to Model Citizens: Indigenismo and the 'revindication' of the Mexican Indian, 1920-40." *Journal of Latin American Studies* 30, no. 2, May (1998): 279–308. doi:10.1017/S0022216X98005057.

Diguet, L. *Por tierras occidentales: entre sierras y barrancas.* México: Centro de estudios mexicanos y centroamericanos, 2005.

Draper, Mario. "The Force Publique's Campaigns in the Congo-Arab War, 1892–1894." *Small Wars and Insurgencies* 30, no. 04–05, 2019.

Fallaw, B, and T. Rugeley. "Redrafting History: The Challenges of Scholarship on the Mexican Military Experience." In *Forced Marches: Soldiers and Military Caciques in Modern Mexico*, edited by B. Fallaw and T. Rugeley, 1–22. Tucson: University of Arizona Press, 2012.

Flores Flores, M. E., and Flores Sánchez. *Memorias políticas de Manuel Flores Flores: la Revolución Mexicana y la Guerra Cristera en el municipio de La Yesca.* México: Arlequín Editorial, 2000.

Friedlander, Judith. *Being Indian in Hueyapan: A Study of Forced Identity in Contemporary Mexico.* New York: St. Martin's Press, 1975.

Friedrich, P. *Agrarian Revolt in a Mexican Village.* Chicago: University of Chicago Press, 1977.

Friedrich, P. *The Princes of Naranja: An Essay in Anthrohistorical Method.* Austin: University of Texas Press, 1986.

Gerhard, P. *La frontera Norte de la Nueva España.* México: UNAM, 1996.

Gitlitz, J. *Administrando justicia al margen del Estado: las rondas campesinas de Cajamarca.* Lima: Instituto De Estudios Peruanos, 2013.

Gómez Canedo, L. "Huicot: Antecendentes Misionales." *Estudios de Historia Novohispana* 9 (1987): 94–145.

Hernández Chávez, A. "*Lozada* no Muere." In *De Cantón de Tepic a Estado de Nayarit, 1810–1940,* edited by J. Meyer, 209–220. Guadalajara: Universidad de Guadalajara, 1990.

Hinton, R.E. "Indian Acculturation in Nayarit: The Cora Response to Mestizoization." In *The Social Anthropology of Latin America: Essays in Honor of Ralph Leon Beals,* edited by Walter Goldschmidt and Harry Hoijer, 16–35. Los Angeles: University of California Press, 1970.

Hinton, T.E "Pre-Conquest Acculturation of the Cora." *Journal of the Arizona Archaeological and Historical Society* 37, no. 4 (Summer, 1972): 161–168.

Hinton, T.E, and P. C. Weigand, eds. *Themes of Indigenous Acculturation in Northwest Mexico,* 9–21. Tucon: University of Arizona Press, 1981.

Hobsbawm, E. *Bandits.* New York: Delacorte Press, 1969.

Hrdlička, A. *Physiological and Medical Observations among the Indians of Southwestern United States and Northern Mexico.* Washington DC: Government Printing Office, 1908.

Hu-Dehart, E. *Yaqui Resistance and Survival: The Struggle for Land and Autonomy, 1821–1910.* Madison: University of Wisconcin Press, 1984.

Jáuregui, J. *Los Coras.* México: CDI, 2004.

Jáuregui, J, and L. Magriñá. "Estudio etnohistórico acerca del origen de los mexicaneros (hablantes del náhuatl) de la Sierra Madre Occidental." *Dimensión Antropológica* 26, (2002): 27–81 September-December.

Jáuregui, J, and J. Meyer, eds. *El Tigre de Alica: mitos e historia de Manuel Lozada.* México: SEP-CONAFE, 1997.

Joseph, G. "Caciquismo and the Revolution: Carrillo Puerto in Yucatán." In *Caudillo and Peasant in the Mexican Revolution,* edited by D. A. Brading, 193–221. Cambridge: Cambridge University Press, 1980.

Joseph, G. "Rethinking Mexican Revolutionary Mobilization: Yucatán's Seasons of Upheaval, 1909–1915." In *Everyday Forms of State Formation: Revolution and the*

Negotiation of Rule in Modern Mexico, edited by G. Joseph and D. Nugent, 135–69. Durham N.C.: Duke University Press, 1994.

Joseph, G, and D. Nugent, eds. *Everyday Forms of State Formation: Revolution and the Negotiation of Rule in Modern Mexico*. Durham N.C.: Duke University Press, 1994.

Knight, A. *The Mexican Revolution, 1910–1920*. Vol. 2. Cambridge: Cambridge University Press, 1986.

Knight, A. "Cardenismo: Juggernaut or Jalopy?" *Journal of Latin American Studies* 26, no. 1, February (1994): 73–107. doi:10.1017/S0022216X0001885X.

Knight, A. "Caciquismo in Twentieth-Century Mexico." In *Caciquismo in Twentieth-Century Mexico*, edited by A. Knight and W. Pansters, 1–48. London: Institute for the Study of the Americas, 2005.

Liffman, P.M. *Huichol Territoriality and the Mexican Nation: Indigenous Ritual, Land Conflict and Sovereignty Claims*. Tucson: University of Arizona Press, 2011.

Lira Larios, R. "De buenos mexicanos, cristianos, soldados y valientes: pueblos coras y huicholes en la configuración de una región, 1840 a 1880." *Historia Mexicana* forthcoming, January–March 2020.

Lomnitz-Adler, C. *Exits from the Labyrinth: Culture and Ideology in the Mexican National Space*. Berkeley: University of California Press, 1992.

López Díaz, P. *Fuego en Nayarit*. Guadalajara: Editorial Gráfica, 1957.

Lumholtz, C.S. *Unknown Mexico*. Vol. 2. New York: Scriber and Sons, 1902.

Macola, G. "Guerrilla Warfare in Katanga: The Sanga Rebellion of the 1890s and Its Suppression." *Small Wars and Insurgencies* 30, no. 4–5, 2019.

Mallón, F. "Reflections on the Ruins: Everyday Forms of State Formation in Nineteenth-Century Mexico." In *Everyday Forms of State Formation: Revolution and the Negotiation of Rule in Modern Mexico*, edited by G. Joseph and D. Nugent, 69–106. Durham N.C.: Duke University Press, 1994.

Meyer, J. "El reino de Lozada en Tepic." In *Esperando a Lozada*, edited by Jean Meyer, 95–109. Zamora: El Colegio de Michoacán, 1984.

Meyer, J. *Esperando a Lozada*. Zamora: El Colegio de Michoacán, 1984.

Meyer, J. *De Cantón de Tepic a Estado de Nayarit, 1810–1940*. Guadalajara: Universidad de Guadalajara, 1990.

Meyer, J. *Manuel Lozada: El Tigre de Álica, General, Revolucionario, Rebelde*. México: Editorial Tusquets, 2015.

Meza Aguirre, R. *Nayarit Memoria Oral: Cachirul, Cabalgando con Lozada*. Tepic: UAN, 1996.

Morris, N, "'The World Created Anew': Land, Religion and Revolution in Mexico's Gran Nayar Region, 1920–1940." Unpublished DPhil Diss., University of Oxford, 2015.

Morris, N. "'¿"Forjando Patria"? Las políticas del Estado revolucionario y el ocaso de los vínculos intercomunales coras en la Sierra del Nayar." *Relaciones: Estudios de Historia Y Sociedad* 39, no. 156, November (2018). doi:10.24901/rehs.v39i156.316.

Neurath, J. *Las fiestas de la casa grande: Procesos rituales, cosmovisión y estructura social en una comunidad huichola*. México: CONACULTA-INAH, 2002.

Nugent, D. *Spent Cartridges of Revolution: An Anthropological History of Namiquipa, Chihuahua*. Chicago: University of Chicago Press, 1993.

Nugent, D, and A. M. Alonso. "Multiple Selective Traditions in Agrarian Reform and Agrarian Struggle: Popular Culture and State Formation in the Ejido of Namiquipa, Chihuahua." In *Everyday Forms of State Formation: Revolution and the Negotiation of Rule in Modern Mexico*, edited by G. Joseph and D. Nugent, 209–246. Durham N. C.: Duke University Press, 1994.

Ohmstede, A., and Escobar Teresa Rojas Rabiela, eds. *La Presencia del Indígena en la Prensa Capitalina del Siglo XIX*. Mexico: Centro de Investigaciones, 1993.

Oosterhout, Van, 'Popular Conservatism in Mexico: Religion, Land, and Popular Politics in Nayarit and Queretaro, 1750–1873'. Unpublished PhD Diss., Michigan State University, 2014.

Pérez González, J. *Ensayo Estadístico y Geográfico de Territorio de Tepic*. Tepic: Imprimador de Retes, 1894.

Pruess, K. T, and J. Jáuregui, eds. *Fiesta, literatura y magia en el Nayarit: Ensayos sobre coras, huicholes y mexicaneros de Konrad Theodor Preuss*. México: Centro de estudios mexicanos y centroamericanos, 1998.

Purnell, J. *Popular Movements and State Formation in Revolutionary Mexico: The Agraristas and Cristeros of Michoacán*. Durham, NC: Duke University Press, 1999.

Quevedo Y Zuvieta, S. *México: Recuerdos de un Emigrado*. Madrid: Estudio tipográfico de los Sucesores de Rivadeneyra, 1883.

Radding. "Peasant Resistance on the Yaqui Delta: An Historical Inquiry into the Meaning of Ethnicity." *Journal of the Southwest* 31, no. 3 Autumn, (1989): 330–361.

Reichel-Dolmatoff, G, and A. D. Reichel-Dolmatoff. *The People of Aritama: The Cultural Personality of a Colombian Mestizo Village*. London: Routledge and Kegan Paul, 1961.

Reina, L. *Las Rebeliones Campesinas en México, 1819–1906*. México: Siglo Veintiuno, 1988.

Rendon, A. *Aldana, Manuel Lozada y Las Comunidades Indígenas*. México: Fondo de Cultura Económica16, 1983.

Rendon, M. Aldana. *La Rebelión Agraria de Manuel Lozada, 1873*. México: SEP 1983.

Rugeley, T. *Yucatán's Maya Peasantry and the Origins of the Caste War*. Austin: University of Texas Press, 1996.

Samaniego, M.A. "La revolución mexicana en Baja California: maderismo, magonismo, filibusterismo y la pequeña revuelta local." *Historia Mexicana* LVI, no. 4 (2007): 1201–1262.

Santoscoy. *Nayarit; colección de documentos, inéditos, historicos y etnograficos, acerca de la Sierra de ese Nombre*. Guadalajara: Tipo-Lit.y Enc. de J. Maria Yguíniz, 1899.

Scott, J.C. "Patron-Client Politics and Political Change in Southeast Asia." *The American Political Science Review* 66, no. 1, March (1972): 91–113. doi:10.2307/1959280.

Scott, J.C. *The Moral Economy of the Peasant: Rebellion and Subsistence in Southeast Asia*. New Haven: Yale University Press, 1976.

Scott, J.C. *Weapons of the Weak: Everyday Forms of Peasant Resistance*. New Haven: Yale University Press, 1986.

Smith, B.T. *Pistoleros and Popular Movements: The Politics of State Formation in Postrevolutionary Oaxaca*. Lincoln: University of Nebraska Press, 2009.

Starn, O. *Nightwatch: The Politics of Protest in the Andes*. Durham, NC: Duke University Press, 1999.

Trueba, Torres. "Nahuat Factionalism." *Ethnology* 12, no. 4 (1973): 463–474. doi:10.2307/3773373.

Weber, M. *On Charisma and Institution Building*. Chicago: University of Chicago Press, 1968. S.N. Eisenstadt (ed.).

Yetman, D. *The Ópatas: In Search of a Sonoran People*. Tucson: University of Arizona Press, 2010.

Guerrilla warfare in Katanga: the Sanga rebellion of the 1890s and its suppression

Giacomo Macola and Jack Hogan

ABSTRACT

This chapter discusses the origins and development of the Sanga insurgency of the 1890s with a view to demonstrating that, contrary to commonly held stereotypes, pre-colonial warfare was neither simple nor unchanging. Its tactics, it is argued here, repay the sort of close analysis commonly reserved for other typologies and theatres of war. The Yeke, against whose exploitative system of rule the Sanga and their allies rose up in 1891, survived the onslaught by entering into a strategic alliance with Lofoi, a newly established station of the Congo Free State, and its limited contingent of regular Force Publique troops. An in-depth examination of the joint Yeke-Force Publique counterinsurgency campaign leads to the conclusion that the novelty of the 'small wars' that accompanied the Scramble for Africa should not be overstated. In southern Katanga and, by implication, elsewhere, these confrontations were shaped by processes of mutual borrowing in which African military practices and even political aims were not necessarily subordinate to European ones.

The study of warfare in Africa remains altogether undeveloped. As a result of this dearth of scholarship, African military history has yet to be fully decolonized, and the sub-field remains 'perhaps the last bastion of the kind of distorted Eurocentric scholarship that characterised African studies before the 1960s.'[1] Richard Reid's remarks apply with special force to the pre-colonial period, whose conflicts have rarely – if ever – been deemed worthy of detailed operational analysis, even in such instances where the available evidence would readily lend itself to this kind of treatment.[2] Drawing on an underutilized body of sources – primarily the records of the Plymouth Brethren (PB) missionaries, in Katanga since 1886, and the personal correspondence of Clément Brasseur, the Lofoi *chef de poste* in 1893–1897[3] – this article examines the dynamics of one specific insurgency on the cusp of the colonial occupation of Katanga with a view to demonstrating that, contrary to commonly held stereotypes, pre-colonial warfare was neither simple nor unchanging. Its tactics and strategies, it is argued here, repay the sort of close analysis commonly reserved for other typologies and theatres of war.

Since the Sanga insurgency was eventually suppressed by a newly-minted Euro-African coalition, this paper also aims to intervene in debates about the military aspects of the Scramble for Africa. Patterns of African resistance to the colonial conquest, of course, were a key concern of the first generation of professional historians of Africa. Their analyses, however, were coloured by the politics of the period and, specifically, the attempt to posit a direct connection between so-called 'primary resistances' and later decolonization movements.[4] For their part, military historians proper have been slower in taking up the study of late nineteenth-century Euro-African confrontations, and their efforts have also been impaired by the tendency to examine these conflicts in isolation from their background and African roots.[5] Thus, our second objective in this paper is to demonstrate the benefits that accrue from placing the 'small wars' that accompanied the Scramble for Africa in a deeper chronological context than is commonly the case.[6] Only by so doing – we contend – can the debate about the ultimate nature and ostensible modernity of these military operations be placed on firmer foundations.

A grassroots rebellion against an oppressive foreign elite, Msiri's Yeke, the Sanga insurgency of the 1890s casts important new light on the rise of warlordism in central Africa in the era of global commerce and the manner in which the unprecedented levels of exploitation it brought in its wake were experienced by its victims. It also illustrates the fragility of these new political formations vis-à-vis determined efforts to sever the external links to which they largely owed their economic and military strength. The workings of Msiri's warlord polity, Garenganze, form the subject of the first substantive section of the present article. From an operational point of view, the Sanga insurgency represents an unusually well-documented instance of guerrilla warfare in an environment still unaffected by direct European politico-military action. Its origins, development and innovative aspects are examined in the second part of the article. A key watershed in the history of the rebellion was the arrival of European-led military forces on the Katangese scene and their alliance with the beleaguered Yeke military. The counterinsurgency campaign jointly carried out by the new partners and the rebels' varied reactions to it are discussed in the third section of the article.

The Yeke warlord state

Between the early 1850s and the late 1860s, Msiri, a caravan leader turned conqueror, succeeded in imposing his and his companions' sway over much of present-day *Haut-Katanga*, a copper-producing district which western Tanzanian traders, including Msiri's own father, Kalasa, had begun occasionally to visit from the 1830s and which had previously formed the western periphery of the Lunda kingdom of Kazembe.[7] Better equipped than longer-established aristocracies to face the forces of the global trade that was then enveloping the Congo basin,[8] Msiri's followers – the Yeke – conquered the Sanga and other pre-existing inhabitants of the region, and spawned an innovative type of polity. Typifying the revolutionary changes in governmentality which were then taking place across vast swathes of what would become the Congo Free State (CFS), Garenganze was a warlord state in which the exercise of political power had less to do with religious sanction and hereditary principles than with its ruler's charisma and the systematic use of commercially-driven violence. Owing their wealth and military strength to trade, Yeke state-builders displayed a consistent willingness to resort to forms of extreme compulsion to obtain the ivory and slaves that fed the long-distance exchanges with both the eastern and, increasingly, the western coasts of the continent.

Arab-Swahili trade in southern Katanga might have reached its high point in the late 1860s. Thereafter, its place in the political economy of the new polity was partly taken over by Luso-African and Ovimbundu caravans from present-day Angola. A direct trading link between Garenganze and the Ovimbundu plateau was inaugurated in about 1870 by a João Baptista Ferreira, who responded to some earlier Yeke entreaties and supplied Msiri 'with powder, guns, and cloth in exchange for ivory.'[9] Yeke and other sources clarify that, unlike in the case of the eastern trade towards the Swahili coast, the Yeke did not merely await the arrival of foreign merchants in their capital, but actively organized independent west-bound caravans to transport the ivory and slaves accumulated by Msiri.[10] Already in 1875, Benguela, the terminus of Ovimbundu trade routes on the Angolan coast, was the principal port of entry for the Portuguese- and Belgian-made muskets and gunpowder which Angolan caravans brought to Msiri capital, Bunkeya, after negotiating Chokwe and Luvale countries and crossing the upper Lualaba to the south of the Upemba depression (see Map 1).[11]

At the apogee of his power in the 1880s, Msiri acted very much like a merchant prince, lording it over Bunkeya – a bustling commercial hub located at the crossroad of several regional and long-distance exchange networks[12] – and exerting tight control over the kingdom's external trade. Central privileges were especially rigidly enforced in the case of the ivory and slave trades. Not only were ivory and slaves exhaustible resources, but they also represented the most highly sought-after items of trade among Ovimbundu gun-dealers. Thus, by centralizing in his hands the commerce in these two commodities, Msiri was also able to impose a practical monopoly over the acquisition and distribution of guns. By 1884, the Ovimbundu-run trade in guns and gunpowder had resulted in Msiri being equipped with an impressive force of '2,000 or 3,000 flintlock muskets'.[13] This arsenal exceeded that of any other regional power, and its 'considerable' size was well-known to neighbouring peoples.[14]

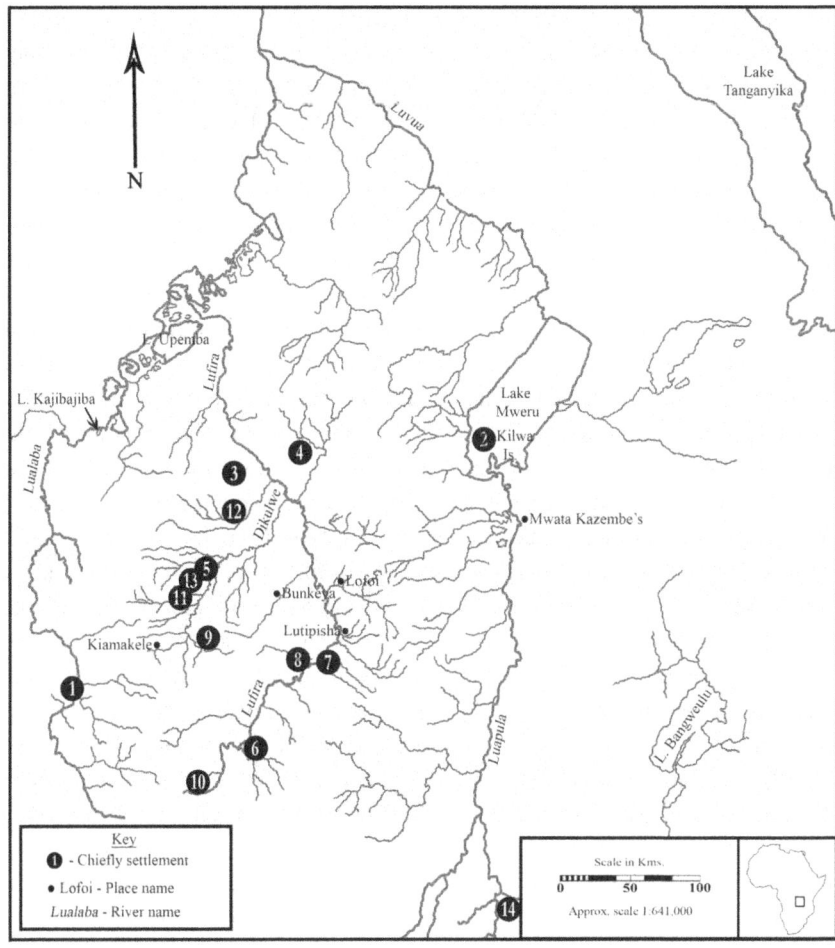

Katanga in the 1890s

1. Kazembe of the Lualaba
2. Shimba
3. Kalera
4. Mukana
5. Mulowanyama (to late 1895)
6. Katanga
7. Mwashya
8. Mutwila
9. Kalunkumia
10. Ntenke
11. Mwenda Mukoshi
12. Kalala Ngombe
13. Kasangula
14. Kiwala

Map 1. Katanga in the 1890s.

To be sure, the creation of dependencies and relations of reciprocity through the circulation of foreign goods was part of Msiri's political repertoire; yet such institutions of rule were largely confined to the Yeke and Yekeized elites. Conversely, the bulk of Garenganze's subject peoples constituted a mass of primary producers to be preyed upon. Resource extraction in the heartland of the polity and even some of the more distant peripheral regions acquired over the course of the 1870s–1880s was ensured by a network of centrally-appointed officials, the *banangwa* (sing. *mwanangwa*). Frequently chosen from within the ranks of Msiri's close patrilineal relatives, Yeke *banangwa* served as tribute collectors and also as military and caravan leaders.[15] Insofar as ivory and slaves were concerned, the Yeke devised a 'system of compulsory production quotas' destined to increase their ruler's access to such indispensable commodities.[16] Msiri's exactions – what Delcommune, the leader of one of the first Belgian expeditions to Katanga, called the 'insatiable avidity of the king' in demanding ivory and slaves for sale to the Ovimbundu[17] – caused real hardship to subject peoples. Chief Mapanda – for instance – complained to PB missionary Crawford that Msiri had once given him only 'a paltry four yards of cloth' and 'no remuneration whatever to his men' in return for a massive tribute of 'forty teeth of ivory'.[18] In another instance, a seeming shortfall in the expected ivory payment late in 1889 led to the execution of an unnamed local leader by the son of Dikuku, Msiri's brother. 'In a day or two' – wrote another missionary – 'the head of this poor unfortunate, with those of all his male villagers, will be presented to Msidi with great pomp and ceremony. The females and children will become the slaves of the Vayeke.'[19]

Besides tribute, raids were the principal means through which captives were obtained for both trading and internal purposes. Carried out by comparatively heavily-armed, semi-professional infantrymen, slave forays punctuated the economic life of Garenganze and accounted for the spread of violence and the overall militarization of society in Katanga in the latter part of the century. The PB missionary Arnot, the first European to spend a long period of time among the Yeke, remarked in August 1886 that Msiri was 'constantly sending out war parties to the countries all round'.[20] These war parties – he explained elsewhere – brought back '[l]arge numbers of slaves' who were 'sold chiefly to Arab traders from Zanzibar and to Ovimbundu traders from Bihé. Strong young men' – he concluded – 'have been sold for ten or twelve yards of cotton cloth.'[21] Women were probably more frequently kept in Bunkeya and surrounding areas with a view to exploiting their labour and reproductive potential.[22] The pervasiveness of slave-raiding activities accounts for the fact that the 'proportion of women to men [was] very unequal' in Bunkeya and for the capital's oft-remarked-upon cosmopolitan nature.[23] For '[w]hat means the babel of tongues heard in this country, if not that Msidi's war parties have brought in from nearly every point of the compass gangs of poor down-trodden mortals, who were swooped down upon in their little hamlets far away and carried off?'[24]

According to Legros' informants, the average Yeke raiding party was led by one or more *mutwale* (who were also frequently *banangwa*) designated and appointed by Msiri. Each *mutwale* commanded a *bulungu*, the basic Yeke military unit. This consisted of a core of about twenty gunmen, whose firearms were personally entrusted to them by Msiri, and a further group of more lightly armed fighters, numbering between sixty and one hundred.[25] Depending on the nature of the particular expedition being prepared, several *bulungu* could operate together. In 1885, for instance, Capello and Ivens encountered a large Yeke war band near present-day Lubumbashi. On this occasion, the overall leader of the 'horde of brigands', the aforementioned Dikuku ('Licuco'), commanded 'about 400 men' equipped with 'firearms, arrows [and] spears'.[26]

Insurgency

How did Yeke militarism and violent forms of governance impact on the social landscape of Katanga? From a purely military point of view, the frequency of Yeke predatory raids and, to a lesser extent, full-blown wars of expansion and their reliance on firearms account for the spread and increasing use of fortifications in the region. Defences around settlements varied a great deal. The most elaborate were probably those surrounding permanent Arab-Swahili bases in the interior. For instance, Shimba's stockaded headquarters on Kilwa Island, Lake Mweru, which neither Yeke nor Congo Free State Forces ever managed to storm in the late 1880s and early 1890s,[27] included 'two bullet-proof blockhouses; [there was] no ditch but a two- or three-metre tall wooden rampart that could not be scaled.'[28] But basic fortifications consisting of a crenelated palisade and encircling ditch were to be found in virtually every Katangese village.[29] These became especially effective when associated with such natural defences as caves and crags. Defensive complexes of this kind enabled a number of mountain communities to retain a degree of autonomy vis-à-vis the Yeke,[30] and would prove difficult to overcome even for European-led forces. In June 1895, for example, *chef de poste* Brasseur had 'one man killed and three wounded' when he rushed the village of Kalera, on the eastern flank of the Mitumba Mountains. 'The natives', after 'defend[ing] themselves like devils', evacuated their village and retreated into a natural tunnel which Brasseur dared not enter 'for I would lose half of my soldiers there. [...]. There must be three or four exits to these tunnels, but no one besides the people of the village knows them.'[31] Three years later, Delvaux, one of Brasseur's successors in Katanga, came to identical conclusions when he took on another naturally-fortified settlement, Mukana's. There was – he argued – simply 'no hope of entering the cave' in which the defenders had sought shelter after pulling out of their village.[32]

Because of the wide distribution of fortifications, most military encounters in late nineteenth-century Katanga appear to have involved more or less prolonged sieges (though, at least in areas of limited visibility, ambushes must have been common as well). Reflecting the prejudices of his time and race – which construed African conflict as either exceptionally savage or, conversely, as exceptionally mild and cowardly[33] – Brasseur was dismissive of this form of warfare and its lethality.

When the natives attack another village, they take position nearby – at a
distance of a few hundred metres – build a stockade and remain there for
months, killing whoever leaves the village and destroying the crops. 'They
should make a sortie', you will say. They are niggers, you see, and then the
attacker is always supposedly stronger than the defender. In short, if there are
two or three fatalities on both sides, the affair was a hot one.[34]

In fact – regardless of whether assaults were truly 'unknown' to local
Africans, another of Brasseur's questionable generalizations[35] – sieges no
doubt entailed significant hardship and constant vigilance. Coupled with the
effects of the adoption of firearms in battle, they are likely to have resulted
in enhanced military discipline for attackers and defenders alike.[36]

   If these were the most common rules of battle at the height of Yeke power – if,
that is, late nineteenth-century Katangese warfare was at least partly static – then
the Sanga-led insurgency of 1891 was a game changer. Coinciding with the
adoption of openly aggressive tactics by the victims of Yeke predation, it
witnessed the emergence of a type of hit-and-run warfare which even coeval
witnesses saw fit to describe under the rubric of 'guerrilla'.[37] The key advantage
of guerrilla methods, of course, is that they enable relatively small, but highly
mobile, group of fighters to avoid set-piece battles while still striking at the heart
of what, at least initially, constitutes a superior enemy in terms of both manpower
and resources. Crucially, the Sanga met three essential preconditions for waging
a successful insurgency: intimate knowledge of local terrain; widespread grassroots support; and a highly motivated leadership.

   The Sanga were 'famous' hunters who had apparently enjoyed a 'monopoly
over elephant hunting' before the inception of Garenganze.[38] No doubt, the
bushcraft skills accumulated during decades of intensive hunting stood them in
good stead once they embarked on the path of armed resistance.[39] Moreover,
to the extent that tactics are shaped by terrain, guerrilla methods were well-
suited to the topography of southern Katanga, a region characterised by the
presence of such difficult and sparsely populated environments as mountain
ranges and flatlands subject to seasonal flooding. The leaders of the rebellion
were Sanga hereditary title-holders, whom the Yeke had subjugated but not
replaced. While our sources do not permit a full discussion of their individual
motivations and commitment to the insurgency, some tantalizing insights into
the background and disposition of at least one rebel leader, Mulowanyama, are
provided by Campbell. Mulowanyama – the PB missionary reminisced in the
early 1920s – was a 'veritable modern Nimrod' whose name, in fact, meant
'game wizard'. Having had 'his ears cropped off by order' of Msiri, his hatred for
the latter 'and his tribe, the Ba-Yeke', was 'implacable'.[40] In this, Mulowanyama
was far from alone, for 'all the Sanga' – if we are to believe Delcommune – hated
the Yeke 'looters with a passion'. The mere mention of the name 'Yeke' –
'detested throughout Katanga' – was enough to 'inflame their eyes with rage'.[41]
This reservoir of popular resentment – itself the product of the divisiveness and

exploitativeness of the late nineteenth-century warlord order – granted the insurgents the grassroots backing and protection that every successful guerrilla movement requires. And if further proof is needed of the popularity of the rebellion, it is provided by the rapidity with which it spread, drawing into its fold not only the bulk of the Sanga, but several non-Sanga communities as well.

The insurgency's spark, as is confirmed by several coeval sources, was the accidental murder of Masengo, a Sanga woman, and the subsequent refusal on Msiri's part to hand over the legal culprit – that is, the Yeke owner of the slave who had fired the accidental gunshot.[42] But the fate of Masengo was a mere 'casus belli', behind which, as the eyewitness Crawford put it, lay 'deep, deep-seated grudges of long-standing against Msidi's tyranny'.[43] Building perhaps on the socially-sanctioned recourse of slighted parties, who became kapondo (assassins) to avenge an individual wrong,[44] the Sanga, who had certainly managed to accumulate some muskets by the time of the rebellion, began a series of nightly attacks against Bunkeya and other centres of Yeke power. The first recorded action of the war took place during the night of 20–21 February 1891, when seven huts in Msiri's own section of Bunkeya, Nkulu, 'were maliciously set on fire and burnt to the ground' under the cover of darkness. Two women and a child were shot in another quarter of the capital, and one more casualty was recorded in Kankofu, the neighbouring village of Ntalasha ('Ndalasia'), Msiri's classificatory brother.[45]

The timing of the attack is interesting: January-February is the height of the rainy season in Katanga, and floods make movement, in general, and military manoeuvres, in particular, extremely problematic for anyone but the swiftest and most experienced of assailants.[46] Far from slowing down the Sanga, rain conditions might have actually favoured them, for their forays increased in intensity and geographical extent in the aftermath of the first raid on Bunkeya. As early as 26 February, 'two or three houses were burned down [...] at different villages', and 'many of the Vasanda [sic]' were already said to have joined the rebellion.[47] Between March and April, the Yeke settlements of both Dikuku ('Lukuku') and the more distant Mulenga ('Molenga') came under attack.[48] An entire village belonging to the former was burnt down during the night and several people shot, 'four of whom died immediately'.[49] By then, Msiri – whom the Sanga insurrection had reportedly left 'bewildered' and 'in a strange lethargic condition'[50] – 'no longer [slept] unprotected, but ha[d] a large bodyguard sleeping around his house every night',[51] while the inhabitants of Bunkeya's outlying quarters were being forced to seek shelter from 'the night raids of the Vasanga' in the central, fortified, areas of the capital, where they built a great many temporary accommodations.[52]

One of the obvious aims of the guerrillas was to generate panic among the inhabitants of Yeke settlements by asserting their ability to strike undetected at even the most closely guarded of targets. In this, they were successful, since by May feelings 'in and around the Capital' were running high, as residents felt unprotected by Msiri and 'liable to be shot down at any time.'[53] The contrast could not have been greater with the situation obtaining only a few years earlier, when the silence and safety of Bunkeya nights had impressed Arnot, prompting him to conclude that 'life and property [...] are safer here than in much-favoured England.'[54] But Sanga attacks were not always random and clearly drew on preliminary intelligence. On occasions, they singled out specific members of the Yeke or Yekeized hierarchy, such as headman 'Muluwe', who had his 'omande' shell – a key symbol of Yekehood – removed from his body before being 'shot through the heart' in the early hours of 10 May.[55] The fact that Muluwe's women and children escaped his burning village and sought refuge at the mission suggests that the seizing of prisoners (which in some cases might have amounted to the liberation of captives of the Yeke) was probably a frequent by-product of Sanga incursions, though the PB missionaries only make explicit mention of it once, in the context of an attack on a Yeke colony near chief Katanga in late May or early June.[56] In this connection, it is probably relevant to note that, in 1895, Mulowanyama was still in the possession of 'a hundred or so women he had abducted from the Yeke' at an unspecified period in the past.[57]

Not the least innovative aspect of the insurgency was its economic dimension. Revealing a good grasp of the political economy of Msiri's warlord state, the Sanga resolved to pair their direct military actions with a policy of strangulation intended to disrupt the lines of communication between the heartland of Garenganze and the Ovimbundu plateau, thereby preventing Msiri from resupplying his military stock with new muskets and, especially, gunpowder, at a time in which his forces were also being tested by the Lunda of Kazembe, on the lower Luapula, and by Shimba, on Lake Mweru. By 'stopping and looting Msiri's caravans' at, or en route to, the Lualaba ferries,[58] the Sanga achieved the twin objective of weakening and impoverishing Msiri while funding their own rebellion. This tactic proved highly effective in restricting freedom of movement in southern Katanga, as attested in the late spring by CFS official Paul Le Marinel, who discovered it was impossible to travel between Bunkeya and the headwaters of the Dikulwe, 'for the Sanga were blocking the routes'.[59] Finding himself constrained by lack of gunpowder, Msiri could not reply decisively to the outbreak of the insurgency and was forced to postpone his counteroffensive for several months.[60]

By the end of May, 'many reports of murders by the Va-sanga' were reaching the PB missionaries, who also learnt that Msiri's own Sanga drummers had fled Bunkeya 'and joined their countrymen in the rebellion.'[61] Over the course of the same month, however, the Yeke supply situation registered a temporary improvement, thanks to the powder which Le Marinel eventually agreed to give Msiri and to the arrival of a well-stocked Angolan caravan which had managed to force the Lualaba blockade.[62] Finally 'in a position to make a big muster of war parties', Msiri dispatched two separate military expeditions in the early summer of 1891. While Dikuku went up the Lufira with a view to protecting the salt pans of Mwashya from attacks by the Sanga of Mutwila, another column, led by Msiri's son, Mukanda Bantu, took a south-westerly direction and confronted the enemy in the 'Va-Sanga stronghold' of Kalunkumia, the holder of the Mpande, the ancient paramount chiefly title of the Sanga. After much firing, Kalunkumia's 'Va-sanga were compelled to evacuate the town and flee south, in the direction of Ntenke's', who despite being a member of Msiri's original following and his representative among the copper-producing peoples of the upper Lufira River, had by then also joined the rebellion.[63] 'No chase seems to have followed, and the war parties returned to the capital without striking any decisive blow.'[64]

Despite this setback, the Sanga and their growing network of allies held out, while the internal situation of the besieged and isolated Garenganze continued to deteriorate. Not only had the 'war [...] reduced Msidi's ivory tribute this year to almost nothing', but ongoing Sanga forays also forced the inhabitants of Bunkeya not to 'venture out far to the fields they were wont to cultivate' before the start of the insurgency and to limit themselves to 'tilling the poor stony soil' of the hill around which the Yeke capital had been built.[65] This had the effect of accelerating the onset of a famine partly brought about by that year 'exceptionally dry condition.'[66] Famine and security concerns, in turn, led to population losses. By October-November, 'more than half' of Bunkeya's population had fled,[67] 'the capital [had] quite a deserted look about it',[68] and Msiri was once again short of powder.[69] Dikuku, meanwhile, had failed to overcome Mutwila,[70] while the Lomotwa of the Kundelungu Range, who 'abhorred' their Yeke *mwanangwa*, Kifuntwe ('Kifumtiè'), had thrown their lot in with the rebels.[71] Running out of options, Msiri sought unsuccessfully to induce both Delcommune and the CFS officials in Lofoi – the station inaugurated by Le Marinel late in May 1891 – to fight the insurgents on his behalf, and might have even contemplated the possibility of shifting the location of his capital.[72] By mid-December, when William Stairs and his 300 men (200 of whom equipped with modern breech-loaders) made their entry into Bunkeya, the famine was 'appalling' and the central symbol of Yeke power a pale shadow of its former self.[73] Msiri's father-in-law, the Angolan trader Coïmbra, told Stairs that the ten quarters into which Bunkeya had been subdivided a mere three years earlier were now reduced to one and that the hills to the south-west of the town, once 'dotted with flourishing villages', were now almost entirely depopulated.[74]

Lack of gunpowder and the Sanga blockade remained Msiri's most pressing concerns to the very end. One last Yeke sortie, against Mulowanyama, took place between November and December, in retaliation for an earlier attack on a party of eastbound traders at the Lualaba ferry. At the time, the missionary Thompson was being detained by Mulowanyama in the village of his Bena Mitumba ally, Mwenda Mukoshi.[75] Stating that 'the Va-Sanga will no longer by the slaves of Msidi', Mulowanyama 'refused to allow our caravan or any other to come to Msidi's'.[76] Eventually, the rebel leader – who appreciated the neutral stance taken by the PB and their unwillingness to assist Msiri in communicating with 'traders to bring powder to him' – was persuaded to give Thompson permission to proceed to Bunkeya, but also insisted that the members of his caravan who carried powder stay put until they had exchanged all of it locally.[77] On 2 December, while negotiations were still ongoing, Thompson witnessed Mukanda Bantu's ambush. 'I saw men and women returning from the fields in hot haste, and making for their villages beside our camp. Immediately I heard the beat of a war drum, and a number of shots were discharged in the bush, about 250 yards from our camp. [. . .]. Msidi's party caught about eleven of the Va-sanga in the fields and forthwith decapitated them, as a *quid pro quo*, I suppose, for those killed at the ford by the Va-sanga.' Thompson's intervention apparently prevented further bloodshed and convinced the Yeke aggressors to decamp without having succeeded in lifting the blockade.[78] Even the Yeke military no longer believed in victory.

Counterinsurgency

Stairs clashed with Msiri over his determination to force the CFS flag on the Yeke warlord and refusal to furnish him with the gunpowder he craved.[79] Tension escalated between 16 and 20 December, the day in which Msiri was shot dead by Belgian officer Bodson and his escort.[80] The assassination of Msiri further compounded the plight of the Yeke and their new leader, Mukanda Bantu, who were now faced with the very real prospect of com-plete annihilation. Mulowanyama, for one, is said to have appealed to Stairs to 'exterminate' all of Msiri's people before his departure.[81] In some cases, these aspirations were being into practice. In June 1892, for instance, a group of Lomotwa assassinated their former *mwanangwa*, Kifuntwe ('Cifuntwi') – an episode which Crawford read as a demonstration of the fact that 'they who once were the oppressors are now in a degree the oppressed.'[82] 'A short time' afterwards, ten followers of Mokembe, Msiri's brother, were murdered by the 'Ve na Mitumba' (Bena Mitumba, or Sanga of the Mitumba Mountains) 'when on a journey of peaceable intent.'[83]

The Yeke's decision to abandon Bunkeya and regroup in and around Lutipisha, to the immediate south of the newly-founded CFS post of Lofoi, in late 1892-early 1893 was a clear sign of the precariousness of their position.[84] In this tense context, Legat, the officer in charge of Lofoi in 1891–1893, and his companion Verdick followed a very cautious course of action, doing little more than (literally) holding the fort with the thirty-odd regular Force Publique soldiers at their disposal. Despite having been enjoined by his superiors to make Lofoi self-supporting and to gather the greatest possible amount of ivory, the most valuable Katangese commodity on the world's market in the 1890s, Legat – as he would openly admit to his successor before leaving the region – never felt strong enough to force the hand of the Sanga and other leaders who kept harrying the

Yeke, paid no taxes and kept well clear of Lofoi.[85] This wait-and-see policy changed with a vengeance from the autumn of 1893, when Lofoi was taken over by a relief column from Lusambo commanded by Lieutenant Clément Brasseur, to whom reference has already been made above.

Determined to maximize the State's – and therefore his own[86] – revenue at all costs, Brasseur inaugurated a reign of terror that made a vivid impression on both the PB missionaries, who would eventually denounce his doings, and local Africans, by whom he was nicknamed *Nkulukulu*. The Nkulukulu – Judge Jenniges explained a few years after Brasseur's death – was a bird 'whose inner wings are bloody red. Now, the natives say, Mr. Brasseur was only happy when he had blood up to his armpits. Then he looked like the bird in question'.[87] Brasseur's extractive methods were especially crude. Bespeaking of an earlier warlord order, they revolved around the deployment of so-called 'sentries' – unsupervised regular soldiers stationed in prominent villages and entrusted with the task of collecting '"all the ivory in the country"' by hook or crook[88] – and the staging of large-scale military expeditions intended both to plunder the surroundings of Lofoi and to impress upon local leaders the need to comply with instructions relating to taxation in kind and labour.

For all of his rapaciousness, racism and penchant for violence, Brasseur, left to his own devises and those of his few regular soldiers, would never have secured what he termed 'the complete subjugation of the country'.[89] Thus, in the further-ance of this objective, he – like countless empire-builders before and after him – set out to exploit local conflicts and manpower,[90] and turned to the beleaguered Yeke. Brasseur himself explained that popular 'hatred against the Yeke suit[ed him] perfectly.'[91] 'Having raided, killed and looted overmuch', the 'remnants' of the Yeke 'were now pitted against every other race'. They were as a result 'devoted to the whites'.[92] But Mukanda Bantu's surviving followers, of course, had a pressing agenda of their own. By reinventing themselves into a rapid deployment, auxiliary force at the service of Lofoi, they aspired to rebuild that regional hegemony which the Sanga rebellion and the death of Msiri were threatening fully to eradicate. The alliance with the CFS, in other words, offered the Yeke the chance to roll back the years and check the assertiveness of the Sanga and other former subjects. In the process, the Yeke military played a critical role in shaping from below a violent counterinsurgency campaign which the Europeans coordinated and benefitted from, but whose ultimate meanings and direction they did not always control or even fully understand.

Beginning in 1894, Mukanda Bantu and his gunmen participated in all of the major expeditions carried out by Brasseur and/or his two deputies. In July of that year, about one-hundred Yeke joined Brasseur's foray against Mutwila, guilty of preventing chef Mwashya from delivering his quota of salt to Lofoi.[93] Mutwila – as we know – had stood up to Dikuku in the summer of 1891 and was regarded by the Yeke as a serious menace and one of the main causes of 'the ruin of Msiri'.[94] The fortified villages of Mutwila and his brother were stormed by the CFS-Yeke force. Numerous children and women were killed during what the missionary Campbell called the 'great slaughter' that followed the attack.[95] Brasseur, his soldiers and auxiliaries returned home with, inter alia, sixty female prisoners and fourteen ivory tusks. Even though, on this occasion, Brasseur was unimpressed by the performance of his Yeke allies – who had apparently behaved cowardly during the assault but as a 'pack of famished wolves scrambling for a corpse' in its aftermath[96] – Mukanda Bantu was naturally delighted at the demise of the old opponent, treating Brasseur to a drinking party in Lutipisha and then escorting him back to Lofoi with a following of 'at least 200' people.[97]

But Brasseur's violence was not always indiscriminate, and his actions suggest that he understood that most basic of counterinsurgency tactics: the sowing of divisions among opponents by means of both threats and inducements.[98] He, in particular, went out of his way to prevent his Yeke auxiliaries from attacking Kalunkumia and Mulowanyama's bases during a joint expedition to Sanga and Bena Mitumba territory in the spring of 1895. In the case of the former, Brasseur's conciliatory policy initially worked. Persuaded to return to his capital, which he had fled as the military column approached, Kalunkumia accepted to house one of Lofoi's outstations and became a regular ivory contributor until the death of Brasseur.[99] Mulowanyama, on the contrary, sought refuge deep in the Mitumba Mountains, whence – as will be seen below – he would spend the last few years of his adventurous life scheming against the new dispensation and Yeke-CFS rule.

Less accommodating or strategically important leaders, on the other hand, could expect little mercy. Thus, over the course of the same 1895 tour, Brasseur, his force Publique soldiers and Mukanda Bantu's warriors (probably numbering between 100 and 200) took the side of local Yeke grandees against autochthonous claimants in the succession dispute that followed the death of one of Msiri's former subordinates, Kalala Ngombe.[100] Both the aforementioned attack on Kalera's and the destruction of Ngonga's settlement late in May are to be set in this context.[101] One year later, the bulk of the porters and irregulars who accompanied Brasseur's on his longest tour yet was still made up of Yeke. Targeting another historical enemy of Msiri, the Luba-speakers of the Upemba Depression and the Luvua River, this expedition – the most murderous of Brasseur's Katangese journeys – did not witness the direct participation of Mukanda Bantu, whose place was taken by *mwanangwa* Kasadi, attracted to the upper Lualaba by the prospect of capturing female slaves galore.[102]

Apart from the direct casualties of massacres – along the Lualaba and Luvua rivers alone, Brasseur's mixed forces were responsible for a minimum of 75 deaths between late June and early August 1896[103] – the human costs of these 'pacification campaigns' were probably magnified by the processes of adaptation and mutual borrowing which underlay them. Mukanda Bantu showed no compunction about imitating Brasseur's textbook infantry tactics.[104] But influence did not flow in one direction only, and the pre-colonial military know-how of Yeke militias could not but have an impact on Brasseur and his Force Publique. Both the practice of living off the land while on campaign and the punitive burning of villages and crops ('a general custom with the Vayeke', according to the missionary Swan[105]) enhanced the lethality of the Yeke-Force Publique counterinsurgency campaign and the sufferings it inflicted on non-combatants. Equally important in psychological terms was the mutilation of the bodies of dead enemies, a practice stemming both from Brasseur's desire to monitor cartridge usage *and* Yeke beliefs in the need to purify the heads of slain enemies.[106] The large-scale taking of prisoners, meanwhile, served both to increase Lofoi's servile workforce and to replenish Mukanda Bantu and the other Yeke chiefs' depleted slave stocks. Early in 1892, in the immediate aftermath of Msiri's death, Mukanda Bantu had cut a powerless figure, surrounded by a mere 'handful of people' and camping among the ruins of Bunkeya, his father's all but deserted former capital.[107] By Jan. 1896, however, Lutipisha was being described as a 'great seething town' harbouring (in Brasseur's conservative estimate) some 3,000 residents.[108] The figure of 5,000 was given in June 1897 by Campbell,[109] who, however, would later downsize his estimate to 3,000. Of these only one fifth were reported to be men,[110] a skewed demographic composition which suggests that the economic role of female slaves was no less important in Lutipisha than it had been in Bunkeya.

Despite – or perhaps because of – Brasseur and Mukanda Bantu's efforts, the enemies of the Yeke kept up a low-intensity guerrilla war throughout the mid-1890s. Now that the nerve centres of the new dispensation – Lofoi and Lutipisha – had been strengthened, the insurgents' forays targeted more remote areas, primarily the Mitumba Mountains, where the inhabitants of Yeke settlements were reportedly being fired upon when going into the woods or the fields in August 1895.[111] It was also in the Mitumbas and, specifically, the Kiamakele network of natural tunnels that Mulowanyana sought shelter after he abandoned his less easily defensible village in October of the same year.[112] Thereafter, Mulowanyama struck an alliance with the then Kazembe of the Lualaba, an important title-holder who controlled a ferry across the Lualaba and whom Mulowanyama appears to have turned against Lofoi in the early part of 1896.[113]

Such worrying developments amongst the Bena Mitumba and along the Lualaba prompted Brasseur to send Mukanda Bantu and a 200-man-strong war party drawn from at least fifty separate villages on a solo expedition to the areas in question.[114] Granted complete freedom of action during the journey, Mukanda Bantu was expected to secure the Lualaba crossing, attack a number of unyielding Sanga and Bena Mitumba chiefs and bring back as much ivory and as many prisoners as possible.[115] Mukanda Bantu – who now reportedly insisted on being called *Nzige* (locust), because '"I have many people, and when the white man sends me somewhere, I raze everything to the ground, like the locusts do"'[116] – and his paternal uncle Mokembe did not disappoint expectations. Not only did they replace the Kazembe of the Lualaba with a man of their choice, kill chief Kasangula and burn down the village and crops of Mwenda Mukoshi.[117] But by the time they returned to Lofoi in January 1897, they had also accumulated 'twenty-odd women, some boys and 26 tusks', as well as 'a great deal of plunder from raided villages and the skulls of the people killed during the journey (31) in baskets.'[118] Since such a raid was an almost exact replica of one of Msiri's *rafles*, it is perhaps not surprising that, at least according to Campbell, "'one of the male prisoners [...] was sacrificed at the grave of old Msiri unknown to the State.'"[119]

For all its immediate effectiveness, Mukanda Bantu's counterinsurgency expedition created more problems than it solved for Brasseur. By as early as March 1897, the chiefs targeted by Mukanda Bantu had regrouped and declared war on the Yeke village of Kalala Ngombe.[120] At about the same time, a Sanga force of which Mulowanyana might have been part seized the village of the Yeke-backed Kazembe of the Lualaba and replaced the latter with a competitor.[121] By April, 'exception made for five or six villages, the Bena Mitumba [had] risen up almost everywhere',[122] attacking isolated Yeke villages in a repeat of the hit-and-run tactics pioneered by their Sanga peers at the beginning of the decade. In the summer, 'hardly a month' went by without 'one chief or another' complaining to Brasseur that 'his village ha[d] been burned down and his people killed.'[123] Eventually, Brasseur came to the conclusion that the rebellion that was engulfing the Mitumbas – one in which he suspected the Sanga paramount, Kalunkumia, of being implicated[124] – necessitated the dispatching of another punitive expedition by Mukanda Bantu. This was tentatively scheduled for after his own return from the Lualaba, where he travelled in August 1897 to thwart the Sanga-supported Kazembe of the Lualaba and officially re-install the Yeke's candidate.[125]

Brasseur's plans for reducing the ongoing insurgency were brought to an abrupt end by his death in November 1897, when he was fatally wounded during the siege of the fortified settlement of the coastal trader Kiwala, on the upper Luapula River. Despite missionary expectations to the effect that his demise might bring about 'a more humane state of affairs in Katanga',[126] the Yeke-Force Publique counterinsurgency effort proceeded unabated, reaching its violent climax in 1899. In 1898 – as Verdick, the new *chef de poste*, discovered during a tour of the south-east – Kalunkumia, 'disgruntled' by 'the protection afforded to the Yeke' by Lofoi,[127] threw moderation to the wind: followed by many of his subjects, he abandoned his village on the upper Dikulwe and settled in Kiamakele, near Mulowanyama's.[128] During the previous two years, Mulowanyama – whose local 'prestige and influence' were 'considerable' – had reportedly 'harassed' some of the outposts of Lofoi and committed 'more than fifty murders'.[129] He was also rumoured to be planning an attack on the missionary George during his upcoming return journey from Angola and to be still bent on controlling the Lualaba ferry with a view to monopolizing communications with Ovimbundu slave and gun traders.[130]

Early in 1899, following some renewed Sanga and Bena Mitumba threats against the Kazembe of the Lualaba and the nearby Yeke village of Nguba, Verdick, accompanied by one hundred Force Publique soldiers, 'about fifty' of Mukanda Bantu's irregulars and a 47mm Nordenfelt cannon, marched against the rebels' headquarters.[131] Mulowanyama's village was bombarded on 20 March. After a heavy exchange of fire – during which a Belgian sub-lieutenant, Fromont, as well as three regular soldiers and two Yeke auxiliaries were killed – the defenders withdrew into their tunnels, which the attackers did not penetrate but sealed from the outside by means of large boulders.[132] Despite Verdick's and, later, Delvaux's attempts to negotiate their surrender – and even though Kalunkumia and several other Sanga and Bena Mitumba chiefs tendered their final submission in the weeks that followed the assault[133] – Mulowanyama and his remaining supporters held out in their redoubt for more than three months. Having run out of food supplies, the survivors – a paltry group of two men, five women and six children – gave up on 25 June; by that point, most of the defenders, including the irrepressible Mulowanyama, had succumbed. Once, after several days' work, he negotiated a way into the main tunnel, part of which had collapsed, the Belgian Delvaux discovered to his 'horror' that it was 'nothing but an immense graveyard'. He counted 178 asphyxiated and decomposing corpses and 142 guns.[134] Before leaving Kiamakele, Delvaux set fire to the place. 'The tunnels caved in one by one, and Mulowanyama's grave was shut away for ever.'[135] The Yeke-Force Publique counterinsurgency had carried the day.

Conclusion

During the violent 1890s, Yeke policies were dominated by the imperatives of surviving the consequences of the Sanga rebellion and of reconstituting their hegemony in tandem with the newly arrived forces of the Congo Free State. Intersecting with European motives, Mukanda Bantu's strategy proved wonderfully effective: not only did the Yeke survive as a corporate group against considerable odds, but they did so as the dominant African ethnicity in colonial southern Katanga. The real victims were those communities who had rebelled against Yeke predation, had briefly regarded the incoming whites as possible saviours,[136] and then found themselves at the receiving end of a nasty counterinsurgency campaign jointly carried out by the same European-led forces and their former Yeke exploiters.

The Sanga rebellion was defeated – and several decades of colonial social engineering and politico-economic change would prove necessary to heal old wounds. Eventually, as Yeke, Sanga and other local ethnicities found a new unity under the banner of an invented Katangese 'autochthony',[137] memory of the Sanga insurgency faded into the background (a process to which the very gruesomeness of the rebellion's ending might have contributed). But though unsuccessful and now largely forgotten,[138] the Sanga rebellion still has some important lessons to teach us. First, the central role played by Yeke forces in the 'pacification' of Katanga implies that historians who stress the innovative character of the 'small wars' of nineteenth-century imperialism – be it on account of the new technologies they utilized or the unprecedented strategic objectives which underpinned them – have invariably tended to overplay their hand.[139] This is not only because – as Walter has recently argued – the imperial wars of the late nineteenth century display a 'striking family resemblance' with all the violent conflicts that have accompanied the territorial expansion of Europe from the sixteenth century.[140] More important is the fact that what southern Katanga (and much of the rest of central Africa as well) witnessed in the 1890s was a process of mutual borrowing and hybridization in which African military practices and even political aims were far from being ancillary to European ones. In fact – as Macola has argued more at length elsewhere[141] – 'creolization' processes went beyond the military sphere and invested the totality of the early colonial dispensation in the Congo basin, which was Africanized, not only through the armed personnel on which it drew, but also in the institutions and practices of rule it adopted and in the symbolism of power it was charged with. In sum, when it is examined from the ground up and from the perspective of the defeated, the conquest of Katanga looks much more like the continuation of an earlier warlord order in disguise than like the external imposition of 'modern' military and political models.

More generally, this well-documented instance of rebellion and counter-insurgency before and during the colonial occupation of Katanga forcefully reminds us that African military history does not begin with the arrival of European forces. Military dynamics were no less vibrant and subject to change in the pre-colonial era than in later periods. And a full grasp of even the most modern of central African conflicts is destined to remain elusive unless these same conflicts are placed in a *longue-durée* perspective – one that must at least make room for the revolutionary transformations experienced by the sub-continent in the nineteenth century and the enhanced militarism and preda-tory style of politics that they brought into being.

Notes

1. Reid, *War in Pre-Colonial Eastern Africa*, 3.
2. For more on this point, see Thornton, "Placing the Military in African History."
3. Macola has recently edited and co-translated into English seventeen of Brasseur's 'Katangese' letters to his brother Désiré: Macola ed., *The Colonial Occupation of Katanga*. All future references to Brasseur's letters come from this edition. A full description of the original collection – housed in the

historical archives of the Musée Royal de l'Afrique Centrale (MRAC), Tervuren, RG 768/81.15 – can be found in G. Macola, "Introduction: Brasseur's Papers and the African Roots of the Congo Free State", in Macola, *The Colonial Occupation of Katanga* (henceforth *TCOK*).

4. The classic statement is Terence Ranger's two-part article: "Connexions between 'Primary Resistance Movements' and Modern Mass Nationalism in East and Central Africa". The long life of this whiggish interpretation is attested, inter alia, by Nzongola-Ntalaja, *The Congo from Leopold to Kabila*, chapters 1–2.

5. See, e.g. Vandervort, *Wars of Imperial Conquest in Africa, 1830–1914*. A more accomplished and ambitious work of synthesis is Walter, *Colonial Violence: European Empires and the Use of Force*.

6. For a critique of the 'presentism' of much recent Africanist historiography, see Reid, "Past and Presentism," 135–155.

7. The timing of Msiri's arrival in southern Katanga is discussed by Legros, *Chasseurs d'ivoire*, 28–29.

8. The incorporation of the central African interior into the Indian Ocean and Atlantic trading frontiers has long formed a subject of scholarly enquiry. Recent overviews of the relevant literature are Gordon, "The Abolition of the Slave Trade and the Transformation of the South-Central African Interior during the Nineteenth Century," 915–938, and Macola, *The Gun in Central Africa*, chapter 1.

9. Arnot, *Garenganze*, 233; Legros, *Chasseurs*, 119.

10. "Lettre de Mwenda II Mukanda Bantu à S.M. le Roi Albert à l'époque, Prince héritier de Belgique. Traduit par A. Mwenda Munongo," 41, 43; and Legros, *Chasseurs*, 121.

11. One of the principal Lualaba crossing points was located to the immediate south of Lake Kajibajiba, among peripheral Luba peoples subjugated by the Yeke in the early 1870s. Another one was to be found further upstream, in the area of the Kazembe of the Lualaba. See, e.g. Le Marinel, *Carnets de route dans l'Etat indépendant du Congo de 1887 à 1910*, 155–156, 159; and Arnot, *Bihé and Garenganze*, 56–7.

12. See, e.g. Arnot's fascinated description in *Garenganze*, 235.

13. Reichard, "Herr Paul Reichard," 119.

14. Giraud was told of Msiri's military might while among the Aushi of the upper Luapula in 1883; *Les lacs de l'Afrique équatoriale*, 318–9.

15. Legros, *Chasseurs*, 80–85.

16. Ibid., 110–111.

17. Delcommune, *Vingt années de vie africaine*, 329.

18. D. Crawford, 'Dec. 1891', in *Echoes of Service* (henceforth *ES*), 273 (Nov. 1892), 257.

19. C.A. Swan, 14 December 1889, in *ES*, 230 (Jan. 1891), 25.

20. F.S. Arnot, 11 August 1886, in *ES*, 184 (April 1887), 59–60.

21. Arnot, *Garenganze*, 243.

22. Ibid., 194.

23. Ibid.

24. Crawford, 'Dec. 1891', 258.

25. Legros, *Chasseurs*, 3–74.

26. Capello and Ivens, *De Angola á contra-costa*, 169, 173.

27. For more details, see Macola, *The Kingdom of Kazembe*, 155–9, 171–2.

28. Macola, *TCOK*, 4.

29. Ibid., 35, 56.

30. Ibid., 33; Arnot, *Garenganze*, 198n.
31. Macola, *TCOK*, 61.
32. Delvaux, *L'occupation du Katanga, 1891–1900*, 70.
33. Smith, *Warfare and Diplomacy in Pre-Colonial West Africa*, 50. See also Reid, "Revisiting Primitive War," 1–2.
34. Macola, *TCOK*, 54. For a consistent, though less stridently formulated, view, see C.A. Swan, 23 January 1888, in *ES*, 202 (Oct. 1888), 317.
35. Macola, *TCOK*, 512.
36. Iliffe, *A Modern History of Tanganyika*, 52.
37. D. Crawford, 18 May 1891, *ES*, 252 (Dec. 1891), 299; Stairs, "De Zanzibar au Katanga," 183.
38. Macola, *TCOK*, p. 32; Campbell, *In the Heart of Bantuland*, 264.
39. Here, my observations echo Storey, *Guns, Race and Power in Colonial South Africa*, 78.
40. Campbell, *Heart of Bantuland*, 26. See also Stairs, "De Zanzibar au Katanga," 191.
41. Delcommune, *Vingt années*, 314–315.
42. See, e.g. C.A. Swan, 10, 21, 26 and 28 Feb. and 1 March 1890, in *ES*, 251 (Dec. 1891), pp. 288–289; and "Lettre de Mwenda II Mukanda Bantu", 47, 49, 51.
43. Crawford, 'Dec. 1891', 258.
44. C.A. Swan, 21 February 1891, in *ES*, 251 (Dec. 1891), 288; Jenniges, *Dictionnaire français-kiluba*, 17.
45. Swan, 21 February 1891, 288; Legros, *Chasseurs*, 84.
46. See, e.g. Macola, *TCOK*, 145.
47. C.A. Swan, 26 February 1891, in *ES*, 251 (Dec. 1891), 289.
48. C. A. Swan, 20 March 1891, in *ES*, 251 (Dec. 1891), p.290; C.A. Swan, 22 April 1891, in *ES*, 253 (Jan. 1892), 9.
49. C.A. Swan, 20 March 1891, in *ES*, 251 (Dec. 1891), 290; "Lettre de Mwenda II Mukanda Bantu," 51.
50. D. Crawford, 18 May 1891, *ES*, 252 (Dec. 1891), 299; D. Crawford, 'Dec. 1891', in *ES*, 273 (Nov. 1892), 257.
51. Swan, 20 March 1891, 290.
52. C. A. Swan, 30 March 1891, in *ES*, 251 (Dec. 1891), 291; Le Marinel, *Carnets de route*, 201.
53. Crawford, 18 May 1891, 299.
54. F.S. Arnot, 11 August 1886, in *ES*, 184 (April 1887), 60.
55. C.A. Swan, 10 May 1891, in *ES*, 253 (Jan. 1892), 10; Le Marinel, *Carnets de route*, 194. On the 'omande' (*mpande*), see Arnot, *Garenganze*, 234.
56. C.A. Swan, 5 June 1891, in *ES*, 253 (Jan. 1892), 11.
57. Macola, *TCOK*, 64.
58. Macola, *TCOK*, 32. See n. 11.
59. Le Marinel, *Carnets de route*, 202.
60. Crawford, 'Dec. 1891', 257.
61. C.A. Swan, 24 May 1891, in *ES*, 253 (Jan. 1892), 10.
62. Le Marinel, *Carnets de route*, 190–1.
63. Crawford, 'Dec. 1891', p. 257. For information on Ntenke, see Delcommune, *Vingt années*, 277–278. See also "Lettre de Mwenda II Mukanda Bantu", 51.
64. Crawford, 'Dec. 1891', 257.
65. Ibid. See also Arnot, *Bihé and Garenganze*, 112.
66. Delcommune, *Vingt années*, 249. See also Le Marinel, *Carnets de route*, 204.

67. A. Legat to A. Delcommune, Lofoi, 13 October 1891, in Delcommune, *Vingt années*, 272.
68. Crawford, 'Dec. 1891', 257.
69. P. Briart (ed. Dominique Ryelandt), *Aux sources du fleuve Congo*, 177, 179.
70. Delcommune, *Vingt années*, 274, 295, 298.
71. Ibid., 243, 296–297.
72. Ibid., 260, 276, 291; Legat to Delcommune, 13 October 1891, in Ibid., 272; and Stairs, "De Zanzibar au Katanga," 183.
73. Stairs, "De Zanzibar au Katanga," 191.
74. Stairs, "De Zanzibar au Katanga," 197.
75. "Lettre de Mwenda II Mukanda Bantu," 53.
76. Arnot, *Bihé and Garenganze*, 111; and Crawford, 'Dec. 1891', 259.
77. Arnot, *Bihé and Garenganze*, 111.
78. Ibid., 111–2.
79. Stairs, "De Zanzibar au Katanga," 191; "De Zanzibar au Katanga," 198; and Crawford, 'Dec. 1891', 260–261.
80. Stairs, "De Zanzibar au Katanga," 199. Bodson himself was wounded by Msiri's gunmen and died shortly thereafter.
81. Mwenda Munongo ed. and transl., "Chants historiques des Bayeke," 64.
82. D. Crawford, 25 June 1892, in *ES*, 281 (March 1893), 56–57.
83. H.B. Thompson, 2 March 1893, in *ES*, 296 (Oct. 1893), 238.
84. Ibid.; D. Crawford, 12 February 1893, in *ES*, 297 (Nov. 1893), 251.
85. Macola, *TCOK*, 3, 5, 382.
86. The payment of premiums on the amount of produce that territorial officials were able to siphon out of their respective regions was a key means through which the CFS's central administration sought to encourage extractive activities in Leopold's colony. See, e.g. Stengers and Vansina, "King Leopold's Congo, 1886-1908," 339.
87. J. Jenniges to Public Prosecutor, Lukafu, 1 March '1903' [*sic*, but 1905], Papiers L. Guebels, MRAC, RG 917.
88. D. Campbell to H.R. Fox Bourne, Johnston Falls, 14 May 1904, in Morel, *King Leopold's Rule in Africa*, 460. See also D. Campbell, 7 May 1895, *ES*, 345 (Nov. 1895), 268, and D. Crawford, 1 September 1895, *ES*, 351 (Feb. 1896), 45–6.
89. Macola, *TCOK*, p. 26. A brief discussion of Brasseur's personality and worldview is to be found in Macola, "Introduction," xx–xxii.
90. Walter, *Colonial Violence*, 96.
91. Macola, *TCOK*, 60.
92. Ibid., 6–7.
93. Ibid., 14.
94. Ibid., 16.
95. D. Campbell, 2 May 1897, in *ES*, 397 (Jan. 1898), 13.
96. Macola, *TCOK*, 15.
97. Ibid., 16.
98. See, e.g. Walter, *Colonial Violence*, 107.
99. Macola, *TCOK*, 54, 55, 67–8.
100. For contrasting estimates of the size of Mukanda Bantu's auxiliary force on this occasion, cf. Macola, *TCOK*, 49 and 52.
101. Ibid., 56, 61.
102. Ibid., 223.
103. Ibid., 227–56.

104. See, e.g. Ibid., 82, 327.
105. C.A. Swan, 17 June 1890, in *ES*, 239 (June 1891), pp. 137–38.
106. Macola, *TCOK*, 229; C.A. Swan, 19 November 1889, in *ES*, 230 (Jan. 1891), pp. 23–4; Legos, *Chasseurs*, 74–5.
107. D. Crawford, 24 February 1892, in *ES*, 274 (Nov. 1892), 269.
108. D. Crawford, 5 January 1896, in *ES*, 364 (Aug. 1896), 250; Macola, *TCOK*, 35.
109. D. Campbell, 6 June 1897, in *ES*, 394 (Nov. 1897), 345.
110. Campbell to Fox-Bourne, in Morel, *King Leopold's Rule*, 460.
111. Macola, *TCOK*, 108.
112. Ibid., 123. For a description of a section of these tunnels, see Ibid., 465, and Arnot, *Garenganze*, 198.
113. Macola, *TCOK*, 145, 160, 195.
114. Ibid., 294, 295.
115. Ibid., 294.
116. Ibid., 326.
117. Ibid., 327, 359.
118. Ibid., 358.
119. Campbell to Fox-Bourne, in Morel, *King Leopold's Rule*, 456.
120. Macola, *TCOK*, 375.
121. Ibid., 390, 403, 460, 461–462, 470.
122. Ibid., 394.
123. Ibid., 448.
124. Ibid., 460–461.
125. Ibid., 451, 471.
126. W. George, 3 January 1898, in *Echoes of Service*, 409 (July 1898), 204.
127. Verdick, *Les premiers jours au Katanga (1890–1903)*, 114.
128. Ibid., 115.
129. Delvaux, *L'occupation du Katanga*, 74.
130. J.W. M'Lachlan, 23 April 1899, *ES*, 438 (Sept. 1899).
131. Verdick, *Les premiers jours*, 130–131.
132. Ibid., 137, 139.
133. Ibid., 141–143; and Delvaux, *L'occupation du Katanga*, 76.
134. Delvaux, *L'occupation du Katanga*, 76. See also W. George, 12 August 1899, *ES*, 448 (Feb. 1900), 57.
135. Delvaux, *L'occupation du Katanga*, 76.
136. W.G. Stairs to D. Crawford, [25 December 1891], in Crawford, 'Dec. 1891', p. 261.
137. Larmer and Kennes, "Rethinking the Katangese Secession," 741–61.
138. Cf., however, Asani bin Katompa, "L'opposition Sanga à Msiri et à l'administration coloniale Belge (1891–1911),".
139. FitzSimons, "Sizing up the 'Small Wars' of African Empire," 63–78. For a classic example of crude technological determinism, see Headrick, *The Tools of Empire: Technology and European Imperialism in the Nineteenth Century*.
140. Walter, *Colonial Violence*, 5.
141. Macola, "Introduction."

Disclosure statement

No potential conflict of interest was reported by the authors.

Bibliography

Arnot, F.S. *Garenganze; Or, Seven Years' Pioneer Mission Work in Central Africa*. London: James E. Hawkins, 1889.

Arnot, F.S. *Bihé and Garenganze; Or, Four Years' Further Work and Travel in Central Africa*. London: James E. Hawkins, 1893.

Asani Bin Katompa, K. "*L'opposition Sanga à Msiri et à l'administration coloniale Belge* (1891–1911)" Mémoire de licence en histoire. Lubumbashi: Université Nationale du Zaïre, 1977.

Briart, P. *Aux sources du fleuve Congo. Carnets du Katanga (1890-1893)*. edited by Dominique Ryelandt. Paris: L'Harmattan. 2003.

Campbell, D. *In the Heart of Bantuland*. London: Seeley, Service & Co., 1922.

Capello, H., and R. Ivens. *De Angola Á Contra-costa. Descripção De Uma Viagem Atravez Do Continente Africano*. 2 vols. Lisbon: Imprensa Nacional, 1886.

Delcommune, A. *Vingt années de vie africaine*. 2 vols. Brussels: Ferdinand Larcier, 1922.

Delvaux, H. *L'occupation du katanga, 1891-1900. Notes et Souvenirs du seul survivant*. Elisabethville: Imbelco, 1950.

FitzSimons, W. "Sizing up the 'Small Wars' of African Empire: An Assessment of the Context and Legacies of Nineteenth-Century Colonial Warfare." *Journal of African Military History* 2, no. 1 (2018): 63–78. doi:10.1163/24680966-00201005.

Giraud, V. *Les Lacs De l'Afrique Équatoriale*. Paris: Librairie Hachette, 1890.

Gordon, D.M. "The Abolition of the Slave Trade and the Transformation of the South-Central African Interior during the Nineteenth Century." *William and Mary Quarterly* 66, no. 4 (2009): 915–938.

Headrick, D.H. *The Tools of Empire: Technology and European Imperialism in the Nineteenth Century*. New York: Oxford University Press, 1981.

Iliffe, J. *A Modern History of Tanganyika*. Cambridge: Cambridge University Press, 1979.

Jenniges, E. *Dictionnaire Français-kiluba*. Brussels: Spineux, 1909.

Larmer, M., and E. Kennes. "Rethinking the Katangese Secession." *Journal of Imperial and Commonwealth History* 42, no. 4 (2014): 741–761. doi:10.1080/03086534.2014.894716.

Le Marinel, P. *Carnets de route dans l'Etat indépendant du Congo de 1887 à1910*. Brussels: Editions Progress, 1991.

Legros, H. *Chasseurs d'ivoire. Une histoire du royaume yeke du Shaba (Zaïre)*. Brussels: Editions de l'Université de Bruxelles, 1996.

Macola, G. *The Kingdom of Kazembe: History and Politics in North-Eastern Zambia and Katanga to 1950*. Hamburg: LIT Verlag, 2002.

Macola, G. *The Gun in Central Africa: A History of Technology and Politics*. Athens, OH: Ohio University Press, 2016.

Macola, G., ed.. *The Colonial Occupation of Katanga: The Personal Correspondence of Clément Brasseur, 1893-1897*. Oxford: Oxford University Press for the British Academy, 2018.

Macola, G. "Introduction: Brasseur's Papers and the African Roots of the Congo Free State." In *The Colonial Occupation of Katanga: The Personal Correspondence of Clément Brasseur, 1893-1897*, edited by G. Macola, ix–xxxviii. Oxford: Oxford University Press for the British Academy, 2018.

Morel, E.D. *King Leopold's Rule in Africa*. London: William Heinemann, 1904.

"Lettre De Mwenda II Mukanda Bantu À S.M. Le Roi Albert À L'époque, Prince Héritier De Belgique. Traduit Par A. Mwenda Munongo." In *Pages D'histoire Yeke*, edited by A. Mwenda Munongo, Lubumbashi, 1967.

Mwenda Munongo, A., ed. "Chants Historiques Des Bayeke." *Problèmes Sociaux Congolais*. 77 (1967): 35–139.

Nzongola-Ntalaja, G. *The Congo from Leopold to Kabila: A People's History*. London: Zed Books, 2003.

Ranger, T.O. "Connexions between 'Primary Resistance Movements' and Modern Mass Nationalism in East and Central Africa." *Journal of African History* 9, no. 3/4 (1968): 437–453, 631–641. doi:10.1017/S0021853700008665.

Reichard, P. "Herr Paul Reichard: Bericht Über Seine Reisen in Ostafrika Und Dem Quellgebiet Des Kongo." *Verhandlungen Der Gesellschaft Für Erdkunde Zu Berlin* 13, no. 2 (1886): 107–125.

Reid, R.J. *War in Pre-Colonial Eastern Africa: The Patterns & Meanings of State-Level Conflict in the Nineteenth Century*. Oxford: James Currey, 2007.

Reid, R.J. "Revisiting Primitive War: Perceptions of Violence and Race in History." *War and Society* 26, no. 2 (2007): 1–25. doi:10.1179/072924707791591677.

Reid, R.J. "Past and Presentism: The "precolonial" and the Foreshortening of African History." *Journal of African History* 52, no. 2 (2011): 135–155. doi:10.1017/S0021853711000223.

Smith, R.S. *Warfare and Diplomacy in Pre-Colonial West Africa*. London: Methuen, 1976.

Stairs, W. G. "De Zanzibar Au Katanga: Journal Du Capitaine Stairs (1890-1891)." *Le Congo Illustré* 2, no. 23/25 (1893): 181–183, 189–191, 197–199.

Stengers, J., and J. Vansina. "King Leopold's Congo, 1886-1908." In *The Cambridge History of Africa. Vol. VI*, edited by R. Oliver and N. Sanderson, 315–358. Cambridge: Cambridge University Press, 1985.

Storey, W.K. *Guns, Race and Power in Colonial South Africa*. Cambridge: Cambridge University Press, 2008.

Thornton, J. K. "Placing the Military in African History: A Reflection." *Journal of African Military History* 1, no. 1/2 (2017): 112–119. doi:10.1163/24680966-00101007.

Vandervort, B. *Wars of Imperial Conquest in Africa, 1830-1914*. London: UCL Press, 1998.

Verdick, E. *Les premiers jours au Katanga (1890-1903)*. Brussels: Comité Spécial du Katanga, 1952.

Walter, D. *Colonial Violence: European Empires and the Use of Force*. London: Hurst, 2017.

Ireland: rebellion and counter-insurgency, 1848–1867

Timothy Bowman

ABSTRACT

The period 1848 to 1867 witnessed what could be regarded as a very small-scale insurgency campaign in Ireland, waged by agrarian groups; the Whiteboys and Ribbonmen. 1848 and 1867 witnessed rebellions by the Young Irelanders and Fenians, which proved to be small-scale and of short duration but the British government had prepared for a nationwide counter-insurgency campaign. The government relied heavily on the militarised Irish Constabulary but in 1848 and 1867 troops were used in large numbers and there were concerns about how they could be best concentrated to meet the envisaged threat.

What might be termed Ireland's 'long nineteenth century' is bracketed by two major and well-researched rebellions; those of 1798 and 1916. The 1798 Rebellion witnessed some examples of insurgency and counter-insurgency, from 1797 to 1803, though, like the rebellion itself, these were confined to localised areas.[1] Those members of the Irish Volunteers and Irish Citizens' Army who survived the Easter Rising of 1916, which was a traditional coup d'etat aimed at capturing key buildings in central Dublin, turned their attention to an insurgency campaign which lasted from 1919 until 1921 and can be seen as one of the first modern insurgencies. Most parts of Ireland saw military action during this period but the majority occurred in Dublin and Cork, with events in Ulster being marked by sectarian rioting as much as an insurgency and counter-insurgency campaign.[2] By comparison, revolutionary activity in the nineteenth century Ireland has been compara-tively poorly served by historians although a number of important academic studies have been completed on the Fenian movement in Great Britain and North America.[3] This chapter focuses on two much smaller rebellions, in terms of casualties and duration, if not necessarily importance; those of 1848 and 1867. Agrarian agitation, which existed throughout the nineteenth century will also be considered. While elements of this were simple 'crime' by any definition of the term, there were enough politicised groups of 'Whiteboys' and more especially, 'Ribbonmen' within this movement for it to be seen as a serious, if low level and highly localised, insurgency.

The 1848 and 1867 rebellions proved to be short-lived, almost comic affairs, compared to the 1798 and 1916 Risings, which were more celebrated in traditional Nationalist historiography. Indeed, academic work on nineteenth century Irish nationalism has tended to focus much more on Daniel O'Connell's campaigns for Catholic Emancipation and Repeal of the Union, which provided a model of a constitutional mass movement, in European, not just Irish or British terms. The revisionist approach to the Fenian movement in this period, by Vincent Comerford, leaves us with the impression of a movement which was more social than revolutionary; indeed, there is an argument that while the British government organised a counter-insurgency campaign, in 1866–7, there were actually very few insurgents to counter.[4]

Policing in Ireland (and even by 1848 Ireland had a long established country-wide police force, which was to be the model for other police forces throughout the British Empire) has been the subject of a number of important works.[5] However, the role of the British Army in acting against rebellion and insurgency has received comparatively little attention in this period. The archival research for this article draws on the largely neglected Kilmainham papers at the National Library of Ireland which, amongst much else, contain the in and out letter books of the Commander in Chief in Ireland, providing a detailed insight into military planning and reaction to outbreaks of trouble.

The 1848 Rebellion was the work of the Young Irelanders, a small, radical group, which had split from the Repeal movement in 1846. They emerged in 1842 led by Thomas Davis, Charles Gavan Duffy and John Blake Dillon. Young Irelanders believed in the promotion of a non-sectarian, cultural nationality and developed their ideas through the *Nation* newspaper, established in 1842. The membership of the Young Irelanders was both Catholic and Protestant and mainly middle class; many being graduates of Trinity College Dublin. Throughout the 1840s the group sought to promote the Irish language and Irish literature. Initially the Young Irelanders were a part of O'Connell's Repeal movement, but they publicly split with O'Connell in 1845 when O'Connell opposed the creation of the Queen's Colleges in Ireland, which offered non-denominational university level education. Thomas Davis, in particular, was a firm advocate of mixed education and supported the establishment of the non-sectarian Queen's Colleges in Belfast, Cork and Galway, while O'Connell saw these as an attack on Catholicism in Ireland. In 1846 the Young Irelanders criticised O'Connell's negotiations with the Whigs. As a result, O'Connell required members of the Repeal Association to renounce the use of force, terms to which the Young Irelanders would not agree and they left the movement.

Young Ireland attempted, in 1847, to transform itself into the Irish Confederation, which was designed to have mass membership and mobilise the middle class in a demand for a separate Irish parliament. However, the results were disappointing with the leadership admitting a failure to mobilise either large numbers of the middle class or the peasantry. News of the revolution in France in February 1848 provided a revolutionary impetus to the movement but the Young Irelanders failed to secure support from either the new French government or the radical Chartist movement in England. Plans for a Rising were developed by early 1848, but, warned by an effective spy network, the British government moved to arrest leaders in April 1848 when they arrested William Smith O'Brien and Thomas Meagher for inflammatory speeches and John Mitchel for sedition. O'Brien and Meagher were acquitted, but Mitchell was found guilty and sentenced to 14 years transportation.

The 1848 rebellion itself was a fairly miserable affair as around 100 armed peasants, led by William Smith O'Brien, a very reluctant insurgent leader, confronted a party of police at Ballingarry, Co. Tipperary on the 29 July 1848. The police took refuge in widow McCormack's farmhouse and the rebellion collapsed as the police fired on the Young Irelanders, killing two of them, and British military reinforcements arrived. O'Brien was arrested and tried but was transported to Australia rather than executed. While the 1848 rebellion was an abject failure; widely parodied as 'The Battle of Widow McCormack's cabbage garden', in strictly military terms, the Young Irelanders influenced later generations of Irish revolutionaries with their promotion of cultural nationalism.[6]

The Fenian movement or Irish Republican Brotherhood, was formed in Dublin in 1858 by James Stephens, a veteran of the Young Ireland rising of 1848. The Fenians were a revolutionary movement, from the outset, and relied heavily on Irish immigrants in the USA for financial support. The aim of the Fenians was to establish a democratic Irish Republic and they were a secret society. Indeed, it is worth noting that, unlike the United Irishmen or the Young Irelanders, the Fenian movement did not experience a constitutional phase, being committed to armed rebellion from their formation. Stephens managed to build a mass movement, though Professor R. V. Comerford has noted that many members seemed more interested in the sporting and social activities available within Fenianism, than in revolutionary activity, though John Newsinger has sought to re-establish the revolutionary credentials of the movement.[7] Fenianism had recruited well amongst Irish communities in Great Britain and amongst Irish soldiers serving in the British army.[8] However, Stephens fell out with his compatriots in the USA, which led to various problems in financing and arming the organisation. The Catholic Church also opposed the Fenian movement and British government spies quickly infiltrated it, leading to a series of arrests, including of Stephens himself.

The main Fenian rising in Ireland occurred in Dublin on the night of 4th to 5 March 1867, with smaller, unco-ordinated actions occurring in Clare, Cork, Kerry, Limerick, Louth and Tipperary during February and March. The plan appears to have been that about 5,000 Fenians from Dublin itself were to meet on Tallaght Hill, outside the city. They would then march on Wicklow, where they would be joined by thousands more Fenians from Counties Wicklow and Wexford. This demonstration, it was believed, would see most British troops leave Dublin City to deal with this threat, allowing for a Rising by the 10,000 Fenians left in the city against the denuded Crown forces. However, these plans were revealed to the authorities by a number of spies and informers within the Fenian movement, who were effectively managed by 'G' Division of the Dublin Metropolitan Police, which meant that the Dublin garrison was strengthened. Indeed, key positions such as the Four Courts, the Royal Exchange and the Amiens Street [now Connolly] railway station received military guards. In addition to this the night of 4/5 March 1867 saw prolonged downpours of snow and sleet which quickly sapped Fenian morale. Ultimately, the 1867 Rising was to prove something of a debacle. A breakdown in command structures and communications meant that many thousands of Fenians assembled, mainly at Tallaght Hill, waiting for orders which never came and then dispersed as they believed the Rising had been called off. No major buildings in Dublin City were attacked, though police barracks at Tallaght, Dundrum, Stepaside and Glencullen were attacked and the policemen at Stepaside and Glencullen surrendered.[9] The Fenian movement was certainly not a spent force after 1867, indeed, it survived as an underground movement and was behind the 1916 Easter Rising, but from c. 1879 the Irish Parliamentary Party and Land League absorbed many of the popular energies that had been devoted to Fenianism.

The Young Irelanders and Fenians left many written sources on their own account and were the focus of much government attention by the police and military. However, what could be termed the 'primitive rebels' or 'resisters' involved in agrarian agitation remain a much more shadowy group.[10] Ribbonmen, developed from the Defender movement of the 1790s which had, at least in some areas, formed an alliance with the United Irishmen. In J. J. Lee's view what drove them were economic concerns, with little sense of national consciousness and nothing that resembled a national organisation. Indeed, he believes that contemporary police concerns about a nationwide conspiracy were fuelled by the reports of informers, who wanted to be seen to have earned their rewards.[11] It should be noted here, that agrarian crime rates in Ireland were low comparative to those in Great Britain and that much of the agrarian crime was low level intimidation, such as the sending of threatening letters. However, other historians have made a distinction between 'Whiteboys', who might be viewed as the most primitive of primitive rebels, focused purely on agrarian agitation and a more politicised 'Ribbonmen' movement. M. R. Beames in his important article on Tipperary, notes that the groups targeted by 'Whiteboys' were landlords and their agents, not policemen, magistrates or soldiers; the more obvious representatives of the British colonial state in Ireland. He thinks that Whiteboys had a wider sense of class consciousness than the term 'peasant' might imply; but concludes that they had a minimal sense of nationalism.[12]

This confusion is understandable, given that contemporaries often were not entirely sure about the place of these movements within a wider Nationalist struggle. A. M. Sullivan a journalist and early nationalist historian, writing in the 1870s, described Ribbonism as a 'Maffia' [sic] which varied its purposes from time to time and place to place. In Ulster it was a Catholic League against Protestant Orangeism, in Munster it combated tithing (taxes raised for the Church of Ireland until the 1830s), in Connaught it resisted rack-renting and in Leinster it was portrayed as an early form of trade unionism. Sullivan believed that there was no National organisation and limited middle class involvement.[13] By contrast Michael Davitt, the famous Land League leader and Fenian veteran saw Ribbonism as a precursor to the Fenian movement; politicised, nationwide and blending religious, nationalist and class sentiment.[14]

Careful work by Beames and Garvin on the Ribbonmen builds up a picture of an organisation which was led by artisans, shopkeepers, publicans and farmers, with very few of the middle or upper class or peasantry in their ranks. This is almost identical to the 'classes above the masses' which Comerford identifies in the later Fenian movement. As Beames puts it:

> A number of social groups or classes among the Catholic population were absent or only weakly represented among the membership. The Catholic middle classes – lawyers, doctors, land agents, merchants – hardly figure at all. Certain trades – generally those with a developed trade-union structure such as bookbinders, builders, plasterers, bricklayers and cabinet-makers – scarcely appeared. Although some farmers are listed among the membership, Ribbonism cannot adequately be described as 'peasant' in character. Finally, it excluded the destitute lumpenproletariat of cities such as Dublin.[15]

Beames sees the Ribbonmen as having a generally nationalist agenda, but looking to others to take the lead. The Ribbon vision of rebellion was that all would rise simultaneously to throw off the English yoke. Organisation was more local and regional than national and the society was shrouded in mystery with oaths and catechisms, which made mobilisation difficult and provided a cover for charlatans. The most recent work on Ribbonism, sees the organisation in Ulster as a much more political force, stating, 'although its politics could adapt to time and place and to political expediency, Ribbonism persisted in carrying forward an important symbolic grievance with Britain and the belief that Catholic Ireland was oppressed by a Protestant Ascendancy.'[16]

The British military commander in Ireland in 1866, Lord Strathnairn, saw the Fenian movement as something which had developed markedly from the Whiteboys and Ribbonmen; interestingly making no distinction between the two:

> Another material consideration presenting itself; the Leaders who would in the event of an outbreak command the Insurgents, are as superior to the leaders in 1848, the Ribbonmen & the Whiteboys as valuable military experience can be to utter ignorance in military matters.

Lord Strathnairn has had the honor at different times to bring to the notice of the Govt. that the American-Irish Fenian Leaders, so many of whom are stated to be here now, are experienced & resolute soldiers, who have won their experience in a very great & important war ... Lord Strathnairn has also had the honor to make known to the late Govt. that all the intercepted plans of operations of the Fenians, shewed [sic.] that their Leaders had acquired in the American War a dangerous knowledge of practical strategy. All these plans pointed as an indispensable [?] commencement & guarantee of success to the capture of Athlone, the Magazine Fort at Dublin, & other excellent Military Positions.[17]

In pursuing counter-insurgency policy, British police and military forces operated within a curious legal structure. Following the Act of Union of 1800, Ireland, supposedly an integral part of the United Kingdom, was to be covered by much special legislation.[18] Some of the most notable acts were; the Arms Act of 1843 which meant that arms had to be registered, subject to a £10 fine; possession of a pike would mean up to 12 months imprisonment; Habeaus Corpus Suspension Acts of 1848 (continued to 1849) and 1866 (continued to 1869), Party Processions Act 1850, which empowered magistrates to order parades with banners or songs likely to cause offence, to disperse; Peace Preservation (Ireland) Act, 1856 which proclaimed districts where firearms must be surrendered. This was all very different from the crude martial law which had been used in 1797–1803 and was to be used again in 1916, but, as David French has noted for later British counter-insurgencies, this did little to enhance the legitimacy of the colonial state.[19]

Such legislation was not always seen as helpful by the strongest supporters of the Government. Lord Strathnairn saw the suspension of habeas corpus in 1866 as, essentially, unworkable, and felt that it would increase tensions as those arrested had a reasonable right of complaint under British law.[20] Certainly there is some evidence to show that military officers were wary of exceeding their powers in a situation where martial law had not been declared. One officer, writing to army headquarters for clarification of his position asked if, in the event of a 'rising or disturbance', he should act on his own judgement or wait for a magistrate to arrive.[21] In 1848 a complaint regarding the absence of any magistrates at Killeshandra received a reply noting that, 'the presence of one single policeman enables the troops to act in case of a riot' which appears to be an odd understanding of the legislation then prevailing.[22] In December 1866 the solution offered was to appoint five field officers in Cork and four in Dublin as magistrates; allowing the appointed officers to make use of troops 'in aid of the civil power' without a formal request from ordinary magistrates or police officers.[23]

When crisis points were reached, it was not always felt that civilian magistrates made sensible or effective use of the military forces available. Lieutenant Langford Leir of the 31st Regiment felt that his detachment had been badly misused in March 1867 when they were called out, by a local magistrate, from their barracks in Templemore. Langford Leir reported that they were with a force of 10 Irish Constabulary when they came upon a group of between 60 and 70 men armed with pikes and guns. The magistrate leading the crown forces ordered the police only to open fire. Major General Bates, commanding troops in Cork, commented, 'had the military been permitted to take a more decided action, the best result would have happened, probably the capture of the whole body of Fenians.' Lord Strathnairn concurred, noting that the magistrate concerned, 'would not appear to be a fit person to be in charge of Troops in these times.'[24] Worse, at least from the point of view of the military, was the behaviour of the magistrates in Macroom. Having called for military support, they refused to accompany the troops (a detachment of the 60th Rifles). But when these arrived the magistrates called upon the troops to open fire on the crowd who had greeted the arriving forces with stones.[25]

The Irish Constabulary (following actions in 1867 against the Fenians to be awarded the prefix 'Royal') was a centrally controlled and armed force from its creation in 1836. Very much at variance with the developing norms of policing in Great Britain which was of a locally raised and administered, and unarmed forces. The Irish Constabulary pattern was much closer to a continental gendarmerie and was to provide the model for Britain's Colonial police forces.[26] However, the military aspects of the force can be overplayed. It reached a high point of 12,358 men in 1850 before settling down to an establishment of 10,000. Throughout the island of Ireland this certainly did not appear as an 'army of occupation'. Distribution was also uneven; the large county of Donegal was allocated just 176 men, while Tipperary had 1,030.[27] Reforms in 1839 and 1845 saw the establishment of a reserve of 200 men in Dublin, housed in the Irish Constabulary Depot in Phoenix Park; a limited 'emergency reserve' if a crisis came and most notably deployed in 1852 to replace the partisan local police force in Belfast which had collapsed in the face of sustained rioting.

The rather grandly entitled 'Police Barracks' were normally ordinary houses in small towns and villages, with often as little as four policemen occupying them. The Irish Constabulary were given all sorts of tasks as the force developed which weakened them as a crime prevention, much less counter-insurgency force; reports on the potato crop in the 1840s, census returns, customs duties, weights and measures inspection. The detective branch of the force was very small and much police intelligence relied on paid informers or local gossip; a system which spectacularly broke down in 1919–21 when the police were boycotted and abandoned many smaller barracks.

The role of the Irish Constabulary as an armed force was also open to criticism. A number of local magistrates claimed that this was a distraction from their 'thief-taking' role and that carrying the Long Enfield Rifle meant that they were easily outpaced by criminals. *The Times*, in a surprisingly optimistic leader of 1864, went so far as to declare that, 'We might as well, when we go partridge shooting, carry spears and rifles for fear of an attack from a mastodon or a swoop from a pterodactyl, as march out our police-men in battle array to catch pickpockets. The English plesiosaurus and the Irish rebel are extinct.'[28] Taking another view of the possible arming of the Irish Constabulary with the new Long Enfield Rifle, the then C in C in 1861 asked, 'are you really prepared to arm them with a weapon which will shoot five or six people at once?'[29] This might be seen as a very early example of the so-called 'minimum force' doctrine in British counter-insurgency. Strathnairn, facing the crisis in December 1866 was concerned that, while the police were armed, they had not been properly, 'trained to arms' and would only be of use as 'good scouts' for the 'first notice of danger'.[30] As Strathnairn was later to appreciate from personal experience, the Irish Constabulary were not always effective scouts. Being led to Tallaght, in the immediate aftermath of the Fenian Rising, Strathnairn soon realised that his police guides did not actually know the way, a major problem as, 'The Country people because disaffected, will not point out the way, and if they did could not be relied upon.'[31] The Dublin Metropolitan Police had been established along the lines of their London counterparts. After 1893, unlike the RIC, the force was not armed, due to the perceived lack of an armed threat in the city. However, prior to this DMP officers were armed and in the immediate aftermath of the Fenian Rising it is clear that 200 modern rifles were supplied to this force, for use in their outstations.[32]

The British army was seen as an uncertain instrument in the hands of successive British governments dealing with political agitation in Ireland, due to the large number of Irishmen in the ranks. By the mid-nineteenth century the British army was disproportionately Irish in its composition; certainly less Irish in its composition than it had been earlier in the century, but still in 1840 the Irish made up 37.2% of the army when they accounted for 30.6% of the UK population; a figure which fell to 28.4% of the army to

20% of the population by 1861.[33] Prior to the localisation and creation of the modern regimental system in 1868–1881 these Irishmen were fairly evenly distributed throughout the army. This, naturally led to concerns of Fenian infiltration of the British army garrisons in Ireland. Fenian leaders claimed to have recruited 8,000 Irish soldiers. The Young Ireland movement was not seen to have infiltrated the army in any similar way. The case of Private Patrick Connor of the 40th Regiment who was imprisoned with hard labour for stating, 'T. T. Meagher would get the better of his object' seems to be an isolated one.[34]

However, no unit disobeyed orders and following investigations only 150 men were tried by court-martial for Fenian activity.[35] Lord Strathnairn's observation that, 'very many, and far too many, cases of individual, but not collective, treason, have occurred amongst the Irish Roman Catholic soldiers' seems entirely accurate.[36] Some of the evidence about Fenian activity in the army also suggests that this was not a well-planned conspiracy, with a surprising number of active Fenians professing their views in public; sometimes to superior officers. So, for example, Francis Quinn, alias Devlin of the 63rd Regiment was reported as saying, 'I have drilled the Fenians, and will drill them again. Damn the Queen.' A private of the 39th Regiment, who was refused service at a public house in Armagh, responded by going out into the street and calling for three cheers for the 'Head Centre'; the leader of the Fenian movement.[37] One of the most concerning examples of Fenian sentiment occurred in the 67th Regiment at Fermoy, where it was reported that some men had said that they would not fire on their 'Fenian brethren' if called upon to do so. However, this does not seem to have been taken very seriously by the military authorities.[38] The Kilmainham papers contain some examples of soldiers giving evidence against those trying to recruit them to the Fenian movement. For example Private David Fox of the 2nd Battalion, 3rd Regiment received an expression of thanks from the Commander of the Forces, 'at his loyal and soldier like conduct in having delivered to Justice a man who had attempted to seduce him from his allegiance.'[39]

The part-time militia, which had been reformed in 1854 in Ireland, was seen as particularly prone to Fenian infiltration and, indeed, in Great Britain there were similar concerns that Fenians had infiltrated some militia and Rifle Volunteer units. In normal circumstances, the militia enlisted recruits for a four month period, following which they were called up for 28 days training each year. During the period 1865–70, directly due to fears of Fenian infiltration, no new recruits were trained and all annual trainings were cancelled in Ireland, though the militia in Great Britain carried on as usual. In 1871 the Irish militia had effectively to be recruited again from scratch.[40]

Similarly, in 1848 there was a reluctance to rely on any locally raised auxiliary units. Following the Young Ireland rising the magistrates in Tralee wanted the 'loyal inhabitants' of the town to be armed and made an application to the officer commanding the depot of the 88th Regiment for arms. This was merely noted by the government, with no further action. A proposal made by Major General Napier to arm 150 'Palatines'; Protestant, German settlers, in Limerick was refused as it was felt that this, 'would make their homes the object of attack by the lawless during the Winter.'[41]

In 1848, that great Irish General, the Duke of Wellington, looking perhaps to Robert Emmet's Rebellion of 1803, Chartist activity in England or continental experiences of revolution, felt that determined efforts had to be made to defend Dublin Castle against a determined attack; including the demolition of some houses close to the Castle. He also believed that the Lord Lieutenant should seek refuge in his Lodge in Phoenix Park, which would be easier to defend than the Castle itself. The Duke of Cambridge, Major General commanding the Dublin District, asked for the withdrawal of detachments at Navan and Trim, presumably to reinforce the Dublin Garrison. However, this was over-ruled by the Lord Lieutenant, who was concerned at the risk of abandoning such relatively important towns.[42]

Lord Straithnairn as Commander in Chief could draw on his experience in command of the Central Indian Force during the Mutiny of 1857–58 and subsequent counter-insurgency operations which lasted into 1859. His perceived success in these operations were rewarded by promotion and appointment as Commander in Chief India from 1860–65 and Commander in Chief, Ireland 1865–70, with elevation to the peerage in July 1866.[43] He was, however, concerned with a Police Commissioners' report of 1866, which suggested that a Rising would take place simultaneously throughout the country:

> These documents shew that in the opinion of the Govt the state of this Country is very critical; that an outbreak, which will take effect, or occur, in all parts of the Country, 'which will not threaten any one point but will affect simultaneously each special locality,' is probable, & that on the Army in Ireland must devolve the principal duty of putting down a revolt which is to be aided by a new & in the opinion of the Commissioners a dangerous weapon.

> Under these circumstances the Commander of the Forces again ventures to draw the attention of the Govt of Ireland, to the positive danger to the Army & therefore to the State, & the Public Peace which the Army protect, which accrues from small detach[ment]ts under young & inexperienced leaders, an expedient which materially affects the discipline, prestige, power & safety of Troops. For it is an axiom that nothing does so much harm to a Regt., even the best, as Detachments.

> Removed from the control of their constituted & experienced superiors, they are placed under officers who commit the shortcomings which are to be expected from those who for the first time exercise two difficult duties which more than any other require experience, command & discipline, &

that in a country where they may be called on at any moment to resist a sudden & treacherous attack or assist the Civil Power under very difficult circumstances, against disaffected [Ruffians?]

The Detachment is numerically too small to possess either self-confidence or to exercise physical & moral influence abroad.

Strathnairn went on to outline his other major concern about the deployment of the troops under his command, namely that these were generally quartered in small outposts which were, in many cases badly situated, being commanded by high ground and were in no sense mutually supporting. This dispersal had been insisted on by the government and it was by no means clear to Strathnairn why detachments had been asked for in certain areas or what purpose they could serve, beyond defending themselves, in the event of an outbreak. He cited some particularly bad examples, noting that in Bantry, which he regarded as 'particularly disaffected', the garrison consisted of a mere company and there were no rail communications, with the nearest military posts being Skibbereen, 16 miles distant on a poor road, and Bandon, 32 miles distant by a good road. In Gort, another isolated company was in garrison, occupying the partially dismantled old cavalry barracks. The nearest military posts were 18 and 27 miles away and, 'The Government in applying for this Detachment do not state any reasons except the request by the Magistrate.' Ballina was seen as a particularly problematic town to garrison and Strathnairn strongly opposed government instructions to send a detachment there. It was noted that the roads to it ran through bogs and narrow passes, making it difficult to withdraw the garrison, under fire, in the event of an uprising. Strathnairn also believed that the proximity of Ballina to the coast made it more suitable for support from the Royal Navy, rather than the army.[44] This was a long running complaint by Strathnairn. In January 1867 he complained that the barracks at Mitchelstown was in an 'unmilitary position' being commanded by Lord Kingston's Deer Park, 170 yards distant. He felt that there was no 'valid reason' why the detachment was there but, if it was to be maintained in Mitchelstown it should occupy the Castle on Kingston's estate.[45] Other quarters were simply seen to be insanitary. Lord Strathnairn inspected the detachment at Tipperary Work House and found the building damp and the sick rate high.[46]

As Fenian outbreaks occurred in Tralee and Templemore, Strahnairn again pushed for isolated detachments to be withdrawn. He asked for those at Cloghan, Mitchelstown, Lismore and Clonakilty, none more than 100 strong, to be concentrated to defend Fermoy, an important town, strategically, which was believed to be vulnerable. He explained;

The precaution should be taken of withdrawing weak detachments in defenceless positions, which so far from being able to repel aggression invite attack;

and it is unnecessary to observe that the moral effect of the surprise or destruction of a detachment of regular troops would be productive of most grave consequences.[47]

Writing shortly after this to the Earl of Erne, a leading Conservative in Ulster, Strathnairn provided a very confident report of the ability of crown forces to deal with the Fenian threat. He noted that troop levels were about 5,000 men above normal and continued stating that this:

> Would render it, I think, an act of insanity, were the Fenians to attempt an outbreak, the more so, as in case of necessity, strong reinforcements of all arms, could proceed into Ireland, at the shortest notice. The Fenian conspirators have neither arms, artillery, cavalry or organisation. How they are to take the field against an army provided with these material indispensibles and in possession of all the strongholds in Ireland I am at a loss to imagine.[48]

The context of Strathnairn's letter to the Earl of Erne is unclear. Strathnairn may have been providing a more optimistic view of the situation than he believed it to be as the amateur forces formed in Protestant Ulster; notably the Yeomanry which existed between 1796 and 1834 had not been seen as a great military asset to the British forces. Formed in small, dispersed troops, poorly trained and sometimes poorly mounted, the force was seen as deeply sectarian in nature; more likely to exacerbate political tensions than to calm them.[49] The Earl of Erne may have been making tentative suggestions about reforming such a force.

Strathnairn was clearly concerned about disposing his troops in small penny packets throughout the country, unable to offer mutual support. He organised four 'flying columns' in early March 1867; two in Munster, one in Leinster and one in Connaught. The columns were to operate in areas where there was no barrack accommodation for troops. To negate some of the legal complications which had been experienced previously, it was made clear that the officers commanding these columns would hold the Commission of the Peace and would also be accompanied by selected magistrates. The importance of these columns was probably as much psychological as practical, in Crossman's view. However, Strathnairn himself, thought that they had accomplished much noting, in mid-April 1867 when they were being stood down, that they had been effective in, 'reassuring the loyal and overawing the disaffected.' He asked for special allowances for officers, who had been forced to pay 'tourist rates' for hotel accommodation and for other ranks, who had often been forced to sleep, fully clothed, in outbuildings.[50]

The period from 1848 to 1867 saw little determined attempt to overthrow British rule in Ireland, with the two risings of 1848 and 1867 almost paling into insignificance compared to the more serious outbreaks of 1798 and 1916. Owen Burne, Lord Strathnairn's ADC in 1867 put it rather well when he stated, 'This rising did eventually come into an active phase, although, fortunately for us, it was so badly organized that the outbreak, instead of being simultaneous in all parts of Ireland, which would have made it very formidable, fizzled, so to speak, here and there like a damp squib, and thus gave us plenty of warning.'[51] However, as Burne's comments show, there were serious concerns about the likelihood of a national rising in 1866–7 and the possibility that some Irish soldiers would defect to the Fenians, or at least, refuse to fire on them. Concerns about 'national uprisings' meant that in both 1848 and 1866–7 British military commanders had difficult decisions to make concerning the dispersal or concentration of their forces, concerned that, as in 1803, a major attempt would be made to capture the seat of British power, in Dublin Castle. Otherwise British forces can be portrayed as being involved in a long-running counter-insurgency against rural insurgents. This was very low-level, and rarely targeted the agents of the British state themselves. For most of this period it was left to the paramilitary Irish Constabulary to deal with this threat.

The threat to the British state of formed and armed 'loyalist' or Protestant forces which were to provide such problems in 1913–14 and 1966–1998 did not exist in the same way in the period covered by this article. However, it is worth noting, especially in the context of the small numbers killed in the rebellions of 1848 and 1867, that sectarian rioting in Belfast in 1857 and 1864 proved to be a major problem for British forces in this period. These riots are best understood as sectarian and intercommunal, rather than directly opposed to the British state but the rioting in Belfast from 8th to 22 August 1864 led to 11 deaths and 316 people being injured. The locally-recruited Belfast Borough Police, a mere 160 strong, unarmed and open to accusations of Protestant bias were disbanded in the wake of these riots with 450 RIC officers being drafted into the town to replace them.[52]

Notes

1. Bartlett, "Defence, counter-insurgency and rebellion," Chambers, *Rebellion in Kildare*; and O'Donnell, *Aftermath*.
2. Barton and Foy, *The Easter Rising*; Hopkinson, *The Irish war of independence*; McGarry, *The Rising*; Townshend, *Easter 1916*; and Townshend, *The British campaign in Ireland, 1919–21*.
3. See, for example, Gantt, *Irish Terrorism in the Atlantic Community*, 23–65; Jenkins, *The Fenian problem*, Newsinger, *Fenianism in mid-Victorian Britain*, Senior, *Last invasion of Canada*; and Steward and McGovern, *The Fenians*.
4. Comerford, "Patriotism as Pastime" and *The Fenians in Context*.
5. Lowe and Malcolm, "The domestification of the Royal Irish Constabulary, 1836–1922,"; Malcolm, Elizabeth, *The Irish policeman 1822–1922*; and Palmer, *Police and protest in England and Ireland*.

6. The standard work on the Young Irelanders remains, Davis, *The Young Ireland Movement*. See also Kinealy, *Repeal and Revolution*.
7. Comerford, "Patriotism as Pastime," *The Fenians in Context* and "Fenianism,"; and Newsinger, *Fenianism in Mid-Victorian Britain*.
8. Ó Cathaoir, *Soldiers of liberty*, 118–139.
9. Ibid: 164–189; and Takagami, "The Fenian Rising in Dublin."
10. Hobsbawm, *Primitive Rebels;* and Townshend, *Political violence in Ireland*, 1–50.
11. Lee, "The Ribbonmen."
12. Beames, "Rural Conflict in Pre-Famine Ireland"; and *Peasants and power*, 89–101.
13. Sullivan, *New Ireland*, 69–95. Cited in Garvin, "Defenders, Ribbonmen and others," 222.
14. Davitt, *The Fall of Feudalism in Ireland*. Cited in Garvin, "Defenders, Ribbonmen and others," 223.
15. Beames, "The Ribbon Societies," 249.
16. Ó Luain, "Popular Collective Action in Catholic Ulster 1848–1867," 98.
17. National Library of Ireland (hereafter NLI), Kilmainham papers, Ms. 1059, memorandum Lord Strathnairn, Commander in Chief, Ireland to Chief Secretary, 6 December 1866.
18. This has been carefully detailed in Crossman, *Politics, Law & Order in 19th Century Ireland*. Appendix F, 199–230.
19. French, "Nasty not nice," 747.
20. Crossman, *Politics, Law & Order*, 111–2.
21. NLI, Kilmainham papers, Ms. 1059, letter Major Mockler, 64th Regiment, Tipperary to Military Secretary, 21 December 1866.
22. NLI, Kilmainham papers, Ms. 1054, letter R. Geaves [Colonel?] to Major General Bainbridge, 11 March 1848.
23. NLI, Kilmainham papers, MS. 1059, letter Sir Thomas Larcom to Commander in Chief, 7 December 1866.
24. NLI, Kilmainham papers, MS. 1059, report from Lieutenant Langford Leir, 5 March 1867; letter from Colonel S. Smyth to Under Secretary, 9 March 1867, notes by Major General Bates and Lord Strathnairn, 7 March 1867.
25. NLI, Kilmainham papers, MS. 1059, letter Colonel S. Smyth to Under Secretary, 14 March 1867 and Ms. 1060, same to same, 9 April 1867.
26. Broeker, *Rural disorder and police reform in Ireland*, 128–159; Malcolm, *The Irish policeman*, 26–44; and Palmer, *Police and Protest*, 316–375.
27. Curtis, *History of the Royal Irish Constabulary*, 88–9; cited in Townshend, *Political violence in Ireland*, 69.
28. *The Times*, 18 April 1864 cited in Townshend, *Political violence in Ireland*, 74.
29. Edward Cardwell to Thomas Larcom, 1 February 1861, NLI Ms 7617, f. 68, cited in Townshend, *Political violence in Ireland*, 75.
30. NLI, Ms. 1059, Kilmainham papers, Strathnairn to Chief Secretary, 3 December 1866.
31. NLI, Kilmainham papers, Ms. 1059, letter Colonel S. Smyth to Under Secretary, 14 March 1867.
32. NLI, Kilmainham papers, Ms. 1059, letter Colonel S. Smyth to Under Secretary, 11 March 1867; and Scanlon, *The Dublin Metropolitan Police*, 10.
33. Hanham, "Religion and Nationality in the Mid-Victorian Army," 162.
34. NLI, Kilmainham papers, Ms. 1054, note of 21 August 1848.

35. Crossman, *Politics, Law & Order in 19th Century Ireland*, 107–8, Ó Cathaoir, *Soldiers of liberty*, 118–38; and Semple, "The Fenian infiltration of the British army."
36. TNA, WO32/6000, General Lord Strathnairn's views on the social and political state of Ireland, June 1867, cited in Butler, *The Irish amateur military tradition in the British army*, 19.
37. NLI, Kilmainham papers, Ms. 1059, letter Colonel S. Smyth to Under Secretary, 10 December 1866 and Colonel L. Curzon to Under Secretary, 20 December 1866.
38. NLI, Kilmainham papers, Ms.1059, letter Colonel L. Curzon to Under Secretary, 21 December 1866.
39. NLI, Kilmainham papers, MS. 1059, letter Major O. S. Burne to Under Secretary, 26 December 1866.
40. Butler, *The Irish amateur military tradition in the British army*, 18–19.
41. NLI, Ms. 1054, Kilmainham papers, notes dated 29th July and 2 August 1848.
42. Kinealy, *Repeal and Revolution*, 161 and NLI, Ms. 1054, Kilmainham papers, memorandum of 29 January 1848.
43. Robson, *Sir Hugh Rose and the Central India campaign 1858*, xviii-xix; and Robson, "Rose, Hugh Henry, Baron Strathnairn."
44. NLI, Kilmainham papers, Ms. 1059, memorandum Lord Strathnairn, C in C, Ireland to Chief Secretary, 3 December 1866.
45. NLI, Kilmainham papers, Ms. 1059, letter Colonel S. Smyth to Under Secretary, 1 January 1867.
46. NLI, Kilmainham papers, Ms. 1059, letter Kenneth D. Mackenzie [Captain?] to Under Secretary, 30 January 1867.
47. NLI, Kilmainham papers, Ms. 1059, letter, Colonel S. Smyth to Under Secretary, 5 March 1867.
48. PRONI, D1939/21/9/9, letter Lord Strathnairn to Lord Erne, 16 December 1866.
49. Blackstock, *An Ascendancy Army*, 232–268.
50. NLI, Kilmainham papers, MS. 1059, letter Colonel S. Smyth to Under Secretary, 10 March 1867 and Ms. 1060, same to same, 10 April 1867; letter Strathnairn to Secretary of State for War, 12 April 1867; and Crossman, *Politics, Law & Order*, 110.
51. Burne, *Memories*, 71.
52. Farrell, *Rituals and riots*, 160; Griffin, *The Bulkies*, 116–42; and Townshend; *Political Violence in Ireland*, 43–44.

Disclosure statement

No potential conflict of interest was reported by the author.

Bibliography

Bartlett, Thomas. "Defence, Counter-insurgency and Rebellion: Ireland, 1793–1803." In *A Military History of Ireland*, edited by Thomas Bartlett and Keith Jeffery, 247–293. Cambridge: Cambridge University Press, 1996.

Barton, Brian, and Michael Foy. *The Easter Rising*. Stroud: History Press, 2011.

Beames, Michael. *Peasants and Power: The Whiteboy Movements and Their Control in pre-Famine Ireland*. Sussex: The Harvester Press, 1983.

Beames, M. R. "The Ribbon Societies: Lower-class Nationalism in Pre-famine Ireland." In *Nationalism and Popular Protest in Ireland*, edited by C. H. E. Philpin, 245–263. Cambridge: Cambridge University Press, 1987.

Beames, M. R. "Rural Conflict in Pre-Famine Ireland: Peasant Assassinations in Tipperary, 1837–1847." In *Nationalism and Popular Protest in Ireland*, edited by C. H. E. Philpin, 264–283. Cambridge: Cambridge University Press, 1987.

Bew, Paul. *Land and the National Question in Ireland 1858–82*. Dublin: Gill and Macmillan, 1978.

Blackstock, Allan. *An Ascendancy Army: The Irish Yeomanry 1796–1834*. Dublin: Four Courts Press, 1998.

Broeker, Galen. *Rural Disorder and Police Reform in Ireland 1812–36*. London: Routledge & Kegan Paul, 1970.

Butler, William. *The Irish Amateur Military Tradition in the British Army, 1854–1992*. Manchester: Manchester University Press, 2016.

Chambers, Liam. *Rebellion in Kildare 1790–1803*. Dublin: Four Courts Press, 1998.

Choille, Mac Giolla. "Breandán Fenians, Rice and Ribbonmen in County Monaghan, 1864–'67." *Clogher Record* 6, no. 2 (1967): 221–252. doi:10.2307/27695596.

Choille, Mac Giolla. "Breandán, 'fenian Documents in the State Paper Office'." *Irish Historical Studies* 16, no. 63 (1969): 258–284. doi:10.1017/S0021121400022148.

Clark, Samuel, and James S. Donnelly. *Irish Peasants: Violence and Political Unrest 1780–1914*. Dublin: Gill and Macmillan, 1983.

Comerford, R. V. "Patriotism as Pastime: The Appeal of Fenianism in the Mid-1860s." *Irish Historical Studies* 22, no. 86 (1980): 239–250. doi:10.1017/S0021121400024925.

Comerford, R.V. *The Fenians in Context: Irish Politics and Society 1848–82*. Dublin: Wolfhound Press, 1998.

Crossman, Virginia. *Politics, Law & Order in 19th Century Ireland*. Dublin: Gill & Macmillan, 1996.

Curtis, Robert. *The History of the Royal Irish Constabulary*. Dublin: McGlashan & Gill, 1869.

Davis, Richard. *The Young Ireland Movement*. Dublin: Gill and Macmillan, 1988.

Davitt, Michael. *The Fall of Feudalism in Ireland*. London and New York: Harper Brothers, 1904.

Desmond, Williams, T., ed. *Secret Societies in Ireland*. Dublin: Gill and Macmillan, 1973.

Emsley, Clive, and Barbara Weinberger, eds. *Policing Western Europe: Politics, Professionalism, and Public Order, 1850–1940*. Westport, CT: Greenwood Press, 1991.

Farrell, Sean. *Rituals and Riots: Sectarian Violence and Political Culture in Ulster, 1784–1886*. Lexington: The University Press of Kentucky, 2000.

Foot, M. R. D., ed. *War and Society: Historical Essays in Honour and Memory of J. R. Western 1928–1971*. London: Paul Elek, 1973.

French, David. "Nasty Not Nice: British Counter-insurgency Doctrine and Practice, 1945–1967." *Small Wars & Insurgencies* 23, no. 4–5 (2012): 744–761. doi:10.1080/09592318.2012.709763.

Gantt, Jonathan. *Irish Terrorism in the Atlantic Community, 1865–1922*, 23–65. London: Palgrave, 2010.

Garvin, Tom. *The Evolution of Irish Nationalist Politics*. New York: Holmes & Meier, 1981.

Garvin, Tom. "Defenders, Ribbonmen and Others: Underground Political Networks in Pre-famine Ireland." *Past & Present* 96, no. 1 (1982): 133–155. doi:10.1093/past/96.1.133.

Griffin, Brian. *The Bulkies: Police and Crime in Belfast 1800–1865*. Dublin: Irish Academic Press in association with the Irish Legal History Society, 1997.

Hanham, H. J. "Religion and Nationality in the Mid-Victorian Army." In *War and Society: Historical Essays in Honour and Memory of J. R. Western 1928–1971*, edited by M. R. D. Foot, 159–182. London: Paul Elek, 1973.

Hobsbawm, E. J. *Primitive Rebels: Studies in Archaic Forms of Social Movement in the 19th and 20th Centuries*. Manchester: Manchester University Press, 1978.

Hopkinson, Michael. *The Irish War of Independence*. Dublin: Gill & Macmillan, 2014.

Jenkins, Brian. *The Fenian Problem: Insurgency and Terrorism in a Liberal State, 1858–1874*. London: McGill-Queen's University Press, 2008.

Kinealy, Christine. *Repeal and Revolution: 1848 in Ireland*. Manchester: Manchester University Press, 2009.

Lowe, W. J., and E. L. Malcolm. "The Domestication of the Royal Irish Constabulary, 1836–1922." *Irish Economic and Social History* XIX (1992): 27–48. doi:10.1177/033248939201900102.

Malcolm, Elizabeth. *The Irish Policeman 1822–1922: A Life*. Dublin: Four Courts Press, 2006.

McGarry, Fearghal. *The Rising: Ireland – Easter 1916*. Oxford: Oxford University Press, 2010.

McGarry, Fearghal, and James McConnel, eds. *The Black Hand of Republicanism: Fenianism in Modern Ireland*. Dublin: Irish Academic Press, 2009.

Murray, A.C. "Agrarian Violence and Nationalism in Nineteenth-century Ireland: The Myth of Ribbonism." *Irish Economic and Social History* XIII (1986): 56–73. doi:10.1177/033248938601300103.

Newsinger, John. *Fenianism in Mid-Victorian Britain*. London: Pluto Press, 1994.

Ó Cathaoir, Eva. *Soldiers of Liberty: A Study of Fenianism 1858–1908*. Dublin: Lilliput Press, 2018.

Ó Luain, Kerron, 'Popular Collective Action in Catholic Ulster 1848–1867'. Unpublished Ph.D. thesis, Queen's University Belfast, 2016.

O'Donnell, Ruan. *Aftermath: Post-rebellion Insurgency in Wicklow, 1799–1803*. Dublin: Irish Academic Press, 1998.

Palmer, Stanley H. *Police and Protest in England and Ireland 1780–1850*. Cambridge: Cambridge University Press, 1988.

Philpin, C. H. E., ed. *Nationalism and Popular Protest in Ireland*. Cambridge: Cambridge University Press, 1987.

Pollard, H. B. C. *The Secret Societies of Ireland: Their Rise and Progress*. London: Philip Allan, 1921.

Robson, Brian, ed. *Sir Hugh Rose and the Central India Campaign 1858*. Stroud: Sutton Publishing for the Army Records Society, 2000.

Robson, Brian. "Rose, Hugh Henry, Baron Strathnairn." *Oxford Dictionary of National Biography*, https://www-oxforddnbcom.chain.kent.ac.uk/view/10.1093/ref:odnb/9780198614128.001.0001/odnb-9780198614128-e-24093?print=pdf

Scanlon, Mary. *The Dublin Metropolitan Police*. London: Minerva Press, 1998.

Semple, A. J. "The Fenian Infiltration of the British Army." *Journal of the Society for Army Historical Research* 52, no. 211 (1974): 133–160.

Senior, Hereward. *The Last Invasion of Canada: The Fenian Raids, 1866–1870*. Toronto: Dundurn Press in collaboration with the Canadian War Museum, 1991.

Spiers, Edward. "George, Prince, Second Duke of Cambridge." *Oxford Dictionary of National Biography*, https://www-oxforddnb-com.chain.kent.ac.uk/view/10.1093/ref:odnb/9780198614128.001.0001/odnb-9780198614128-e-33372?print=pdf

Steward, Patrick, and Bryan McGovern. *The Fenians: Irish Rebellion in the North Atlantic World*. Knoxville: University of Tennessee Press, 2013.

Sullivan, A. M. *New Ireland*. New York: Peter F. Collier, 1878.

Takagami, Shin-Ichi. "The Fenian Rising in Dublin, March 1867." *Irish Historical Studies* 29, no. 115 (1995): 340–362. doi:10.1017/S002112140001186X.

Townshend, Charles. *The British Campaign in Ireland, 1919–21: The Development of Political and Military Policies*. Oxford: Oxford University Press, 1978.

Townshend, Charles. *Political Violence in Ireland: Government and Resistance since 1848*. Oxford: Clarendon Press, 1988.

Townshend, Charles. *Easter 1916: The Irish Rebellion*. London: Penguin Books, 2006.

Vaughan, W. E. *Landlord and Tenants in mid-Victorian Ireland*. Oxford: Clarendon Press, 1994.

'The extraordinary successes which the Russians have achieved' - the Conquest of Central Asia in Callwell's *Small Wars*

Alexander Morrison

ABSTRACT

Charles Callwell's *Small Wars* (1896, 1899, 1906) is widely considered both an *ur-text* for modern counter-insurgency studies, and a primer for the racialized late-Victorian approach to war against 'savages': either way it is usually only considered within a British context. Alongside the numerous examples Callwell used from British colonial campaigns, he frequently referred to those of other European powers – notably the Russian conquest of Central Asia. This chapter will seek to analyse Callwell's views of Russian colonial warfare, establish the sources on which he relied, and evaluate his accuracy and the effect which the Russian example had on his thinking.

Charles Edward Callwell (1859–1928) is perhaps the best-known British military theorist of the 19th century, and his *Small Wars, Their Principles and Practice* (1896, 1899 & 1906)[1] is by far his best-known text, echoed in the title of the *Small Wars* journal and described by Hew Strachan as a 'minor classic of compression and sophistication'.[2] Widely regarded as the originator of the idea of 'counter-insurgency' as a distinct form of warfare (though he never used the term), Callwell is much cited by both military historians and security strategists today.[3] His work has recently undergone considerable scholarly reappraisal, much of which has focused on how far *Small Wars* is still useful and relevant for modern students and practitioners of counter-insurgency.[4] David Betz has suggested that Callwell still has a good deal to teach them, particularly in his focus on understanding the characteristics of the enemy, and his warnings against dogmatism.[5] Daniel Whittingham also argues for Callwell's continued usefulness, while noting that his emphasis on what he called the 'moral effect' of short, sharp actions translated into brutality and forms of collective punishment that are very much at odds with modern counter-insurgency doctrine.[6] Sibylle Scheipers also considers that Callwell devised particular rules of engagement for 'uncivilised' opponents, though insists that the effects of this should not be overemphasised, as military laws of engagement were not wholly absent from colonial conflicts.[7] Dierk Walter notes how Callwell's 'evident obsession with the cult of the offensive' and the practical recommendations he made for colonial warfare were perpetuated throughout much of the 20th century.[8] The most blistering critique has come from Kim Wagner, who focuses on Callwell's racialised understanding of warfare, reflected in the unsuccessful British attempt to have soft-nosed 'Dum Dum' bullets recognised as legitimate under the laws of war when used exclusively against 'savage' opponents. Both Wagner and Douglas Porch dismiss entirely the notion that *Small Wars* should be used as a guide to current doctrine or action.[9]

As Whittingham has argued, Callwell needs to be treated as a historical figure and understood in his contemporary context, rather than as a guru whose work can still be mined for tips.[10] This chapter is not concerned with whether any of his recommendations for fighting 'Small Wars' against what he referred to as 'Savage Races' might still be useful – instead my interest is in understanding how the text of *Small Wars* was produced, and what it reveals about its author and the late Victorian military culture that he sprang from. Despite the extent to which Callwell's text has been treated as a practical guide to counter-insurgency, until recently there has been relatively little attention paid to the man himself. As Whittingham notes, this is partly because he did not leave behind a compact collection of private papers, while correspondence from him in those of other figures is sparse and mostly relates to the end of his career.[11] There has also been little exploration of how Callwell actually wrote and compiled *Small Wars* and his other theoretical texts, something which he studiously avoided referring to in his autobiographical writings.[12] Which sources did he use? How far was he working from personal experience and oral anecdote, and how far from published accounts? How and why did he choose particular examples to illustrate his points, and how accurate was his presentation of them? How reliable is he, in short, as a military historian?

Callwell is often understood as a primarily British figure, and the prize-winning essay which first made his name as a theorist of the 'Small War' only considered British colonial campaigns.[13] While overlooked by Michael Howard in favour of Julian Corbett, he is widely seen as the originator of a distinctively 'British way of warfare', and perhaps the only significant British military theorist of the 19th century.[14] However Callwell's range of reference in *Small Wars* is much wider and more cosmopolitan than this would suggest, reflecting a knowledge of French and German that was characteristic of military intellectuals of his day, and a healthy respect for the achievements and abilities of continental armies.[15] Thomas Bugeaud

(1784–1849) and Mikhail Dmitr'evich Skobelev (1843–1882) feature at least as prominently in *Small Wars* as Garnet Wolseley (1833–1913) or William Lockhart (1841–1900), and rather more so than Herbert Kitchener, whose role in the Sudan campaign is overshadowed by that of Hector Macdonald (1853–1903) in Callwell's account. *Small Wars* contains examples from Mexico, the American West, Algeria, Tonkin, Madagascar and the Russo-Turkish War alongside numerous British colonial campaigns in India and Africa, but the foreign case-study Callwell returns to most consistently is that of the Russian conquest of Central Asia from the 1850s to the 1880s, which he supplements with occasional references to Russian campaigns against Shamil and the Caucasus mountaineers in earlier decades.[16] Though they have since sunk into relative obscurity in Anglophone scholarship, these conquests were the subject of lively interest in Britain at the time when Callwell was writing, largely because of the widely-held but illusory belief that they were the prelude to a Russian invasion of British India, the so-called 'Great Game'.[17] To Callwell's credit, he does not appear to have shared this paranoia, which helped to generate a range of (often rather alarmist) publications about Russia's Central Asian campaigns. Although Callwell did not provide *Small Wars* with references or a bibliography, it is easy enough to establish which sources he used. As he did not read Russian he was reliant on a small body of writings translated into English, and often published in the *RUSI* journal. These usually derived from official military histories or *Voennyi Sbornik* and *Russkii Invalid*, the journals respectively of the Russian Main Staff and Ministry of War, though in some cases the material was a good deal more second-hand. Callwell's knowledge of Russian campaigns was inevitably fragmentary and partial, but this paper will argue that they played an important role in his work. I seek to establish the sources he used, the gaps in his knowledge and inaccuracies which resulted, but also to test whether he identified a distinctively 'Russian' approach to small wars which could be contrasted to a British way of war, or whether he instead advanced a more universal formula for warfare between 'civilized' and 'savage' peoples.

I The Russians and their opponents in small wars

Callwell's evident admiration for the Russian record in wars against 'savage peoples' was shared by many British contemporaries, who saw them as unconstrained by the liberal hand-wringing of Britain's parliamentary system and therefore more decisive – as G. N. Curzon put it: 'A greater contrast than this can scarcely be imagined to the British method, which is to strike gingerly a series of taps, rather than a downright blow; rigidly to prohibit all pillage or slaughter and to abstain not less wholly from subsequent fraternisation. But there can be no doubt that the Russian tactics, however deficient they may be from the moral, are exceedingly effective from the practical point of view'.[18] Curzon here was referring to Skobelev, who was an object of fascination for the British: becoming well-known internationally as a result of the dashing role he played in the Russo-Turkish War, his apotheosis came at the siege of the Akhal-Teke Turkmen fortress of Denghil-Tepe, whose storming in January 1881 was followed by the massacre of 14,000 Turkmen, women and children included. As he told the journalist Charles Marvin later that year, in a turn of phrase that became proverbial:

'I hold it as a principle that in Asia the duration of peace is in direct proportion to the slaughter you inflict on the enemy. The harder you hit them, the longer they will be quiet afterwards.'[19]

By 1882 Skobelev was dead, expiring of a heart attack in the arms of a Moscow prostitute, but his celebrity was if anything enhanced by this, with the first English biography appearing in 1883,[20] while Russian accounts of his final campaigns continued to be published regularly over the next two decades.[21] Callwell quoted Skobelev directly and approvingly more often than any other foreign commander mentioned in Small Wars: 'Do not forget that in Asia he is the master who seizes the people pitilessly by the throat and imposes upon their imagination' (51) – possibly a paraphrase of his words to Marvin quoted above. Callwell was able to draw upon contemporary publications of some of Skobelev's reports and orders,[22] and on the weighty official history of his final campaign against the Turkmen by General N. A. Grodekov (1843–1913), published in four volumes in 1883–4 and almost immediately translated into English for the Indian General Staff.[23] This campaign – the last major conquest undertaken by the Russians in Tsarist times– provided the bulk of Callwell's Central Asian examples.

Russia's Central Asian campaigns accorded well with Callwell's general emphasis on the importance of 'moral force', i.e. of rapid blows to cow and overawe an inferior enemy[24]:

'In the Russian campaigns in Central Asia it has generally been the same. Energy and resolution have been the watchword. The procedure has been rather to overawe the enemy by a vigorous offensive than to bring against him a mighty force, and the result speaks for itself. Prestige is everything in such warfare. It is the commander who recognizes this, and who acts upon it, who conquers inferior races absolutely and for good' (58).

However Callwell's admiration for the Russians stemmed less from their speed and ruthlessness in crushing their enemies than from an appreciation of how those enemies had chosen to fight. To put it simply, the Russians had been lucky in their opponents in Central Asia, who while they often fought valiantly had generally preferred open confrontation on the battlefield or the defence of fortified points to guerrilla tactics, thus offering Russian regular forces a clear objective and enabling rapid decisive battles: 'The Russians in their gradual extension of territory beyond the Caspian have often had to deal with armies, ill armed and organized of course, but nevertheless armies' (6). This reflected Callwell's wider argument that the most pressing problem in fighting small wars was bringing the enemy to decisive battle, rather than allowing the enemy to engage in attritional guerrilla warfare. While the Russians had suffered severely from the latter during their Caucasus campaigns, even here their enemy had eventually obliged them: 'The great Circassian [sic] leader Schamyl kept the Russians at bay for years with guerrilla tactics; his cause declined when he formed his followers into armies and weighed them down with guns' (76).[25] Thus, rather than despising 'irregulars', Callwell simply saw such tactics as a rational and effective response by 'savage' opponents to the superiority regular armies enjoyed in weapons and technology.[26] When a 'savage' opponent chose instead to fight Europeans on their terms, but with armies that were 'a travesty of regular organization' and encumbered with inferior artillery, it made the task of the conqueror far easier: 'The Russians in their campaigns against Khokand and Bokhara had to deal with armies standing on a somewhat similar footing' (10). Callwell drew a parallel, not with the frontier campaigns which characterised Anglo-Indian warfare in his own day, but with the Anglo-Sikh Wars of the 1840s: 'Runjeet Singh was a respected ruler who could dispose of organized forces completely at his command; the Amir of Bokhara stood on a similar footing during the campaigns which ended in the annexation of his khanate to the Russian Empire' (14).[27] In other words, what made the Russian campaigns distinctive was not their willingness to use exemplary violence, even against civilians (of which Callwell clearly approved), but the fact that usually their opponents had obligingly preferred to stand and fight.

Callwell expanded on this theme further when discussing the battle of Irjar, fought between Russian and Bukharan forces at a crossing of the Syr-Darya south of Tashkent in 1866:

'The Russians in Central Asia have been very fortunate in finding their opponents inclined for decisive conflicts. At Yedshar [sic] in 1866, a very large army from Bokhara marching on Tashkend in the hope of recovering that city was confronted by a far inferior Russian force, and a severely contested action ensued in which the latter was completely victorious; two years later a decisive battle was fought under the walls of Samarcand; these two great battles decided the fate of Turkestan, the capture of Tashkend having given the Russians a firm footing in the country to start with. Minor engagements have been conspicuous by their absence in Central Asia, almost every episode in the campaigns which brought the Cossacks to Bokhara and the sources of the Sir Daria was an important operation of war, and to this may be attributed the extraordinary successes which the Russians have achieved' (79–80).

Callwell's source here was an 1873 account of this campaign by Friedrich von Hellwald, an Austrian military writer, whose book on the Russians in Central Asia was translated into English in 1874. It does not appear to have been based on any direct knowledge of the region or of Russian campaigns.[28] Contemporary accounts from both the Russian and the Bukharan side emphasise the Bukharan ruler Amir Muzaffar's foolishness and overconfidence in challenging the Russians directly at Irjar in 1866, though the 1868 engagement before Samarkand, fought on the Chupan-Ata heights, came after several attempts to make peace.[29] In both battles superior Russian artillery and small arms were the decisive factor in bringing about an easy victory. Callwell was at least partially aware of this:

'General Romanovski's decisive victory over the Uzbeg army at Yedshar is worthy of mention in this connection. The Emir of Bokhara had about 40,000 men, the Russian force consisted of only 3,000; but in spite of disparity of numbers General Romanovski attacked without hesitation. The enemy enveloped the Russian force and made desperate attacks upon the baggage which was guarded by only a few companies, and these were at times in great peril. But the Russians resolutely pressed on, the baggage escort meanwhile repulsing the hostile onslaughts as best it could and pushing on whenever it had a moment's respite. The enemy could not stand against the determined advance of General Romanovski and the fire of his guns, and at last became panic-stricken and fled' (163).

However the Russians were not as heavily outnumbered at Irjar as Callwell believed, nor were they the first to attack. In his own account of the battle D. I. Romanovskii (1825–1881) estimated that of the forces facing them only about 5,000 were regular *sarbaz* infantry, with the remaining 35,000 made up of irregular cavalry, who suffered terribly as they attempted a frontal assault on the Russian position.[30] Another celebrated incident in which Russian forces were heavily outnumbered occurred in 1864, when a small force of Ural Cossacks was ambushed by a much larger contingent of Khoqandi cavalry at the village of Iqan between the towns of Turkestan and Chimkent. Callwell gives the following account:

'In 1864, in the early days of the Russian operations against Khokand, a detached sotnia of cavalry with a gun was surrounded by an immensely superior force of Khokandians at Ikan. For two days the Russians defended themselves against overwhelming odds, inflicting great loss upon the enemy. Finally, they managed to escape. The moral effect inspired by the fight made by this detachment was very great, and although it was almost the only conflict of the year it appears to have so gravely impressed the Khokandians as to have materially assisted the Russians next year in their successful attack upon Tashkend' (79).[31]

Callwell's admiring tone here is characteristic of his descriptions of the Russian campaigns – and the emphasis on the moral effect of this victory (which was of no strategic significance) also serves his broader argument well. In this case he may well have had a point – in the *Ta'rikh-i 'Aliquli*, a local history of the Russian campaigns, the author recalled the Khoqandi commander at Iqan, Alimqul, lamenting that 'the surprising thing is that such an army as ours cannot overpower two hundred Russians!'[32]

Callwell goes on to use the capture of Tashkent by General Mikhail Grigor'evich Cherniaev (1828–1898) in 1865 as a key illustration of the effects of 'moral force' and decisiveness in overawing a numerically superior Asiatic enemy:

'The capture of the great walled city of Tashkend by General Tchernaieff in 1865, is a case in point. The place contained more than 100,000 inhabitants and it was defended by 30,000 men. It was one of the great commercial centres of the east, its name was known from Stambul to the Yellow Sea. Its perimeter was about 16 miles. Its ramparts were of stout design, and its battlements sheltered a respectable artillery. The Russian General arrived before it with 2,000 men and 12 guns, and determined upon a *coup de main*. An entrance was surprised at dawn of day at two points by storming parties, and these thereupon opened the gates; the guns upon the battlements on one side were seized and spiked. So great was the effect produced within the city by this daring feat of arms that its notables surrendered at once, although in street fighting the Russian troops could not have successfully coped with the great numerical superiority of their adversaries. And Tashkent was incorporated in the Tsar's dominions from that time forward' (59).

This account, which like that of Irjar was taken from Hellwald, reflects the legend which Cherniaev himself sought to project about the capture of Tashkent (he would later compare it to the exploits of Cortez and Pizarro), but is deficient in many respects.[33] Tashkent did not have 30,000 defenders – that figure would have corresponded to the entire male population of fighting age, and the city was deeply divided between factions that favoured Khoqand, Bukhara and Russia. The city did not surrender at once, and Cherniaev and his men did in fact spend two days fighting through Tashkent's streets.[34] These facts are referred to in numerous contemporary English publications, but having found a version of events that accorded with his thesis Callwell does not seem to have wanted to look any further.[35] This is characteristic of the rather partial and selective way in which he used Russian examples.

II Russian decisiveness and clarity of objectives

For Callwell another reflection of the obliging nature of their Central Asian opponents was that the Russians had always had a set of clearly-defined objectives for their campaigns, after the capture of which serious resistance ceased:

'... the advantage of having a well-defined objective even for a time can scarcely be over-rated, and the Central Asian campaigns of Russia illustrate this vividly. Turkestan was territory inhabited largely by nomads, but studded also with historic cities many of which had been for ages the marts of oriental commerce. The invaders went to work with marked deliberation. They compassed the downfall of the khanates by gradually absorbing these cities, in many cases by very brilliant feats of arms. The conquests were not achieved by any great display of mighty force. The objectives were always clear and determinate. The capture of one city was often held sufficient for a year; but it then became a Russian city. The troops had always an unmistakable goal in front of them, they went deliberately to work to attain that goal, and when it was attained they rested on their laurels till ready for another coup. Such is the military history of the conquest of Central Asia. Desultory operations were throughout conspicuous by their absence. But such conditions are very seldom found in small wars ... ' (16–17).

By contrast:

'The Russian failures in the Caucasus were mainly due to the objectless character of their campaigns. They would assemble a great force and march through the forest and over the hills to capture some stronghold. They would find this abandoned. Then they would march back again harassed all the way by the warlike Circassians, Georgians [sic][36] and Chechens, and would settle down into cantonments till the time came for undertaking some similar spasmodic enterprise'(76).

The lengthy resistance offered by Shamil and the Caucasus mountaineers to Russian imperial expansion was a celebrated theme in Russian romantic literature of the golden age – for instance the disastrous ambush of General Golosoyev's force by the Chechens on the river Valerik in 1840, which Callwell refers to (220), features in Lermontov's *A Hero of Our Time*. Meanwhile the slightly later accounts of the conquest of Central Asia tended to be written by military professionals and to exclude minor or unsuccessful engagements, focused instead on set-piece battles and the capture of cities, which is why they figured so prominently in Callwell's narrative. However in Central Asia too there was a long history of precisely the kind of desultory and purely punitive operations which Callwell deplored – an appropriate comparison with Goloseyev's ambush in 1840 would be the complete destruction of Colonel Rukin's patrol by the Kazakhs of the Adai tribe on the Mangishlaq peninsula in 1870 – but no account of this ever seems to have appeared in English at the time.[37]

Throughout the 18th century and into the 1850s the Russian fortified lines on the steppe were regularly raided by Kazakhs and Turkmen, who were then raided in turn by Cossack forces, the capture of livestock and people being the main aim in both cases.[38] Callwell was in fact aware of this, writing that: 'Fighting the Kirghiz [Kazakhs] and other nomads of the steppes the Russians have always trusted largely to carrying off the camels and flocks of the enemy' (19). He expanded on this further by noting Russian imitation of nomadic tactics:

'On the steppes also the mobile columns are formed of mounted Cossacks and irregulars; The Kirghiz [Kazakh] freebooter Kutebar, in the days when Russia was preparing for the incorporation of Khokand, defied all efforts to effect his capture and his subjection for a decade, and while he was at large his meteoric appearances and his impudent hardihood made him at once a peril to Russian progress and a scourge to the clans which had welcomed the arrival of the Cossack. But, on the other hand, in the desultory warfare in Central Asia against the nomads whose sole wealth consists in flocks and herds, the Russians have themselves employed raids with conspicuous success' (110–111).

Callwell considered such raiding for livestock to be a legitimate tactic of war against a nomadic enemy, though it clearly contradicted his advocacy of decisive operations against a clear objective (116, 210). Referring to the Turkmen, he excused the decade of inconclusive operations which preceded Skobelev's final victory at Denghil-Tepe in 1881 by noting that 'The Russian expeditions against the Tekke Turkomans were partly punitive, but they were undertaken mainly to suppress this formidable fighting nomad race; and the final campaign became a campaign of conquest' (9).

Curiously enough, the first Russian attempt to mount a major military operation in Central Asia – the Khiva expedition of 1839–40 – is not referred to anywhere in *Small Wars* despite the availability of an account in English.[39] While this campaign certainly had a clear objective, it failed miserably owing to the death from winter cold of almost all the 10,000 camels needed to carry supplies. Callwell's failure to mention this is probably another sign of selectivity, since it was a campaign which obeyed all his principles: decisiveness, aggression, clarity of objective – but which nevertheless was unsuccessful owing to a failure to take account of local environmental conditions (and indeed to listen to the advice of the Kazakh camel-drivers, though Callwell could be forgiven for not knowing this).[40] Following this debacle the Russians spent almost twenty years, between 1845 and 1864, attempting to maintain a line of fortifications along the Syr-Darya as a new steppe frontier, which proved another costly failure. The victorious campaigns against Khoqand and Bukhara in the 1860s, Khiva in the 1870s and the Turkmen in the 1880s which Callwell so admired came in the aftermath of numerous disappointments, some of which he must have been aware of, given other examples from that period which he cites – such as an attack on

Fort Perovskii, the former Khoqandi fortress of Aq Masjid on the Syr-Darya, which the Russians captured in 1853:

'A good example of the moral effect of capture of artillery is afforded by the following incident. In 1854 the Khokandians assembled in great force before Fort Perovski, the Russian advanced post on the Sir Daria, and were practically blockading it. The commandant resolved on a bold stroke, and sent out such troops as he could spare to attach the enemy unexpectedly. This small force however soon found itself in a critical situation, threatened from all sides; the enemy in enveloping it left their artillery almost without protection of other troops. Perceiving this the officer in command delivered a vigorous attack upon the guns, and captured them. The effect was immediate. The Khokandians took fright and fled in wild disorder, leaving many trophies in the hands of the insignificant Russian force. Their rout was complete' (133).[41]

Far from being immediately decisive, Russia's war against Khoqand, which had begun in 1853 with the fall of Aq Masjid, only ended in 1876, when the khanate's last remaining territory in the Ferghana valley was annexed. Callwell only made one reference to this final campaign, noting Skobelev's use of carts to move two infantry companies across Ferghana at high speed, something made possible by the relatively good roads of the region (209).[42] This was the campaign in which Skobelev first made his reputation as an independent commander. It was marked by indiscriminate brutality against the civilian population and prolonged guerrilla-style resistance, particularly in the Alai valley in the surrounding highlands.[43] It attracted less attention in Britain than other campaigns as it represented the annexation of an existing protectorate rather than a significant expansion of territory, which may be why Callwell largely overlooked it – certainly it complicated his vision of Russia's Central Asian campaigns as being free from desultory or purely punitive operations.

The only Russian failure that Callwell considers in any depth is that of General N.I. Lomakin's expedition against the Akhal-Teke Turkmen in 1879, which culminated in a humiliating reverse for Russian forces below the walls of the fortress of Denghil-Tepe.

'Denghil Tepe, for instance, became the stronghold in which practically the whole military power of the Tekke Turkomans concentrated itself in 1879 and 1880; the Russians failed in their first campaign through mismanagement, but the objective never was in doubt. In their second venture, the formidable nomad race, which might have taken years to subdue, was with the fall of the fortress crushed for good and all.' (17–18)

Callwell's account of this campaign was based largely on Charles Marvin's translations of the official reports in *Russkii Invalid*, and articles by the correspondent of *Novoe Vremya* who had accompanied the expedition.[44] Echoing contemporary Russian criticisms, Callwell noted that the preceding years had seen a damagingly indecisive approach against the Akhal-Teke which had emboldened the Turkmen:

'The small Russian columns sent against the Tekke Turkomans in 1876–77, afford illustration of the evil of desultory, indecisive operations, although their Asiatic wars have generally been conducted in a very different spirit. Detachments too weak to effect any good purpose were sent out with no very clear object in view and were driven back, the result being merely to damage Russian prestige and to confirm the Turkomans in their hostile attitude' (77).

The 1879 expedition was plagued with difficulties from the outset, not least the sudden death of its initial commander, General Lazarev, from carbunculum.[45] Lomakin only succeeded to command temporarily, and it was his haste to join battle before a replacement could be despatched which proved his undoing: owing to a shortage of camels only 1,400 of the original force of 16,000 made it to Denghil-Tepe, and even they were desperately short of supplies. Callwell wrote that 'General Lomakin [sic – it was Lazarev who made this error] assembled his troops immediately at Chikishlar, and they ate up the supplies as fast as they were disembarked. As a consequence the large force was for months detained in an unhealthy locality' (45–6).[46] Callwell also noted Lomakin's foolishness in cutting off the Turkmen retreat at Denghil-Tepe, and thus rendering them determined to fight to the last: 'The Turkoman chiefs, it transpired afterwards, had contemplated surrender, but when they found themselves hemmed in and saw their families being thus driven back into the shell-swept encampment they concluded that extermination was in store for them, and they resolved to fight to the last. Then just at the wrong moment General Lomakin ordered the assault' (84–5). This echoed contemporary Russian judgements,[47] and broader Victorian assumptions about warfare against inferior or 'savage' races: as James Belich has argued in the case of the Maori, it was generally assumed they would not have the technical ability to construct effective fortifications, or the determination and moral fibre effectively to defend them. A successful defence thus necessarily indicated that they had been driven to desperation.[48] Callwell did however acknowledge that 'the developments to the enceinte of Denghil Tepe added by the Tekkes in anticipation of General Skobelef's campaign, give examples of earthworks of the modern type' (442), though these were added after Lomakin's repulse. Denghil Tepe finally served as a warning of what could happen when European forces forfeited their aura of invincibility by being seen to retreat in the face of a 'savage' enemy.

'At Denghil Tepe in 1875 [sic], the Turkomans who had been utterly disheartened by the Russian bombardment, and who had only manned the ramparts of their fortress in despair when they found themselves hemmed in, no sooner saw the assaulting columns falling back in confusion, than they charged out furiously after the Russian troops. Their counter-attack was delivered with tremendous force. Had it not been for the guns, Lomakin's little army might have been not only defeated but destroyed. And yet up to this moment the Turkomans had shown little inclination to meet their antagonists in battle' (185).

Callwell did thus make use of one Central Asian example to drive home lessons about what *not* to do in a colonial campaign, but this was partly in order to contrast Lomakin's haplessness with the genius of Skobelev whom he so admired.

III Russian tactics, logistics and supply

Callwell devoted considerable attention to Russian tactics and logistics in Central Asia, whether this was their order of march, of battle or of assault. His examples mostly came from the Turkmen campaigns of 1879–81, although his discussion of the technicalities of marching formations clearly comes from the translation of L. F. Kostenko's work on the subject.[49] Callwell cited Skobelev approvingly when advocating the exclusive use of infantry in close formation – either columns or squares – when up against an 'unciv-ilised enemy'[50]:

'It is a striking fact that so skilled and experienced a leader as General Skobelef should have been strongly opposed to anything like dispersed formations in Asiatic warfare. "We shall conquer," he wrote in his instruc-tions prior to the attack on Yangi Kala [...], "by means of close mobile and pliable formations by careful, well – aimed volley firing, and by the bayonet which is in the hands of men who by discipline and soldier-like feeling have been made into a united body – the column is always terrifying." And again, "The main principle of Asiatic tactics is to preserve close formations." These maxims are not quoted as conclusive – on the contrary, they appear to have been enunciated under a mistaken estimate of the Tekke powers of counter-attack and of the fighting qualities of the Turkoman horse; but they are none the less interesting and instructive as the views of a great leader who understood war and who never failed in what he undertook' (333).[51]

For the purposes of defence Callwell advocated the use of the square, noting in its support that 'The Russians have sometimes adopted it in Central Asia, notably during the suppression of the Turkomans in the Khanate of Khiva after the occupation of the oasis in 1874' (235), and that:

'This elastic square formation was employed largely by Prince Woronzoff in his operations against the Chechens amid the extensive forests on the northern slopes of the Caucasus, to cover the working parties which slowly hewed clearings through the woods. In such fighting the arrangement is advantageous at times even when the enemy's attacks are merely of a desultory kind' (310).

Such squares could also be supported by using camels as improvised fortifications, although the example Callwell gives is one where this was used against the Russians successfully by the Turkmen:

'It should be noted that very small parties cannot form a square of any defensive strength round their camels, in such a case the camels must be used as a parapet, the men inside – a plan which the Turkomans used very successfully on one occasion, shortly before General Lomakin started for Denghil Tepe in 1879. This incident gives a remarkable illustration of camelry operations ... ' (374–5).

He also noted and approved the fact that in the conquest of Central Asia the Russians made the company rather than the battalion the key infantry unit, remarking that 'General Skobelef laid particular stress on this' (334), and that it gave greater flexibility and reduced problems of supply.

Such small units in close formation would then require the closest possible artillery support. When the Turkmen counter-attacked after the Russian repulse at Denghil-Tepe, Lomakin's force escaped total disaster thanks to close support from its artillery:

'At Denghil Tepe in 1879, General Lomakin was obliged owing the smallness of his force to deliver the assault with practically no reserves. When the stormers found it impossible to penetrate into the defences and they fell away under the heavy fire poured into them by the Tekkes, there were no reserves to lean upon. Fortunately the guns afforded a refuge to the Russian infantry as this was swept back by the defenders, who charged out over their battlements in great force and with much determination. The retreating infantry masked the artillery for a while but cleared the front in time to allow the guns to deliver some rounds of case into the Turkoman swarms. These sufficed to drive the Tekkes back into their stronghold in confusion.' (158–9).

Here Marvin's account (and hence Callwell's) accords exactly with Lomakin's report of the action.[52] Callwell uses this as an illustration of the principle that artillery's primary role in small wars is to provide close infantry support, something he finds further evidence for in Skobelev's instructions on close artillery support in the successful Turkmen campaign the following year 'The main principle of Asiatic tactics is to preserve close formations'; 'The artillery must devote itself to closely supporting its comrades without the slightest regard for itself' (160).[53] This was also true of picquets and fortified positions: 'In the defence of isolated posts guns are of course most valuable. General Skobelef in forming the advanced depots on the line his troops were to follow towards Denghil Tepe told off several guns to each, the infantry garrisons being very small' (383).

When it came to cavalry tactics Callwell's preference here as elsewhere was for boldness and decision. He seems to have been in thrall to the romantic ideal of the Cossack as the perfect skirmisher and guerrilla horseman, harassing the enemy's lines of communication as they had done in the Napoleonic campaigns:

'The Cossacks, when they were purely irregulars in the Russian service, were wonderfully skilled in the art of luring on an enemy and they practised these manoeuvres with equal success upon the splendid cavalry of Napoleon and upon the Tartar horsemen of the steppes' (216–7).

'Moors, Tartars, and some Asiatics of the steppes fire from horseback, and the Cossacks adopted the plan of firing mounted in the days when their guerilla tactics made them so formidable' (361).

Given this Callwell found it hard to understand why Skobelev did not assign the cavalry a more active role in his instructions for the Turkmen campaign:

' ... it is only right to notice that the views expressed above as to the desirability of cavalry acting with great boldness in warfare of this nature had an opponent of undoubted authority in the person of General Skobelef. His instructions to his cavalry in the Turkoman campaign throughout breathed the spirit of caution. "As long as the enemy's cavalry is unshaken and is not in an unfavourable position, *e.g.*, with an obstacle in rear, in a hollow, &c., our cavalry must not enter on a combat with it. Pursuit of a retreating Turkoman cavalry is useless, as it only breaks up the tactical formations – our one strong point and sheet anchor."[54] Such were his orders, and they sound strange enough; for in a word General Skobelef taught his cavalry to be afraid of the Turkoman horse. It must, however, be remembered that the Russian cavalry operating beyond the Caspian was not well adapted for shock tactics, and that with mounted troops of a different class at his disposal the general might have held other views' (363).

In fact the reasons for this were clear enough. As Mikhail Terent'ev, author of the standard history of the Russian conquest, put it 'The local Asiatics do not rate Cossacks at all, if they are not accompanied by rockets or cannon; they regard the infantry with great respect. A platoon of Russian sharpshooters is far more alarming to them than three *sotnias* of Cossacks ... .'[55] This in turn was because the Cossacks came out of the same tradition of steppe raiding and nomadic warfare as the Turkmen and Kazakhs themselves (in Russian the word is the same – *Kazak*). As such they held few terrors for nomadic opponents, however formidable they might have been for regular European forces. Frequently in Central Asia Cossacks fought most effectively when dismounted, as at Iqan in 1864. Callwell cites a further incident which demonstrates both this and their ultimate reliance on infantry to secure a position, although he does not seem to draw either of these inferences from it:

The affair of 'Petrusvitch's Garden,' near Denghil-Tepe in 1880, is an admirable illustration of this sort of work and of its dangers. The enclosure was held by the Tekkes in some forces. At dawn the whole of the cavalry and some guns moved in this direction under General Petrusvitch in obedience to orders to that effect. When at about 180 yards from the enclosure the general ordered his men to dismount and attack, the horse-holders retiring some distance, while a mounted troop remained in reserve. The dismounted men cleared the enemy out with the bayonet (the cavalry had bayonets), but General Petrusvitch was mortally wounded at their head and there was some confusion in consequence. The Tekkes now issued out round the flank and threatened the horse-holders, but a portion of the reserve troops dismounted and assisted by the guns repulsed this offensive movement. Very severe fighting continued about the enclosure for some time, but the cavalry managed to hold their own till infantry reinforcements, which had been urgently asked for, hurried up and secured what had been won (369).[56]

Overall then Callwell had a tendency to exaggerate both the consistency and the success of Russian tactics, even when they did not accord with his own preconceptions about the continuing importance of cavalry in small wars.

On logistical questions Callwell noted at various points that small wars were as much about the conquest of nature as of indigenous enemies (38),[57] remarking that 'in the steppes Russian forces have similarly disappeared, victims of the enormous distances in which such a territory must be traversed to achieve a military object' (187). In Central Asia neither water transport nor railways (until the 1880s) were of any use to European armies. These might be equipped with the most modern breech-loading rifles and artillery, but were still wholly reliant on animals – principally camels – when it came to moving around.[58] While modern firearms provided an immediate tactical advantage, the consumption of ammunition demanded by modern weapons also worsened problems of supply: 'The Russian infantry during General Lomakin's disastrous attack on Denghil Tepe fired 246,000 rounds, or considerably over 100 rounds per man actually engaged' (347). This translated into a need for thousands of camels to sustain even a very small expeditionary force: 'When the Russians conquered Khiva in 1874 [sic], the column from Tashkend, consisting of 5,500 men, was accompanied by a supply column of 8,800 camels' (40).[59] Callwell also noted that Colonel Markozov's column from Krasnovodsk had been forced to turn back because of inadequate transport (27). He praised the use of four separate invading columns for the Khiva campaign as a means of maximising the chances of success, since it was unlikely that all of them would find their routes impracticable (90–91). In fact, this decision was partly dictated by logistical constraints (a single large column could not have moved over such barren country) and partly by the vanity of the Turkestan Governor-General, General Konstantin Petrovich von Kaufman (1818–1882), who was determined to have a column under his personal command set out from Tashkent, rather than leaving the glory of the expedition to the columns from Orenburg and the Caspian. In the event the Tashkent column very nearly foundered in the Qizil-Qum desert, and was only saved by the skill of its Kazakh guides.[60]

When it came to logistics for Callwell Skobelev was once again the exemplar, both in his insistence on the need for a small, flexible attacking force, and in the balance he was able to achieve between the fighting force and those protecting his communications in his final campaign against the Turkmen in 1880–1:

General Skobelef, when engaged upon his campaign against the Turkomans in 1880, which will be referred to later, was constantly in fear that the Russian government would take alarm at the slowness of his progress and would send reinforcements across the Caspian; weakness of force to him was under the circumstances a source of strength (41).

General Skobelef made great efforts to get every available man up to the front for his operations against Denghil Tepe in 1880, the detachments on the communications being reduced to the lowest possible limits consistent with the safety of the various posts; and yet while the army at the front mustered only about 8,000 men, the troops on the communications mounted up to about 4,500, or over one third of the whole force (97).

What Callwell did not recognise was that this army of over 12,000 men, while small in absolute terms, was in fact the largest single force used in any of the Central Asian campaigns. Skobelev also insisted on the construction of the first railway in Central Asia (from the Caspian shore to the oasis of Qizil-Arvat), anticipating Kitchener's use of railways in the Sudan by fifteen years. This was reflected in the cost of the campaign, at 11 million roubles more than double that of Lomakin's failed attack the previous year.[61] In other words Skobelev was not quite the master of economy of force that Callwell imagined – he was given more men and funds than would usually have been available for such an operation precisely because of the need to wipe out the memory of Lomakin's previous failure.

Conclusion

Callwell's work is refreshingly free from the hypocrisy that attended many British accounts of the Russian campaigns in Central Asia, which accused them of waging war with a peculiar brutality, in implied contrast to the more humane approach of British forces in Asia.[62] Skobelev had come in for particular censure for the massacre at Denghil-Tepe, and in particular for the admission in his report that Turkmen 'of both sexes' (*oboego pola*) had been cut down by his men when fleeing the fortress.[63] Skobelev was in fact a particularly brutal and sadistic commander, and was seen as such by many of his Russian contemporaries,[64] but his actions at Denghil-Tepe and elsewhere were certainly not of a different order of violence to the British campaigns during the Indian rebellion of 1857, on the North-West Frontier or at the Battle of Omdurman, and Callwell tacitly recognised this. For him, as for many other British Indian soldiers and officials, the British and the Russians were engaged in an entirely comparable and indeed complementary colonial enterprise in Asia – they were primarily colleagues rather than enemies, a fact which is often overlooked in the focus on 'Great Game' rivalry.[65] The Russian campaigns were a particularly successful example of colonial expansion through small wars which could be readily mined for tips, and Callwell cherry-picked suitable examples to support his overall thesis about the importance of decisiveness and the use of 'moral force' to intimidate a 'savage' enemy.

For this reason it would be a mistake to view *Small Wars* as 'a reliable portrayal of all colonial war around 1900' in a handily compressed and synthesised form.[66] Looking at the Central Asian example we see that Callwell omitted a good many cases which did not fit his thesis, such as the 1839 Khiva expedition, and misrepresented others such as the fall of Tashkent. This was probably not deliberate, but an unconscious result of the single-minded pursuit of his argument. As he did not know Russian all his information was relayed at second hand, and he did not make full use even of all the sources that would have been available to him in English. *Small Wars* is in the end the work of a professional soldier seeking to produce a manual, not of a military historian, and it needs to be under-stood as such. Its value as a historical document is what it reveals about the late Victorian military mentality – it provides fascinating evidence of the constant flow of information between the elites of European empires in the 19th century, which amounted to a common imperial repertoire of techniques of rule and power which went well beyond military tactics.[67] Callwell was firmly embedded in that world, and *Small Wars* drew on the full range of European colonial experience. If, as Wagner suggests, Callwell's work has been used to argue for a form of British exceptional-ism in the waging of colonial wars – a softer, 'hearts and minds' approach – then this is a misreading.[68] Callwell's evident approval, admiration and advocacy of the tactics employed by Russia's most notor-iously violent general, Skobelev, and his clear acceptance of complete comparability between British, French and Russian colonial campaigns suggests that he was not claiming the existence of or advocating a peculiarly *British* way of war. Instead *Small Wars* is about the common experience of warfare between the armies of 'civilized' states (of which

Russia's was one) and a range of more or less 'savage' opponents – conflicts in which the tactics, weapons and forms of violence used were different from those employed when these states fought each other.

Notes

1. Callwell, *Small Wars*, 1st ed. 1896, 2nd ed. 1899, 3rd ed. 1906 – references throughout are to the 1899 edition.
2. Strachan, "Colonial Warfare," 76.
3. Laqueur, "Origins of Guerrilla Doctrine," 361–3; and Porch, *Counterinsurgency*, 50–1.
4. Gates, "Small Wars," 381–2; Beckett "Another British Way in Warfare"; and Whittingham "Warrior-scholarship in the age of colonial warfare."
5. Betz, "Counter-insurgency, Victorian Style."
6. Whittingham, "Savage warfare."
7. Scheipers, *Unlawful Combatants*, 177–83.
8. Walter, *Colonial Violence*, 248–9.
9. Wagner "Savage Warfare"; Porch, *Counterinsurgency*, 50–7, 76.
10. Whittingham, *Charles Callwell*.
11. Whittingham, *Charles Callwell*; Moreman "Callwell, Sir Charles Edward (1859–1928)". Letters from Callwell can be found in the Jellicoe Papers Vol.XLIX BL Add. MS 49037 f.130 (1919); Campbell-Bannerman Papers Vol.XLVII BL Add. MS 41252 f.173 (121); Callwell to Sir Eyre Crow 22/08/1914 TNA FO 800/102/77 ff.246–248; Callwell's letters to Lord Kitchener in 1915–1916 are in TNA WO159 & PRO30/57.
12. Beckett, "Another British way in Warfare," 92.
13. Callwell, "Lessons to be Learnt."
14. Howard, "The British Way in Warfare"; Strachan, "Colonial Warfare"; Beckett, *Victorians at War*, 189; Beckett, "Another British way in Warfare"; Whittingham, *Charles Callwell*.
15. The first edition of *Small Wars* was translated into French in 1899, and Callwell drew upon some of the observations of his French translator when revising it for the second edition: Finch, *A Progressive Occupation?* 29.
16. The classic account of the Caucasus campaigns in English is Baddeley, *Russian Conquest of the Caucasus*. Published in 1908, the book cannot have been Callwell's source for his Caucasian examples, but Baddeley, the St Petersburg correspondent for the *London Standard*, also published articles on the subject which Callwell may have drawn upon.
17. Morrison, "Killing the Cotton Canard"; and Morrison, "Beyond the 'Great Game'".
18. Curzon, *Russia in Central Asia in 1889*, 86.
19. Marvin, *The Russian Advance Towards India*, 98–9.
20. Novikova, *Skobeleff and the Slavonic Cause*.
21. Maslov, "Rossiya v Srednei Azii"; Shakhovskoi, "Ekspeditsiya protiv Akhal-Tekintsev"; V.N.G.: "Ocherk Ekspeditsii v Akhal-Teke" Kuropatkin *Zavoevaniya Turkmenii*. See further Rogger "The Skobelev Phenomenon."
22. *Siege and Assault of Denghil-Tépé*; Dalton, "General Skobeleff's Instructions."
23. Grodekov, *Voina v Turkmenii* translated as Grodekoff, *The War in Turkumania* [*sic*].

24. Strachan, "Colonial Warfare," 81–2.
25. Shamil was an Avar from Daghestan, not a Cherkess (Circassian).
26. Scheipers, *Unlawful Combatants*, 226–7.
27. Strachan, "Colonial Warfare," 77.
28. von Hellwald, *The Russians in Central Asia*, 157–9.
29. Terent'ev, *Istoriya Zavoevaniya Srednei Azii* I, 345–9; Danish *Risala ya Mukhtasari*, 47–8; and Donish *Istoriya Mangitskoi Dinastii*, 47–8.
30. Romanovskii, *Zametki po Sredneaziatskomu voprosu*, 61 – translated as Romanovski *Notes on the Central Asiatic Question*, where the account of Irjar is on pp.lx – lxiv.
31. Callwell's source here is probably Kostenko 'Turkestan', 912, a partial translation of Kostenko *Turkestanskii Krai*.
32. Beisembiev, *The Life of 'Alimqul*, 64.
33. On this see Morrison, "The Turkestan Generals," 168–72.
34. There is a vivid account in the report of Lt Soltanovskii, who led one of the storming parties: 'Zapiska o deistviyakh vzvoda strelkovoi roty 7 Zapadno-Sibirskago bataliona pri shturme gor. Tashkenta 15, 16 i 17 iiunya 1865g' 22/06/1865 in Serebrennikov, *Turkestanskii Krai*, 215–9.
35. Hellwald, *The Russians in Central Asia*, 146–7. Much more complete accounts can be found in Romanovski, *Notes on the Central Asiatic Question*, 13; Anon, "Russian Advances in Central Asia," 408. Aberigh-Mackay *Notes on Western Turkistan*, 27; Schuyler, *Turkistan* I, 114–5. Callwell was clearly familiar with the latter text as he made use of it when writing of the 1875–6 campaign against Khoqand (see below).
36. In fact Georgian officers and men fought on the *Russian* side in the Caucasian campaigns.
37. See Potto, "Gibel otryada Rukina"; Terent'ev, *Istoriya Zavoevaniya* II, 62.
38. See Khodarkovsky, *Russia's Steppe Frontier*.
39. *Russian Military Expedition to Khiva*, which is a translation of Ivanin/Golosov "Pokhod v Khivu v 1839 godu."
40. On this see Morrison, "Twin Imperial Disasters."
41. Callwell's source here and elsewhere when referring to the wars against Khoqand in the 1850s is almost certainly Michell, *The Russians in Central Asia*, here 363–5.
42. Callwell's source is Schuyler *Turkistan*, II, 292.
43. Michell, "The Russian Expedition to the Alai and Pamir"; The standard account of the final Khoqand campaign in Russian was Serebrennikov, "K istorii kokanskogo pokhoda."
44. Marvin, *The Eye Witnesses' Account of the Disastrous Russian Campaign*; Callwell also seems to have used Delmar Morgan, "The Tekkeh Expedition of 1879."
45. Terent'ev, *Istoriya Zavoevaniya* III, 14; and O'Donovan *The Merv Oasis* I, 136–7.
46. Marvin, *The Eye Witnesses' Account*, 77.
47. Shakhovskoi "Ekspeditsiya protiv Akhal-Tekintsev, 171.
48. Belich, *The New Zealand Wars*, 311–21.
49. Kostenko, "Turkestan."
50. Strachan, "Colonial Warfare," 85–6.
51. Dalton, "General Skobeleff's Instructions," 714–5.
52. Marvin, *The Disastrous Russian Campaign*, 253–5; "Zhurnal zanyatii i voennykh deistvii Akhal-Tekinskogo ekspeditsionnogo otryada" 01/09/1879 Russian State Military-Historical Archive F.1300 Op.1 D.80 I.137*ob*; this also accords almost

exactly with the account given by Demurov 'Boi s tekintsami pri Denghil-Tepe' no.3, 620.

53. Callwell's source is Dalton, "General Skobeleff's Instructions," 714–5.
54. A slight paraphrase of Dalton, "General Skobeleff's Instructions," 716.
55. Terent'ev, *Istoriya Zavoevaniya*,II, 177.
56. Callwell takes this episode from Grodekoff *The War in Turkumania* III, 661–2.
57. Beckett, *The Victorians at War*, 4.
58. Morrison, "Camels and Colonial Armies."
59. Khiva actually fell in 1873. Callwell's source here is Trench, "The Russian Campaign against Khiva"; Trench in turn was relying on accounts in *Russkii Invalid* and on the articles about the campaign written for the *Norddeutsche Allgemeine Zeitung* by Lt. Hugo Stumm of the Prussian army, subsequently published as Stumm *Der russische Feldzug nach Chiwa* & translated as Stumm *The Russian Campaign against Khiva*.
60. Terent'ev, *Istoriya Zavoevaniya*, II, 169–182.
61. Heiden to Miliutin 06/02/1881 in Il'yasov (ed.) *Prisoedinenie Turkmenii k Rossii*, 484.
62. See for instance Curzon *Russia in Central Asia in 1889*, 84–6.
63. Skobelev, "Osada i Shturm kreposti Dengil-Tepe (Geok-Tepe)", 52 translated as *Siege and Assault of Denghil-Tépé*; the phrase about 'both sexes' here occurs on p.54.
64. Nalivkin, "Moi Vospominaniya o Skobeleve", 535–8; and Campbell, "Violent acculturation."
65. Campbell, "Our friendly rivals".
66. Walter, *Colonial Violence*, 249; and Gates, "Small Wars," 381–2.
67. See Mackenzie, "European Imperialism", and other essays in the same volume.
68. Wagner, 'Savage Warfare', 218–220.

Acknowledgements

My thanks to Mark Lawrence for his helpful comments and to Daniel Whittingham for sharing the manuscript of his forthcoming book.

Disclosure statement

No potential conflict of interest was reported by the author.

Bibliography

Aberigh-Mackay, G. A. *Notes on Western Turkistan*. Calcutta: Thacker, Spink & Co, 1875.

Anon. "Russian Advances in Central Asia." *Quarterly Review* 136 (April 1874): 395–433.

Baddeley, J. F. *The Russian Conquest of the Caucasus*. London: Longmans, 1908.

Beckett, Ian. *The Victorians at War*. London & Hambledon: Hambledon Continuum, 2003.

Beckett, Ian F. W. "Another British Way in Warfare: Charles Callwell and Small Wars." In *Victorians at War: New Perspectives*, edited by Ian F. W. Beckett, 89–102. Warminster: Society for Army Historical Research, 2007.

Beisembiev, T. K., ed. & trans. *The Life of 'alimqul. A Native Chronicle of Nineteenth Century Central Asia*. London: Routledge, 2003.

Belich, James. *The New Zealand Wars and the Victorian Interpretation of Racial Conflict*. Auckland: Penguin, 1988.

Betz, David. "Counter-insurgency, Victorian Style." *Survival* 54, no. 4 (2012): 161–182. doi:10.1080/00396338.2012.709395.

Callwell, C. E. "Lessons to Be Learnt from the Campaigns in Which British Forces Have Been Employed since the Year 1865." *RUSI Journal* 31, no. 139 (1887): 357–412.

Callwell, C. E. *Small Wars, Their Principle and Practice*. London: HMSO, 1899.

Campbell, Ian. "'Our Friendly Rivals': Rethinking the Great Game in Ya'qub Beg's Kashgaria, 1867–77.'." *Central Asian Survey* 33, no. 2 (2014): 199–214. doi:10.1080/02634937.2014.915613.

Campbell, Ian. "Violent Acculturation. Alexei Kuropatkin, the Central Asian Revolt, and the Long Shadow of Conquest." In *The Central Asian Revolt of 1916. Rethinking the History of a Collapsing Empire in the Age of War and Revolution*, edited by Aminat Chokobaeva, Alexander Morrison, and Cloé Drieu, 191–208. Manchester: Manchester University Press, 2019.

Curzon, George Nathaniel. *Russia in Central Asia in 1889 and the Anglo-Russian Question*. London: Longmans, Green & Co., 1889.

Dalton, Captain J. C. "General Skobeleff's Instructions for the Reconnaissance and Battle of Geok-Tepe on the 17th July and 30th December, 1880." *RUSI Journal* 25, no. 112 (1881): 712–716.

Danish, Ahmad. *Risala Ya Mukhtasari Az Ta'rikh-i Saltanat-i Khanadan-i Mangitiya*, edited by, Abulghani Mirzoev. Stalinabad: Nashriyat-i Daulati-i Tajikistan, 1960. trans. I. A. Nadzhafova as *Istoriya Mangitskoi Dinastii*. Dushanbe: Donish, 1967.

Demurov, G. "Boi S Tekintsami Pri Denghil-Tepe, 28 Avgusta 1879 Goda" [The Fight with the Turkmen at Denghil-Tepe, 28th August 1879." *Istoricheskii Vestnik* 3 (1881): 617–622.

Finch, Michael. *A Progressive Occupation? the Gallieni-Lyautey Method and Colonial Pacification in Tonkin and Madagascar, 1885-1900*. Oxford: Oxford UP, 2013.

Gates, John H. "Small Wars, Their Principles and Practice. By C.E. Callwell. Review." *Journal of Military History* 61, no. 2 (1997): 381–382. doi:10.2307/2953992.

Grodekoff, N. *The War in Turkumania [sic] Skobeleff's Campaign of 1880-81*. Vol. 4. Simla: Government Central Branch Press, 1884–1885.

Grodekov, N. I. *Voina V Turkmenii: Pokhod Skobeleva V 1880-1881gg* [The War in Turkmenia. Skobelev's Campaign in 1880-81] 4 Vols. St Petersburg: Tip. V. A. Balasheva, 1883-4. translated by J. M. Grierson as Major-General

N. Grodekoff *The War in Turkumania* [sic] *Skobeleff's Campaign of 1880-81* 4 Vols. Simla: Government Central Branch Press, 1884-5.

Howard, Michael. "The British Way in Warfare: A Reappraisal." In *The Causes of Wars and Other Essays*, edited by M. Howard, 169–187. London: Temple Smith, 1983.

Il'yasov, A., ed. *Prisoedinenie Turkmenii K Rossii. Sbornik Arkhivnykh Dokumentov* [The Uniting of Turkmenia to Russia. Collection of Archival Documents]. Ashkhabad: Izd. AN Turkmenskoi SSR, 1960.

Ivanin, M., and D. Golosov. "Pokhod V Khivu V 1839 Godu Otryada Russkikh Voisk, Pod Nachal'stvom General-Ad'yutanta Perovskago." [The Expedition to Khiva in 1839 of a Force of Russian Troops, under the Command of General-Adjutant Perovskii] *Voennyi Sbornik*29, no. 1 (1863): 3–72. No.2: 309–58; No.3: 3–71. Translated as *A Narrative of the Russian Military Expedition to Khiva under General Perofski, in 1839. Translated from the Russian for the Foreign Department of the Government of India*. Calcutta: Office of the Superintendent of Government Printing, 1867.

Khodarkovsky, Michael. *Russia's Steppe Frontier. The Making of a Colonial Empire 1500 – 1800*. Bloomington, IN: Indiana University Press, 2002.

Kostenko, L. F. *Turkestanskii Krai. Opyt Voenno-Statisticheskogo Obozreniya Turkestanskogo Voennogo Okruga* [The Turkestan Region. Results of the Military-Statistical Survey of the Turkestan Military District]. 3 Vols. St Petersburg: Tip. A. Transhelya, 1880. Partially translated by Major F. C. H. Clarke as L. T. [sic] Kostenko "Turkestan", *RUSI Journal* 24, no.108 (1881): 898–921

Kuropatkin, A. N. *Zavoevaniya Turkmenii (pokhod V Akhal-Teke V 1880 – 1881gg) S Ocherkom Voennykh Deistvii V Srednei Azii S 1839 Po 1876g* [The Conquest of Turkmenia (the Expedition to Akhal-Teke in 1880-1881) with an Account of Military Actions in Central Asia from 1839 to 1876]. St Petersburg: V. Berezovskii, 1899.

Laqueur, Walter. "The Origins of Guerrilla Doctrine." *Journal of Contemporary History* 10, no. 3 (1975): 341–382. doi:10.1177/002200947501000301.

Mackenzie, John M. "European Imperialism: A Zone of Co-operation Rather than Competition?" In *Imperial Co-operation and Transfer 1870 – 1930. Empires and Encounters*, edited by Volker Barth and Roland Cvetkovski, 35–56. London: Bloomsbury, 2015.

Marvin, Charles. *The Eye Witnesses' Account of the Disastrous Russian Campaign against the Akhal Tekke Turcomans*. London: W. H. Allen & Co., 1880.

Marvin, Charles. *The Russian Advance Towards India. Conversations with Skobeleff, Ignatieff, and Other Distinguished Russian Generals and Statesmen, on the Central Asian Question*. London: W. H. Allen & Co, 1882.

Maslov, A. "Rossiya V Srednei Azii. Ocherk Nashikh Noveishikh Priobretenii" [russia in Central Asia. An Account of Our New Acquisitions]." *Istoricheskii Vestnik* 5 (1885): 372–423.

Michell, John Robert. *The Russians in Central Asia. Their Occupation of the Kirghiz Steppe and the Line of the Syr-Daria. Their Political Relations with Khiva, Bokhara, and Kokan*. London: Edward Stanford, 1865.

Michell, Robert. "The Russian Expedition to the Alai and Pamir." *Journal of the Royal Geographical Society of London* 47 (1877): 17–47. doi:10.2307/1798737.

Moreman, T. R. "Callwell, Sir Charles Edward (1859–1928), Army Officer and Writer" (2008) *Oxford Dictionary of National Biography*. Accessed January 7, 2019 http://www.oxforddnb.com/view/10.1093/ref:odnb/9780198614128.001.0001/odnb-9780198614128-e-32251

Morgan, E. Delmar. "The Tekkeh Expedition of 1879." *RUSI Journal* 23, no. 103 (1879): 985-1010.

Morrison, Alexander. "Introduction: Killing the Cotton Canard and Getting Rid of the Great Game: Rewriting the Russian Conquest of Central Asia, 1814–1895." *Central Asian Survey* 33, no. 2 (2014): 131–142. doi:10.1080/02634937.2014.915614.

Morrison, Alexander. "Camels and Colonial Armies: The Logistics of Warfare in Central Asia in the Early 19th Century." *Journal of the Economic and Social History of the Orient* 57, no. 4 (2014): 443–485. doi:10.1163/15685209-12341355.

Morrison, Alexander. "Twin Imperial Disasters. The Invasions of Khiva and Afghanistan in the Russian and British Official Mind, 1839–1842." *Modern Asian Studies* 48, no. 1 (2014): 253–300. doi:10.1017/S0026749X13000036.

Morrison, Alexander. "Beyond the 'great game': The Russian Origins of the Second Anglo–Afghan War." *Modern Asian Studies* 51, no. 3 (2017): 686–735. doi:10.1017/S0026749X1500044X.

Morrison, Alexander. "The Turkestan Generals and Russian Military History." *War in History* 26, no. 2 (2019): 153–184. doi:10.1177/0968344517723373.

Nalivkin, V. P. "Moi Vospominaniya o Skobeleve" [My Reminiscences about Skobelev] *Russkii Turkestan* (1906) No.119, reprinted in *Polveka v Turkestane. V. P. Nalivkin: biografiya, dokumenty, trudy* ed. T. V. Kotiukova, 535–538. Moscow: Izd. Mardzhani, 2015.

Novikova, O. A. *Skobeleff and the Slavonic Cause*. London: Longmans, Green & Co, 1883.

O'Donovan, Edmund. *The Merv Oasis Travels and Adventures East of the Caspian, 1879-80-81*. Vol. 2. London: Smith, Elder & Co, 1882.

Porch, Douglas. *Counterinsurgency. Exposing the Myths of the New Way of War*. Cambridge: Cambridge UP, 2013.

Potto, V. A. "Gibel Otryada Rukina V 1870 Godu." [the Doom of Rukin's Force in 1870]." *Istoricheskii Vestnik* 7 (1900): 110–135.

Rogger, Hans. "The Skobelev Phenomenon: The Hero and His Worship." *Oxford Slavonic Papers* 9 (1976): 46–78.

Romanovskii, D. I. *Zametki Po Sredneaziatskomu Voprosu* [Notes on the Central Asian Question]. St Petersburg: Tip. Sobst. E.I.V. Kants, 1868. translated by Robert Michell as D. Romanovski *Notes on the Central Asiatic Question* Calcutta: Office of the Superintendent of Govt. Printing, 1870.

Scheipers, Sibylle. *Unlawful Combatants: A Genealogy of the Irregular Fighter*. Oxford: Oxford University Press, 2015.

Schuyler, Eugene. *Turkistan. Notes of a Journey in Russian Turkistan, Khokand, Bukhara and Kuldja*. Vol. 2. London: Sampson Low, Marston, Searle & Rivington, 1876.

Serebrennikov, A. G. "K Istorii Kokanskogo Pokhoda." [Towards a History of the Khoqand Campaign] *Voennyi Sbornik* 9, (1897): 5–28; (1899) no.4: 211–226; (1901) no.4: 29–55; nos.7–11: 29–55, 69–96, 37–74.

Serebrennikov, A. G., ed. *Turkestanskii Kral. Sbornik Materialov Dlya Istorii Ego Zavoevaniya. 1865g*. Chast I. [The Turkestan Region. A Collection of Documents for the History of Its Conquest. 1865 Part I]. Tashkent: Tip. Shtaba Turkestanskogo Voennogo Okruga, 1914.

Shakhovskoi, V. "Ekspeditsiya Protiv Akhal-Tekintsev V 1879 – 1880 – 1881gg. Posvyashchaetsya Pamyati M. D. Skobeleva." [The Expedition against the Akhal-Tekke in 1879 – 1880 – 1881. Dedicated to the Memory of M. D. Skobelev] *Russkaya Starina* XLVI, no. 4 (1885): 161–286. No.5: 377–410; No.6: 531–558

Skobelev, M. D. "Osada I Shturm Kreposti Dengil-Tepe (geok-tepe)." [Siege and Storming of the Fortress of Denghil-Tepe (gok-tepe)] *Voennyi Sbornik* 4, (1881): 3–64 translated by Lt. J. J. Leverson as *Siege and Assault of Denghil-Tépé. General Skobeleff's Report*. London: HMSO, 1881.

Strachan, Hew. "Colonial Warfare, and Its Contribution to the Art of War in Europe." In *European Armies and the Conduct of War*, edited by H. Strachan, 75–88. London: Unwin Hyman, 1983.

Stumm, Hugo. *Der Russische Feldzug Nach Chiwa*. Berlin: Mittler, 1875. translated by F. Henvey & P. Mosa as Hugo Stumm *The Russian Campaign against Khiva in 1873*. Calcutta: Foreign Dept. Press, 1876.

Terent'ev, M. A. *Istoriya Zavoevaniya Srednei Azi* [History of the Conquest of Central Asia]. Vol. 3. St Petersburg: Tip. A. Komarov, 1906.

Trench, F. Chenevix. "The Russian Campaign against Khiva, in 1873." *RUSI Journal* 18, no. 77 (1874): 212–226.

V.N.G. "Ocherk Ekspeditsii V Akhal-Teke 1879-1880." [Account of the Expedition to Akhal-Teke in 1879-1880] *Voennyi Sbornik* 5, (1888): 204–230. no.6: 396–426.

von Hellwald, Frederick. *The Russians in Central Asia. A Critical Examination down to the Present Time of the Geography and History of Central Asia*. trans Lt.-Col. Theodore Wirgman. London: Henry S. King & Co., 1874.

Wagner, Kim A. "Savage Warfare: Violence and the Rule of Colonial Difference in Early British Counterinsurgency." *History Workshop Journal* 85, no. 1 (2018): 217–237. doi:10.1093/hwj/dbx053.

Walter, Dierk. *Colonial Violence. European Empires and the Use of Force*. London: Hurst, 2017.

Whittingham, Daniel. "'savage Warfare': C.E. Callwell, the Roots of Counter-insurgency, and the Nineteenth Century Context." *Small Wars & Insurgencies* 23, no. 4–5 (2012): 591–607. doi:10.1080/09592318.2012.709769.

Whittingham, Daniel. "Warrior-scholarship in the Age of Colonial WarfarE. Charles E. Callwell and Small Wars." In *The Theory and Practice of Irregular Warfare: Warrior-scholarship in Counter-insurgency Ed*, edited by Andrew Mumford and Bruno Reis, 18–34. London: Routledge, 2013.

Whittingham, Daniel. *Charles Callwell and the British Way in Warfare*. Cambridge: Cambridge University Press, Forthcoming.

General Zuo's counter-insurgency doctrine

Kenneth M. Swope

ABSTRACT

This paper examines the career of one of the most notable of Chinese state officials in the mid-nineteenth century, General Zuo Zongtang (1812–1885) at a time when the Qing Empire was beset by enemies on all sides along with the massive Taiping Rebellion (1851–65). The paper explores how Zuo created a coherent and flexible counter-insurgency doctrine that underpinned his suppression of the Taipings, the Nian, Muslim rebels of the northwest and the defeat of a massive Sufi revolt in Central Asia. The paper also shows how Zuo helped convince the Russians to return territory they occupied in Central Asia, thereby preserving the territorial integrity of the Qing Empire in the region.

Imagine if at the height of the Civil War, the United States was simultaneously confronted with massive, semi-linked Indian uprisings throughout the country, religious motivated uprisings in the west, territorial encroachment by Canada, an attack on the capital by the combined forces of Great Britain and France, and widespread mounted banditry across the Midwest? Such a state of affairs is essentially what confronted the Qing Dynasty (1644–1911) in China in the mid-nineteenth century. While the Qing would eventually crumble before revolutionary nationalist forces on the eve of World War I, it amazingly not only weathered the storms outlined above, but managed to initiate significant governmental, technological and military reforms, and even extended and consolidated its political control over much of Central Asia, an achievement whose repercussions are still being manifested today by virtue of Muslim separatist movements in Xinjiang province in the People's Republic of China. This was largely due to the efforts of a few prominent officials and statesmen, one of the most important of whom was General Zuo Zongtang (1812–1885).[1]

Map 1: Map of the Qing Empire, ca. 1820

Known to Americans as the namesake of the ubiquitous General Zuo's (or General Tso's or General Cao's) Chicken dish one finds in Chinese restaurants in the United States, Zuo Zongtang's life and career offers a fascinating prism through which one can view the tumultuous nineteenth century in China. Zuo witnessed China's humiliating defeat in the Opium War (1839–42) and by virtue of personal connections and his own talents entered government service shortly thereafter. Zuo never attained the highest civil service degree and possessed an admittedly prickly personality, upon entering government service when he was already middle aged. Zuo quickly rose to positions of influence in his native Hunan province and was a key figure in government campaigns to suppress the Taiping rebels, whose challenge to the Qing has gone down as the single bloodiest civil war in human history. After the final suppression of the Taipings, recognizing the significance of naval power for modern states, Zuo embarked upon the construction of China's first modern shipyard, known as the Fuzhou Naval Yard.

But before he could make much headway on this project Zuo was recalled to military service and tasked with the suppression of massive Muslim revolts in northwest China that had dragged on for decades and spread into provinces near China's interior. Yet before he could take up his post there, Zuo was redirected to assist in suppressing the Nian bandits, who had ravaged northeast and central China since 1851, sometimes in conjunction with the Taipings or the Muslims. Zuo managed to stamp out the Nian by the end of 1868 and turned his attention towards the Muslims. Over the next several years Zuo alternated between fighting the rebels and rehabilitating the people and devastated lands of the northwest. Among other things he established cotton and wool mills and prohibited opium use amongst his soldiers while redistributing land and resources. As the rebel remnants fled into Central Asia and joined forces with a local leader there named Yakub Beg, Zuo pressed the court for permission and resources to complete the job. After much debate that included the negotiation of foreign loans to purchase grain for his troops, Zuo launched a campaign that proved spectacularly successful owing to the meticulous planning of its commander.

Zuo's success proved critical in enabling the Qing to press Russia for the return of lands they had recently occupied in Central Asia, ostensibly because the Russians feared disorder along their common frontier. Together with territory recently retaken from Yakub Beg, these lands were formally reincorporated into the Qing Empire as the province of Xinjiang. This was critical in restoring a measure of pride and prestige to the beleaguered Qing by demonstrating that it could still compete in the scramble for empire being waged on a global scale. Near the conclusion of negotiations with Russia, Zuo was summoned to Beijing to serve as an adviser to the throne. He would later be placed in charge of all the military affairs of the empire and commander of Fujian province in the southeast when war with the French broke out in 1884. Zuo would die in 1885 shortly after the war ended.

Despite his many achievements, and despite the plethora of surviving primary sources pertaining to his activities, Zuo has received relatively little attention in the English language scholarship of the period.[2] The only full-length English language biography of Zuo was written in 1937 by an American naval officer serving in China at the time.[3] There are more recent studies that examine certain aspects of Zuo's career, but none looks at his whole life or treats it within the full context of nineteenth century imperialism in China. Zuo's life and experiences offer a perfect lens through which readers can experience the trials and tribulations of China in the nineteenth century. Against a general backdrop of failure and defeat, Zuo played a pivotal role in the late Qing's most noteworthy successes. While being an ardent Chinese patriot, he recognized the value of Western ideas and inventions and sought to incorporate them into his own campaigns and projects, albeit on his own terms. While it ultimately failed, his Fuzhou Navy Yard turned out some of China's most impressive modern ships and helped the Qing build one of the largest navies in the world in just a couple decades.

Zuo's conquest of Xinjiang highlighted the tensions between Han and Muslim groups within the empire even as he created innovative solutions towards easing these tensions. Most germane to the subject at hand, his methods of counterinsurgent warfare were way ahead of his time and reveal both a keen mind and a man who was sincerely concerned with the plight of the common folk, a quality absent from many of his more literarily inclined colleagues. In the remainder of this chapter I will trace the origin of Zuo's ideas and discuss their implementation with respect to the suppression of various rebel groups, most notably the Muslims of China's northwest, as that is where Zuo had the greatest freedom to put his ideas into action, unfettered by meddling fellow officials. It was also where he was able to translate his book knowledge into concrete policies for the suppression of rebellion and the improvement of living standards for local populations. Nonetheless, as evidenced by the current unrest in Xinjiang, Zuo's policies did not end the tension between Han Chinese and Muslim minorities in the region and in fact, the contemporary strife can be considered part of Zuo's legacy so far as it was due to his efforts that Xinjiang was formally incorporated as a province of the empire.

Zuo was born in the village of Xiangyin, in Hunan province in south-central China. His father had been a scholar and teacher and his grandfather had also had scholarly leanings, though neither had held official posts. Zuo was an adept student and passed the lowest level civil service examination, for the *shengyuan* degree at the age of fourteen, subsequently passing the second examination, for the *juren* (recommended man) degree in 1832 at the age of twenty. But he was never able to pass the highest examination to earn the coveted *jinshi* (presented scholar) degree. Nevertheless, he was a dedicated scholar of not only the Confucian classics, but also more practical and (for a traditional scholar) unusual topics such as strategic geography, military arts, and agriculture.[4] These interests would serve him well in his later career as Zuo became familiar with contemporary issues and even got to meet some of China's most famous officials. Following their lead, he would seek to put his ideas about practical statecraft into practice.[5]

Zuo's parents died when he was still fairly young and he was in straitened financial circumstances, forcing him to take a variety of low-paying teaching jobs to make ends meet as he had married shortly after passing the *juren* exam. This impacted him as well and he subsequently stressed the values of frugality and simple living upon his children. He also emphasized the importance of setting an example for the common people and even with his own burdens. Zuo's family regularly contributed to famine relief and other local efforts to aid the populace in times of hardship and disorder, among other things establishing mutual aid societies for growing and sowing grain.[6] He also cultivated some of his own crops, again putting his book knowledge to practical use. Meanwhile, he took advantage of the generosity of friends to plumb their libraries for classical and modern texts to expand his educational horizons. It was when lecturing in the Lujiang Academy in Hunan that he made the acquaintance of Tao Zhu, then a powerful official serving as Viceroy of Liang-Jiang. Tao was so impressed by Zuo's acumen that he made him tutor to his son.[7] It was during the late 1830s that Zuo made use of his position and access to good libraries to study many works of statecraft and military strategy, particularly works pertaining to Northwest China and Central Asia. Zuo would serve as a teacher to the Tao family for over eight years, even after the death of Tao Zhu himself in 1839. He would also be forced to hide out in the nearby mountains on multiple occasions between 1840 and 1850 owing to natural disasters, banditry and secret society activity in the region.

This period encompassed China's shocking and humiliating defeat in the first Opium War (1839–42) and further galvanized Zuo's desire to put his knowledge to use.[8] He compiled at least one military manual in this period and in various letters to his friends he talked of his plans for renovating China's defenses. He called for an integrated system of defenses involving the construction of new ships, new training methods, better cannons, improved bullets and other technological advances. In particular he was impressed by the strength of Western firearms, and saw those as key to Britain's victory. Yet despite his appreciation for their firepower, the war caused Zuo to despise the English for the rest of his life and he never employed them, though he would work with other Europeans later in his career, most notably French, Russians, and Germans.[9] He also realized that the Qing court's longtime policy of simply overawing potential foes was misplaced. Finally, he realized that he might have the chance to realize his life's aspirations without earning the final civil service degree.[10] But all this might not have happened had Zuo not also met an official named Hu Linyi at the time. Hu was then serving in Guizhou province in the southwest and he and Zuo regularly communicated regarding military events and issues of strategy and tactics. In fact, Hu had wanted Zuo to join his staff, but Zuo was reluctant at the time to leave his family, still hiding out in the mountains.[11] Hu was well connected and his recommendation would launch Zuo upon his official career in 1852.

Zuo's first position was as a secretary to Zhang Liangji, who was then serving as the provincial governor of Hunan, threatened by the outbreak and rapid expansion of the Taiping Rebellion, which now had over 50,000 followers in Hunan alone.[12] Knowing of Zuo's reputation, Zhang put him in charge of all military affairs, a testament to the esteem in which Zuo was held despite his lack of formal government experience.[13]

Though they had been driven out of their previous base in Guangxi province, the Taipings advanced rapidly through Hunan, approaching the city of Changsha by the seventh month of 1852. Zhang and Zuo went to the city and navigated through the siege to enter it in early October and assume control over its defense. One key problem facing the Qing at this time was that there were simply too many officials and military commanders with hazy jurisdictions and overlapping responsibilities. Nonetheless, Changsha withstood the Taiping siege for over two months and a combination of land and river operations managed to lift the siege. In the wake of this experience Zuo saw the wisdom of creating new smaller, mobile units, identified as militia (*tuanlian*) in the sources.[14] He also created a new system of titles and rewards and emphasized the importance of personal relationships between commanders and their troops, something which had always been regarded with suspicion in China. In fact, Zuo and other successful Qing commanders were drawing upon doctrines pioneered by the famous late Ming general, Qi Jiguang (1528–88), in this respect.[15]

Shortly thereafter, in January of 1853, the Taipings attacked the important commercial city of Wuchang, northeast of Changsha at the confluence of the Han and Yangzi Rivers. In under a month, however, the efforts of Zuo and his superior had dislodged the rebels, sending them east, and enhancing Zuo's status and reputation. Unfortunately, Zhang was transferred a few months later to Shandong province in the northeast and though he wanted to take Zuo with him, Zuo decided not to go, fearing for the safety of his family in war-torn Hunan and he returned to his home in Xiangyin. Meanwhile, the Taipings occupied the great city of Nanjing, further down the Yangzi River, making it their 'Heavenly Capital'. They would hold the city for over a decade. From Nanjing, the Taipings sent huge armies out in all directions, threatening Beijing, as well as areas to the south and west. This prompted the new governor of Hunan, Luo Bingzhang, to convince Zuo to return to service in his native province.[16]

Zuo's first major assignment was to head a gun manufacturing bureau. There he invented the *dishanpao* (mountain-splitting cannon), which was used on sampans for riverine warfare.[17] Around this time Zuo's fascination with steamers and naval vessels increased, particularly in light of the environment in which he was fighting wherein riverine and lake warfare played crucial roles. He also recalled the spectacular success of the British gunboat *Nemesis* during the recent Opium War and thought it was critical for China to acquire and deploy similar technology. He would make increasing use of steamers in his subsequent campaigns against the Taipings, even if early efforts to manufacture their own steamers from scratch were less than satisfactory.[18]

When the Qing commander Zeng Guofan (1811–1872) retook Wuchang from the Taipings again in late 1854 (they had held it for almost 4 months this time), he selected Zuo Zongtang to be in charge of coordinating all armies and supplies in Hunan, raising him to the fourth grade of officialdom (out of 9) in the Ministry of War.[19] As part of this duty assignment Zuo centralized the salt monopoly and other local revenues pertaining to commerce and transit taxes so as to ensure a steady flow of revenue without imposing undue burdens upon an already sorely pressed populace. Indeed, probably because of his own somewhat impoverished background, Zuo was always acutely sensitive to preserving the livelihood of the people under his purview.

Not unlike his later fellow provincial, Mao Zedong, Zuo further held that the army should serve the people, as rehabilitation was essential to long-term security. To this end he had his troops build and repair canals, streets, and farms, both for their own use and for the locals.[20] Zuo's efforts in this respect were well appreciated, for the defense of Hunan, an important strategic and commercial region, was deemed critical to the survival of the empire as a whole. As one of his biographers put it 'The Empire could not survive a day without Hunan and Hunan could not survive a day without Zuo Zongtang'.[21] He would show the same concern for the masses in later operations in northwest China.

Because he never had a large number of troops and their strength was being dissipated after fighting so many engagements, Zuo temporarily advocated assuming a more defensive position and relying on firearms. These units would be supplemented by mobile units that could assist in sieges and participate in search and destroy operations. He called this his 'from guest to master' strategy.[22] Zuo proved especially adept at securing military supplies from disparate sources, manipulating the tax systems and calling in favors from friends and allies.[23] Where possible, Zuo even lowered taxes on the peasants, diverting tribute grain and other exactions for military use and attempting to curb the skimming that so many officials regarded as their right. Among other things, modern scholars credit him with curbing landlord abuses and assuring that tax burdens were properly assessed in areas under his jurisdiction.[24] Zuo also sought to improve and standardize the pay for his units, hoping to thereby decrease the possibility that the troops would rob and plunder the very people they were supposed to be protecting. This included providing extra cash for maintaining firearms.[25]

Zuo served under Luo Bingzhang and Zeng Guofan for the next five years in Hunan and the power of the Taiping rebels waxed and waned. He appears to have discharged his duties well, but his prickly personality and bluntness got him into trouble and he was impeached on charges of corruption and insubordination, though he was cleared of all of them and hoped to take the highest civil service examination one more time.[26] However, before he got to Beijing he was summoned to Zeng Guofan's headquarters in Anhui province and charged with raising a new volunteer force of 5000 to assist in stemming a resurgent Taiping tide in early 1860. This became known as the Chu Army.[27] Zuo assembled his army at Changsha in the summer of 1860 and they went east to Jiangxi in September. As other commanders assumed defensive positions to keep the Taipings from breaking west, Zuo led his small army through Jiangxi, capturing several important cities. In the process Zuo also helped relieve Taiping pressure on the city of Qimen, which was a gateway to the west, and fought major battles of Leping and Lake Poyang. This was despite the fact that the number of troops under his direct control rarely exceeded 8000 and he was constantly requesting reinforcements from neighboring provinces, both for combat operations and to hold captured territory. As a result of his actions the strategic position of the central Taiping armies was disintegrating and Zuo was promoted Commander-in-Chief of government forces in Zhejiang province.[28] He would be made provincial governor there a month later, when the important city of Hangzhou, located near the coast, fell to Taiping forces. Nearly all of Zhejiang was in rebel hands.

Zuo was wary of being flanked or encircled and having his supply lines cut. So he decided to spread his forces out and both threaten the rebels and hold key points so as to avoid providing them with a static line of defense to attack and ceding the initiative to the Taipings.[29] Within a year he had pacified all of eastern Zhejiang. As various Qing commanders clamored for the right to claim victory and recapture Hangzhou, Zuo cautioned that the city was still heavily fortified and there were other rebel positions that could still threaten the Qing. So he favored starving the Taipings out in Hangzhou and picking off stragglers and remnants elsewhere. Zuo sent his trusted subordinates to root the rebels out of eastern Zhejiang, directing matters from his headquarters in Yanzhou, in the centre of the province. Their progress may have seemed a bit slow to the Qing court, but it symbolized Zuo's thoroughness and desire to get the job done right.[30] The Qing were steadily advancing upon Hangzhou while preventing the Taipings from launching major westward or southern campaigns, essentially Zuo's 'march to the sea'. In the process Zuo impeached and dismissed incompetent officials and military commanders, promoting others for their achievements.[31] In the spring of 1862 he was made Governor-General of Zhejiang and Fujian provinces.

Zuo's forces were stalled at the city of Fuyang, southwest of Hangzhou, by a combination of stout resistance and a sickness that spread through the troops. Zuo brought in more troops from his older Zhejiang forces and recruited some foreign mercenaries to bolster his strength.[32] Fuyang would fall in the late summer of 1862 to a combined land-water attack. Owing to heavy fighting across the front as the desperate Taiping forces sought to save their capital at Nanjing, the siege of Hangzhou did not begin until autumn of 1863. Hangzhou was attacked via sapping, cannonfire, and efforts to induce its defenders to surrender. All bore fruit in their own ways, and the city would be captured in a rainstorm on 1 April 1864.[33] It was reported that allied European troops started looting upon entering the city, but when Zuo personally arrived he prohibited rape, looting, and pillaging and prescribed the death sentence for violators, though his orders were binding only to those troops under his direct command. Upon reporting the victory Zuo was promoted to Junior Guardian of the Heir Apparent and awarded a golden riding jacket by the emperor.[34] In the wake of this triumph Zuo reopened markets and remitted taxes to one third of their normal quotas for Hangzhou and its environs so as to succor the people.

Shortly thereafter, around the beginning of June, the Taiping leader Hong Xiuquan, committed suicide by taking poison, as the 'Heavenly Capital' of

Nanjing finally fell to the Qing after a siege that had lasted intermittently for over a decade. But the Taiping armies were still formidable and they broke out in all directions. Though Zuo quickly secured control of Zhejiang (and was promoted to Earl of Kejing for his efforts), some rebel remnants, estimated at as many as 200,000 strong, escaped south into Fujian province.[35] So Zuo was again appointed Governor-General of Fujian, Guangdong, and Jiangxi provinces and he took up the pursuit of the rebels, all the while trying to rebuild devastated areas and curb the excesses of his own troops.[36] They were soon driven out of Fujian and into neighboring Guangdong province, with Zuo and his subordinates taking care to keep the main Taiping commanders divided and using both land and sea forces to contain them.[37] After a series of battles, the final Taiping remnants were crushed at Jiaying, Guangdong in February of 1866.[38] In this battle some 16,000 were killed, 60,000 surrendered and 734 leaders were captured and executed.[39]

At this point Zuo had hoped to turn his attentions towards the construction of a small naval yard near Fuzhou, in Fujian province. The recent war with the Taipings and ongoing struggles with European imperialists had impressed upon him the urgent need for China to develop its naval capabilities. Moreover, Zuo was convinced that simply buying ships from Westerners or having them build them for China was not sufficient. China needed to develop the capability to build its own ships. He also wanted to establish a school for teaching navigation and the requisite sciences to prospective Chinese naval officers.[40] Unfortunately, just as Zuo was preparing to throw himself into this important project, he was recalled to military service in the north. He had to pass along management of his naval yard project to his friend Shen Baozhen (1820–79), who had also served under Zeng Guofan and participated in the suppression of the Taipings.[41]

Zuo was appointed Governor-General of Shaanxi and Gansu provinces in September of 1866. The region was then in the midst of a massive series of inter-related Muslim uprisings that had arisen independently and spread from Central Asia. On top of this, loosely organized rebel bands known as the Nian, who had occasionally worked with or even accepted military commissions from the Taipings, were still active across north-central China and were foraging into these northwestern provinces raising the possibility that these two movements would become closely linked.[42] For his part, Zuo did not leave Fuzhou until the end of 1866, first proceeding to Hankou, where he started assembling troops for his new operations. By this point there were rumors that the Muslim insurgents numbered over one million and that they were in fact working with the Nian so Zuo was instructed to first deal with the Nian and subsequently proceed to the northwest.[43]

Zuo advanced to the city of Wuchang and sent a missive to the throne which read

When your minister made war in the southeast advantage lay in boats. In the northwest, advantage lays in horses. The Nian and Muslims rush about the Central Plain on horseback, but if we try to meet them with government troops on foot, then surely we will not have success. But when it comes to talking of strength in horses, the north is not the same as the west in terms of resources. The Nian rely on northern produce/resources. Therefore, the Nian are actually more fierce than the Muslims. But right now your minister has only 6000 troops. So now I wish to purchase northern horses and train cavalry units and combine them with mobile battle carts. From Xiang[yang] and Deng [zhou] we'll advance from Jinxing Pass and traverse through Shangzhou to Shaanxi. We'll establish military farms (tuntian) so as to facilitate long-term campaigning. Therefore, when we advance our troops to Shaanxi, we'll first clear the areas beyond the pass of bandits, then we will advance to Gansu; once we advance to Gansu we'll first clear Shaanxi of bandits. When we station troops at Lanzhou we'll first clear the various roads of bandits. Afterwards we'll be able to move armies and supplies without obstructions.[44]

Exercising his usual caution, Zuo added that it would take several months to make all his preparations and begged the emperor to indulge him so that he could bring the campaign to a successful conclusion. Again hearkening back to the instructions of the Ming general Qi Jiguang, Zuo recognized the importance of mixed units combining firearm-wielding infantry, battle carts, and cavalry to deal with mobile, mounted enemies. He prioritized his foes, regarding the Nian as more dangerous, not only because of their mobility and experience, but because of their proximity to the capital. Strategically he wanted to follow his usual practice of gradually securing key points and expanding control to encompass his foes. In the spring of 1867 he would lead his initial force of just 12,000 to the west.[45] By this time one provincial governor of Shaanxi had already been replaced and his successor had requested permission to pull out so Emperor Tongzhi (r. 1862–74) had appointed the Manchu Governor of Ningxia, Mutushan, to take charge, with Zuo as special imperially-deputed overseer of military affairs.[46]

Zuo endeavored to bring many of his trusted subordinates with him and corresponded with old friends about the strategic situation. In a memorial to the throne prior to his departure, Zuo stressed the importance of rehabilitating the region economically once the military campaigns were over.[47] Upon his arrival in the northwest, he realized the degree of mobility the Nian possessed and he decided to muster more battle carts, cavalry, and firearms. Concerning the latter, Zuo procured Krupp guns from the Germans and hired foreign machinists to teach his own weapon manufacturers.[48] He hoped that his carts could stymie the mobility of the Nian cavalry and allow him to concentrate their ranks into his artillery fire.[49]

Zuo warned the court that defeating the Nian would not necessarily be fast, but if his troops were properly trained and equipped, they could get the job done.[50] He estimated that the Eastern Nian still had over 100,000 members and the Western, around 30,000. At least a third of these were mounted. To account for these unfavorable odds, Zuo put his stock in training and weaponry. Every brigade of 500 men had 38 'short mountain-cleaving guns' mounted on battle carts that were supported with Western guns. He had extra shipments of gunpowder brought in from his old base in Zhejiang.[51] Even as Zuo's forces were en route, the Western Nian threatened the city of Xi'an and several Qing forces were defeated. There was increasing danger that the Eastern and Western Nian would link up, but they were thwarted in the spring of 1867 by Zuo and others and most were driven back east.[52]

The essential problem when it came to fighting the Nian was the fact that they were mounted bandits without clear goals, bases, or even leaders. Even if some leaders and groups were wiped out, it did not mean that it would have much effect on the rest. They could also seamlessly blend into the local populace and often find shelter in small villages.[53] Throughout north China small communities would build earth wall stockades and appoint militia captains for local defense in times of strife and disorder. These communities sometimes simply morphed into Nian bands with leaders taking titles such as 'lodge master' or 'banner captain'. The community would pledge mutual support, go out on raids and divide their booty.[54] They would even hold operas and festivals prior to going out on raids, further cementing their ties with local communities. Their depredations enabled and inspired other local warlords who operated independently or took titles from the Qing in a situation not unlike that which prevailed in the same region during the Ming-Qing transition in the seventeenth century.[55] The rugged terrain of north-central China, which consists primarily of forested mountains running through plains, also aided the rebels. In fact, this rugged inter-provincial area was known as 'the region of the three no-governs' as regional and provincial authorities generally denied responsibility for dealing with events in these overlapping jurisdictions.[56]

Prior to Zuo, Zeng Guofan and others had tried for nearly two decades to stamp out the Nian. And despite his preparations, Zuo was initially frustrated as well, his efforts to contain the Nian within the boundaries of Shaanxi failing miserably. He had hoped to hem the Western Nian in by defending several points and launching probing attacks to get them to split their forces, then close in on these isolated units from several directions simultaneously. But he was still short on troop numbers himself and illness further cut into the number of battle-worthy soldiers at his disposal.[57]

Even after Zuo and his colleague (and rival) to the east Li Hongzhang (1823–1901) scored some victories over the Nian, they were hampered by the Nian's mobility, their decentralized structure, and the Qing commanders' own refusal to fully cooperate.[58] Still, Zuo believed that the separated Nian forces could be picked off. After all, in most of the set-piece battles, the Nian had been handily defeated when Zuo could bring his firepower to bear.[59] And the Nian lines were long and susceptible to being severed. Plus, autumn rains would swell the rivers and hamper their movements further. This proved to be a false hope, as the Nian broke out of his attempted encirclement and drove north.[60]

When the Western Nian leader Zhang Zongyu escaped their grasp and raided through the provinces of Shanxi, Henan, and Zhili, even threatening the capital itself, both Zuo and Li Hongzhang were briefly deprived of their official ranks. Matters intensified further when some Muslim units joined the Western Nian and Zhang sent a letter to Lai Wenguang, leader of the Eastern Nian, hoping to link up again.[61] Moving east, the Western Nian split into three groups, striking into Henan as well as Zhili, advancing all the way to Baoding, 160 kilometers southwest of Beijing. The court offered Zuo and the others a chance to redeem themselves and they took advantage. Zuo and his commander Liu Songshan drove the Nian from Baoding as Li Hongzhang exerted pressure from the east. Li now wanted to complete an encirclement campaign whereas Zuo favored combined offensive and defensive operations to make sure the Nian did not escape.[62] Zuo said that the Qing should steadfastly defend the west bank of the Yellow River and let the troops in the east wipe out the units or drive them towards the river, gradually diminishing their base of operations. He argued that he lacked the troop numbers to simply pursue all the remaining Nian forces.

Zuo feared the suppression campaign might drag on for quite awhile, but in fact the situation changed remarkably fast. The Nian suffered a series of major defeats in the spring of 1868. They attempted to flee south into Henan, were thwarted and moved back northeast. After they were repulsed in an attempt to attack Tianjin, they swept back into Shandong. Zuo then posted his units on the west bank of the Grand Canal, which brought grain and supplies to the capital. The Qing court then put a 1 month ultimatum on crushing the Nian, sending another Manchu commander into the field, which irked Zuo, who complained that they were mucking things up with too many overlapping commands and jurisdictions.[63] Still, the government forces were making progress as Zuo and Li hemmed the Nian in, using gunboats on the rivers and even diverting waterworks to hamper Nian movements. Zuo even employed locals to repurpose the very forts the rebels had formerly used for government and community use.[64] Denied resources and hampered by rainy spring weather that negated their mobility, the Nian found their ranks diminishing throughout 1868. They were finally wiped out on 28 June 1868. Zhang Zongyu, however, was never apprehended, and was presumed drowned.[65]

Shortly after the victory Zuo was summoned to Beijing for a court audience. At this meeting he discussed the ongoing Muslim revolt with the Dowager Empresses, Cixi (1835–1908), and Ci'an.[66] Zuo warned them that the disturbance was complicated and he could not expect to bring the matter to a conclusion in less than five years.[67] While not pleased, the empresses knew that Zuo was known for his frank and honest assessments and that he could be counted upon to devote his full energies to the task at hand. Zuo requested 3.5 million *taels* of silver per year for military operations, plus more funds for rehabilitation efforts such as famine relief and funds for rebuilding destroyed homes and cities, though the Ministry of Revenue initially gave him just one million.[68] So he was dispatched to the west and the Qing armies reached Xi'an in the tenth month of 1868. At the time it was estimated that there were perhaps 300,000 rebels in Shaanxi and Gansu. The most prominent leaders were the Muslim Ma Hualong, based at Fort Jinji (Jinjibao), in Gansu, and Bai Yanhu, who operated in southwestern Shaanxi province.

Ma was identified with the so-called 'New Teachings' or 'New Sect,' a Sufi sect that had its roots in anti-Qing activities in the previous century.[69] Founded in the 1760s by one Ma Mingxin, the New Sect was West Asian in nature and its beliefs included the veneration of Sufi saints, meditating at tombs and wonder working.[70] It also prominently featured vocalization rituals that apparently clashed with the practices of the other, 'Old Sect' Muslims.[71] Sufis in Central Asia, like elsewhere, used magic and the curing of illnesses to attract followers and gain recognition as saints. These wandering dervishes had the charisma to unite groups and spread their teachings, especially in the absence of a strong central political or religious authority. Some saints' tombs became holy sites or shrines associated with curing diseases. Sufi brotherhoods and saintly leaders also provided protection and other benefits for their members.[72] Factions would coalesce around *khojas*, or religious leaders, and these would sometimes ally with other political or military groups, occasionally getting drawn into wider conflicts.

As for Ma Mingxin, he had traveled westward to Bukhara and possibly Samarkand prior to founding his New Teaching sect of Sufism. Central to his teaching was vocalization and his group was called the Jahniyya, though there are disputes amongst scholars concerning the precise nature and origins of this group. In any case, a dispute arose between followers of the New and Old Teachings in 1781. The Manchus intervened and arrested Ma Mingxin. This prompted his followers to attack Qing units and besiege the city of Lanzhou, where Ma was being held, prompting the Qing to execute him.[73] Another rebellion in 1784 caused Emperor Qianlong to condemn the New Teachings and ban Muslim teachers from entering China proper. Therefore the sect became akin to a secret society wherein leadership of the sect was transmitted down through the generations. Ma Hualong kept his feet in multiple worlds, purchasing a Qing military title, but also doing fortune telling, curing illnesses, restoring fertility for the barren and receiving tribute and gifts from adherents as far away as Yunnan province.[74]

There was a flourishing tradition of religious leaders and mystics in Central Asia, many of whom could trace their lineages back to various Sufi masters.

Some of these religious figures had been involved in anti-Qing activities in the past so those descended from them had a certain degree of local charisma.[75] Each locality had religious and political figures and many villages were dominated by mosques and religious figures that excluded outside groups. The Qing authorities had difficulty extending their influence in these places.

The reasons for the outbreak of these rebellions involved a complex amalgamation of long-term and more immediate triggers and grievances. The Qing had first conquered the vast regions of Central Asia in the early to mid eighteenth-century, effectively ending the time-old threat of steppe nomads for sedentary China-based empires. And though they now controlled territory vaster than any government in Chinese history, they lacked the manpower to properly administer these far-flung regions. They were also not that interested in local governance or resources, beyond extracting some tribute and taxes. The region had long been populated by peoples from all over and Islam had become the dominant faith. The Qing decided to adopt indirect government procedures, appointing Manchu and Mongol overlords to administer vast tracts of territory that were locally managed by an array of officials and figures, most of whom were Muslim and of different ethnic groups.

As might be expected, Qing rule was not particularly benign; there was ample opportunity for graft and corruption amongst the ruling stratum, many of whom were regarded as mere Qing collaborators.[76] For many of the underlings of the appointed leaders were in fact relatives with no practical government experience who saw the posts as simply a means for enriching themselves. Furthermore, because these posts were not especially desirable, people often bribed their way out of them, relegating them to less well-connected, but possibly even more unscrupulous officials. One commander of the city of Urumqi stayed in his position for seven years, allegedly never setting foot in the countryside and amassing a personal fortune of some 3 million taels.[77] Lower level officials were accused of crimes such as ravaging Muslim women and forcibly taking them as wives or concubines.[78]

Additionally, and more relevant for the situation in Shaanxi and Gansu provinces, the steady population expansion of the high Qing era had encouraged more and more Han Chinese settlers to move to these areas in hopes of starting fresh or improving their lot. Though groups sometimes lived amicably, there were increasing confrontations between these immigrants and local Muslims, variously known as Hui or Tungans in the sources.[79] Ethnically many of these Muslims were of Turkish stock, but not all. As disputes increased, local investigators were often charged with favoring one group or another, though usually the non-Muslims. There were frequent vendettas and even massacres of one side or the other. It was suggested that Han officials used the government directive to create militia to guard against the Nian as a pretext for assembling gangs to massacre Muslims.[80] The Muslims then returned the favor in a series of incidents from 1862–66. For example, when Ma Wenlu captured the city of Suzhou in 1865 only 1100 old and weak people out of an estimated 30,000 in the city survived his massacre. In 1866 dozens of Han dwelling in the city of Taozhou were slaughtered by Muslim rebels.[81]

Moreover, rumors often circulated about impending massacres of Muslims by Manchus or Han Chinese. Mongols also tended to be hostile to Muslim groups. This in turn sometimes provoked preemptive massacres by the aggrieved group. The fact that a recent rebellion in southwest China, known as the Panthay Revolt (1856–1873) had been provoked by such a massacre lent credence to such rumors.[82] These tensions were exacerbated by the fact that the Qing government was understandably preoccupied with the massive rebellions in the interior and the rising challenge of the Western imperialist powers, which of course included the 'Great Game' machinations of the Russians and British in Central Asia. So it was easy for a spark to set off the powder keg and engulf the entire region.

Upon his arrival at Xi'an in the middle of the tenth month of 1868, Zuo Zongtang followed his usual procedures. He evaluated possible supply routes and ordered supply depots and warehouses built at critical junctures. He made use of multiple supply routes to deliver more supplies and men safely. He carefully checked weather patterns and local geography and personally supervised loading, unloading, and transport of supplies whenever he could. He also continued to pursue various avenues for funding, including calling in personal favors and securing transit taxes from southeastern provinces.[83] In terms of troop recruitment, Zuo tried to have his personally known and trusted subordinates recruit and lead the troops. He strongly discouraged looting and other damaging activities by his troops, noting the importance of maintaining positive relations with local populations. As noted above, he was keen on bringing lots of firearms and heavy artillery into the field and he hired many foreigners as trainers and technicians. And he decided to pursue his familiar divide and conquer strategy as opposed to the 'kill them all' approach favored by his Manchu predecessor Duolongga.[84]

Zuo estimated there were perhaps 200,000 troops in some 18 battalions with a central base at Fort Zhenjing from whence they could strike easily at locations across three provinces. Zuo invited his commanders to Xi'an to discuss war strategy and assigned jurisdictions.[85] Zuo's commander Liu Songshan started the campaign off impressively, smashing a large rebel force at Suide, killing 6000 and getting one of their leaders to surrender.[86] Some tried to flee south into Sichuan province, but they were crushed by another Qing commander. Around this time, however, Liu Songshan's units experienced supply difficulties as they battled their way north. They were forced to kill their draft animals for food. Nonetheless, the Qing forces defeated a Muslim army estimated at 30,000–40,000 in early 1869 and turned the tide once more.[87]

Zuo then advanced to Qianzhou and sent more units forth, killing over 30,000 Muslims in battle during his initial offensive as he retook Qingyang, Zhenyuan, and several other towns.[88] Allegedly some 100,000 Muslim rebels quickly surrendered, some serving the Qing loyally for decades thereafter. But the loess gullies of northern Shaanxi were notoriously hard to attack and easy to defend. Still, by the end of 1869, Shaanxi was largely free of Muslim rebels. Zuo therefore moved his headquarters to Pingliang, Gansu. Ma Hualong fled north, but the destruction was heavy. Zuo observed, 'For 1000 *li* all was a wasteland. White bones filled the landscape as far as one could see and there were no cooking fires'.[89] He blamed the Muslims for this state of affairs. Zuo attempted to rehabilitate the region by digging wells and setting up fields for agriculture. Famine relief was distributed and he resettled some 170,000 people dislocated by the recent military operations.[90]

Meanwhile, Ma Hualong evinced a desire to surrender and cease hostilities. But Zuo did not trust Ma and told the court in a secret letter to Liu Songshan that he was just pretending to accept suasion in order to protect his wealth and position and to buy time to recruit more followers to his side.[91] Zuo instructed Liu to deal with Ma by military means. Thus, in the fifth month of 1869, Zuo ordered a three-pronged advance upon the rebel stronghold at Jinjibao (Fort Jinji). This main stronghold was in fact surrounded by several hundred smaller fortified positions arrayed around the loess gullies and hills. The Chinese maintained that Ma used his followers to loot the Han folk, take their produce and ravage their daughters from these strongholds.[92] This provoked even more indignation on the part of Zuo and his subordinates. So Zuo devised his multi-pronged attack strategy while rebuffing further offers from Ma to surrender.

But as the Qing were mobilizing, Liu Songshan was forced to deal with a mutiny amongst his troops, which he quelled, but it slowed his progress as did an impeachment for allegedly sanctioning wanton looting and killing.[93] Liu was cleared of all charges relatively quickly. He would occupy Fort Zhengjing in the seventh month and his forces closed in on Lingzhou, another lynchpin of Muslim defenses, in the eighth month of 1869. The Muslims fought hard but Liu reduced their strongholds to ashes. For his part, Ma Hualong continued to send missives to Zuo asking to submit even as he sent reinforcements to Lingzhou to fight the Qing.[94] The Manchu official Mutushan was inclined to believe Ma and consistently referred to Ma as a 'good Muslim,' enraging Zuo, who heartily distrusted anyone associated with the New Teachings. This provoked Zuo into stepping up his own assaults with the aim of overawing the enemy.[95]

Liu's army occupied Lingzhou shortly thereafter.[96] The Qing then steadily advanced on Jinjibao burning down dozens of stockades and outlying fortresses. Things slowed down as the Qing got closer to the main objective, with the Muslims constantly harassing the attackers with sallies even as they dug in at other positions. Zuo was aided by a timely shipment of supplies from the south and he attempted to bring his superior firepower into play wherever possible. For their part, the Muslims tried to cut Qing supply lines. Liu Songshan was killed in battle in the first month of 1870, which allowed Ma Hualong to once again secure the strategic throat to Jinjibao along the Yellow River. At this juncture, the other four Qing armies pulled back and their strategic advantage collapsed. Zuo was eager to maintain the offensive so he strenuously pushed for the appointment of Songshan's nephew, Liu Jintang, to his command, arguing that the personal relationship the younger Liu enjoyed with the troops would be vital in maintaining momentum and securing victory. In the meantime Zuo decided to regroup, resupply and build up strength again for a renewed assault.[97]

This was a valid strategy but it allowed the Muslim forces to quickly go back on the offensive and take some outlying towns. It also prompted Ma Hualong to once again request permission to submit. Zuo refused this request and the Muslim forces went about raiding for supplies with little success. By the summer of 1870 the Qing were ready to renew their offensive and Liu Jintang was again closing in on Jinjibao in September. It was soon encircled by Liu Jintang's units to the east and south; those of Jin Yunchang to the north; Xu Wenxiu to the northwest; Huang Ding to the west, and Lei Zhengguan to the southwest. The Qing took Hanbobao near the end of the year and by this point only 5 out of 570 outlying forts had not been attacked. As starvation and cannibalism started setting in, Muslim units surrendered to the Qing in large numbers. On 6 January 1871 Ma Hualong and his top allies surrendered their arms to Liu Jintang and others submitted peacefully. But Zuo still distrusted Ma. On 2 March 1871 Liu pulled some 1200 guns out of Jinjibao and Ma Hualong was executed along with 1800 of his followers.[98] According to Chinese accounts Ma had refused Zuo's request to write letters to other Muslim leaders encouraging them to surrender.[99]

Some 12,000 survivors were resettled in Guyuan and elsewhere. Among other things, Zuo offered draft animals, seeds, and tools to the settlers and endeavored to keep them separate from Han Chinese to limit the potential for conflict.[100] Afterwards Zuo commented on the ferocity and casualty figures of the campaign, noting that they were greater than those incurred in more than a decade of fighting the Nian. Perhaps his previous estimate about the greater ferocity of the Nian was misplaced.

Zuo next turned his attention towards another Muslim rebel force based at Hezhou, far to the west in Gansu province. Controlled by an Old Sect Muslim leader named Ma Zhan'ao, Hezhou was also defended by stockades and a system of forts. In preparation for this campaign Zuo again mustered supplies for months but then his commander, Zhou Kaixi, died of illness, delaying the start of the expedition for a couple months. Zuo's armies finally advanced in the seventh month of 1871.[101] Zuo himself directed matters from his base at

Jingning. And even as he was doing this, an unfolding crisis in the Ili Valley near Russia forced Zuo to send troops there just as he dispatched his trusted commander Xu Zhanbiao with 6000 troops to Suzhou, which was one of the cities under Muslim attack.

Zuo's forces advanced steadily towards Hezhou through the summer of 1871, crossing the Tao River with pontoons and taking many outlying fortresses.[102] By late autumn they were dislodging rebels from their bases, prompting others to attempt to cut Zuo's supply lines. They achieved some success in this regard and a pair of Qing generals were killed in the first month of 1872 in a battle near Taizi Temple. Zuo moved quickly to replace them and ensure control of their armies. The Qing still kept up their pressure, killing one of the chief Muslim leaders, Ma Duosan, and obtaining the surrender of Ma Zhan'ao so that Hezhou was pacified in short order. Zuo moved his head-quarters to Lanzhou, northeast of Hezhou, in the autumn of 1872.

He would have an arsenal and woolen mill built there to both supply his troops and stimulate the local economy. The arsenal was under the direction of Lai Chang, a native of Guangdong who had served Zuo back in Fujian. He had worked with Prussian engineers making breach-loading rifles and field guns. Zuo highlighted his success in these endeavors in a letter to the Zongli Yamen (Foreign Affairs Office), saying proudly that the Qing could now manufacture copper caps and fuses and large and small exploding shells, as well as imitating Prussian-style 7-shot breach-loaders. Workmen were brought in from south China to work alongside a few Europeans and some Western observers were impressed at Zuo's efforts. This arsenal proved critical in supplying Zuo's armies in their subsequent campaigns, though it would sadly be shut down in 1882 after Zuo left, illustrating the haphazardly personal dimension of many late Qing modernization efforts.[103] At the same time he ordered the establishment of a printing press in Xi'an for the expressed purpose of distributing edifying Confucian texts to the poor people of the northwest.

Meanwhile as troops were heading northwest to deal with the Ili crisis, yet another Muslim leader, Ma Wenlu, raised the flag of rebellion, taking advantage of the fact that the Qing presence in the region had diminished. [104] Ma gained support from Bai Yanhu, another leader who, as the reader may recall, had previously been involved in the disturbances in Shaanxi and Gansu. Liu Jintang was dispatched by Zuo to deal with this new threat. The rebels entrenched themselves in the mountains, gathered supplies, and dug in as starving refugees streamed into the countryside. At the same time there were renewed uprisings back east, in Shanxi province. But Liu Jintang was experienced in this kind of fighting and his troops were well trained and loyal to him. He defeated most of the rebel groups in the first few months of 1873, obtaining the surrender of many key leaders.[105] Among other things, Zuo sent out reinforcements with more powerful cannon to help reduce fortress walls.[106] Ma Wenlu held out longer, but lost hope when Zuo Zongtang personally came to inspect the besieging army outside the city of Suzhou where Ma was entrenched. His offer to surrender was refused, though Zuo pledged to spare the old and young when the city was taken. Ma personally surrendered to Zuo on 4 November 1873.

Zuo had all the prominent commanders executed eight days later, along with 1573 Muslim soldiers. It was estimated that another 5400 people were killed by Qing armies when they took the city.[107] In fact, virtually no able-bodied males were spared and barely 900 old men, women, and children were said to have survived the bloodletting. Zuo personally regretted this but claimed he was unable to prevent it.[108] Certainly Zuo would later intensify his efforts to restrain his troops during their operations in Central Asia so perhaps he learned something from this incident. In any event, he received more promotions and honors from the court as a result of his successes. On the negative side, the notorious Bai Yanhu had escaped to Hami in the west.

Zuo turned simultaneously to rehabilitating the devastated regions and preparing for potential future operations in Central Asia. In addition to establishing the factories noted above, he engaged his soldiers in produc-tive labor when they were not training. They were enlisted in agricultural and engineering pursuits that including digging wells, clearing roads and planting trees along the latter. Opium sales and use were strictly prohibited amongst his troops, though modern scholars debate the scope and efficacy of these prohibitions. He stressed maintaining positive relations with the local people, claiming that the coming of the army should be 'like a spring rain' for the local populace.[109] He also sought to diminish tensions between Muslims and non-Muslims, saying, 'Whether you are Han or Muslim does not matter. There are only good people and bad people'.[110] Nonetheless, he generally tried to keep Muslim and non-Muslim groups separate and he was consistently distrustful of the New Sect, which was banned. Surrendered Muslims had to yield their horses, grain, and weapons and assist in repairing damaged buildings. They then had to provide information on their native places for later segregation.[111] When relocated they received land, housing, tools and animals and were divided into *baojia* (mutual security) organiza-tions. Other relief measures included setting up soup kitchens, enforcing lower interest rates at pawn shops and establishing distribution centers or food, tools, animals, and clothes after natural disasters.[112] Coinage was revitalized and transit taxes were introduced as used in the southeast. Cotton and sericulture were also introduced and people were encouraged to reclaim fallow lands. Renewed investigative apparatuses were set up to evaluate low ranking officials and guard against corruption.

All these efforts also served practical purposes in that Zuo's troops still needed food and supplies for themselves and given the climate and recent devastation in the area, caused in part by their own activities, the Qing army could hardly expect their needs to be met. Finally, Zuo realized that the armies could not occupy the region forever and he hoped to build an infrastructure for long term peace and prosperity. Incidentally, in terms of the scale of the destruction, it was estimated that out of 700,000–800,000 Muslims who had been living in Shaanxi prior to the uprisings, no more than 300,000 were left, possibly fewer.[113]

Moreover, Zuo encountered a fair degree of resistance to his directives from local gentry, Han officials and the like. Some claimed resettled Muslims caused trouble and requested 'extermination troops' to remove Muslims in their midst. Others simply ignored Zuo's directives and tried to prevent Muslims from receiving their allotted aid, tools, grain, and land. So Zuo had to continually send out his most trusted subordinates to ensure compliance. These were usually men from his old Chu Army units. Zuo's efforts were also sometimes compromised by the Muslims themselves, who frequently chafed at living in new places under a variety of rules and restrictions and resented being forced to alter their customs.[114] These rules, incidentally, included restrictions on the size and materials that could be used for building mosques, since Zuo did not want them to be used as defensive structures in the event of another uprising. Zuo also set up a system of appointed *ahong*, or *imams*, hoping to sever the charismatic links between certain teachers and followers and presaging efforts under the People's Republic.[115] Despite these problems, overall Zuo's methods can be judged to have been effective from the standpoint that the agriculture and economies of Shaanxi and Gansu were restored to pre-rebellion levels within just 5–6 years.

In 1875 Zuo would be placed in charge of military affairs for Xinjiang, where a Muslim uprising had been raging since 1864, under the leadership of a man known as Yakub Beg.[116] In the course of his state-building efforts Yakub Beg had courted relationships with both Russia and Great Britain (via India) and had even managed to convince the Ottoman sultan to name him a vassal. At the same time, the distractions caused by Yakub had allowed the Russians to pursue their interests in the Ili Valley as noted above.[117] It was believed that they also wanted to take control of Urumqi. Thus there emerged at the Qing court a debate over whether Zuo should be authorized to take Xinjiang back for the Qing with many arguing that it was too expensive to maintain and strategically unimportant whereas China should instead focus on building up its naval defenses to counter the threats of the British, French and other imperialist powers. Some even argued in favor of Yakub Beg's request, conveyed through the British, to establish his own tributary state, serving as a Qing vassal but acting as a buffer to the Russians and British.[118]

The main opponent of a campaign in Xinjiang was none other than Zuo's old rival, Li Hongzhang. After reminding the court and his critics that he himself had supervised the construction of a naval yard demonstrating that he understood full well the importance of that theatre, Zuo proceeded to explain why Xinjiang could not simply be 'thrown away'. His arguments were both sentimental and strategic. On one level he appealed to the court's filial piety in noting that the hard-won achievements of their forebears, who had been the first rulers in China's long history to conquer and hold Central Asia, should not just be disrespected and cast aside. More to the point, he argued for the strategic value of Xinjiang as a buffer between China and Russia and the British in India. He argued that if Xinjiang fell, then Mongolia was imperiled. Once Mongolia fell, Beijing would be next. He further added that while the Russians and British in India had proven they were after territorial aggrandizement, the Westerners along the coast were generally merchants who desired only trade. There would still be ample time to build up the navy for a future confrontation. Finally Zuo argued that funds earmarked for naval developments should have no impact on a vitally important war in the northwest. His troops had already demonstrated their mettle. It was time to finish the job and end the threat to Qing sovereignty on this frontier at least.[119]

Among the major sources of opposition to Zuo's plan was the projected cost of the operation. As before, Zuo proved to be creative in searching for funds. He proposed negotiating a loan with European powers through the Hong Kong and Shanghai Bank for some of the funds. He would also supply his armies via military farms and he worked out another series of grain purchases from the Russians at prices that were far above what Russian peasants might get at home, but low enough for Zuo's purposes. Then there was his usual method of pressing friends and allies for *lijin* and other revenues. In the end, Zuo's impassioned pleas and detailed plans convinced enough members of the Qing court, most importantly Cixi, and he got the go-ahead for his campaign.[120] Zuo got his loan in 1876 and moved his headquarters to Suzhou in anticipation of beginning the campaign. Zuo ended up receiving over 52 million taels (8.5 million from the Hong Kong Shanghai Bank Co.) for his campaign, which also used 29,000 camels, 5000 wagons and 5000 donkeys and mules. His troops were equipped with European repeating rifles, Krupp needle guns along with twelve and sixteen pound steel cannon.[121]

He eventually amassed 178 battalions and devised a five-phase plan for the re-conquest of Xinjiang. The total number of troops he had is still open to debate, but was probably around 60,000. Zuo was concerned that many of his battalions were under strength and Yakub Beg had a reputation as a formidable adversary so Zuo stressed additional training with his new weapons.[122] After reorganizing and outfitting the army, they would mass in western Gansu. From Suzhou other bases would be set up, radiating west along both the northern and southern routes into Xinjiang, which is bisected by the Tianshan range. Next his forces would occupy the cities of Hami, Barkul, and Gucheng, using them as advance bases. Once the supply lines were ready and supplies were in place the army would go from Barkul and Gucheng to take Urumqi, then head south to Manas and then Turfan and Korla, converging on Yakub's base of operations at Kashgar.[123] In particular Zuo stressed that he wanted a fast, multi-pronged operation to both cut down on the time his troops spent in the unforgiving wastes of Central Asia and to overawe his presumptuous challenger with the majesty of the Qing.[124] In fact, the outlying cities of Hami, Barkul, and Jincheng were all taken in 1875, in advance of the full-scale campaign. Yakub Beg dispatched some units to counter the Qing, but they do not appear to have done much actual fighting.

Training lasted into early 1876 and it apparently impressed some European observers. One Englishman noted, 'This Chinese Turkestan army has the appearance of one of the strongest European armies'.[125] Such praise from Europeans was always a point of pride for Zuo despite his supposed anti-foreignism. Zuo also sent spies and informants into Xinjiang and gathered as much information about the situation as he could. He was especially concerned about the positions and interests of Britain and Russia. Once he acquired this intelligence, he decided that he could take Urumqi first so that this quick victory would serve to spread word of the awe of his army.[126] They also hoped to apprehend Bai Yanhu, who was reportedly in the vicinity of Urumqi.

Once the campaign began in earnest, it proceeded better than even Zuo could have envisioned. The Qing embarked upon the route north of the Tianshan range first. In the fifth month Liu Jintang and the Manchu commander Jinshun took Huangtian en route to Urumqi, avoiding a possible enemy ambush along the way and putting Bai Yanhu to the run.[127] A couple months later they secured Manas and the northern route was complete. Bai had apparently hoped to lure Qing forces to pursue him into the steppe to no avail. Yakub Beg had not bothered to put up any kind of stout defense, even though some troops had been dispatched to reinforce Urumqi.[128] They attacked Gucheng in the sixth month of 1876 and intercepted a relief force of 1000 troops sent by Yakub Beg. They assaulted Gucheng with heavy artillery and infantry and took the city within five days, killing several thousand inside.[129] The Qing decided to rest for the winter and renew their offensive in the spring of 1877 as Yakub moved his headquarters to Kurla. At this point he still had about 28,500 troops stationed at various points south of the Tianshan.[130]

In April of 1877 Liu Jintang led some 14 battalions south from Urumqi to Dabancheng, taking the city in five days. Turfan fell to another Chinese force less than a week later, with another 10,000 people surrendering to the Qing.[131] Yakub Beg died soon thereafter, though sources differ as to whether he took poison in despair, was assassinated, or had a stroke.[132] Combined with the rapid Qing victories, Muslim morale was sapped and city after city fell into Qing hands. Bai Yanhu kept fleeing, eventually escaping to Russia. Though Zuo hoped to capture him and bring him to justice, the throne felt that there were more important things to deal with such as the rehabilitation of the newly recovered lands. There was also the matter of the ongoing Ili Crisis with Russia and Zuo's victorious army could play an important role, even if only as a deterrent, in that affair. For his part, Zuo immediately started pushing for the reclassification of Xinjiang as a regular province of the empire, a dream which would become reality in 1884 and still has ramifications today.

In assessing Zuo and his achievements, it is hard not to be impressed by all that he accomplished, particularly given his upbringing and the particular circumstances of his life. While he had a reputation for bonding with his men and using troops well, that was not his strongest characteristic. He recognized the importance of winning hearts and minds and saw that the keys to this were simple: food and security. He particularly excelled in planning and supplying his men and was creative in feeding both his troops and the ordinary subjects under his care. He was also innovative in the areas of strategy and tactics and very interested in adopting and adapting new technologies. His ability to apply the old lessons of Qi Jiguang within the contemporary context by using new weapons like steamships and Krupp guns is a testament to this. He also made excellent use of his book knowledge in devising methods for fighting in vastly disparate environments ranging from the riverine areas of south and central China to the coast to the mountains of the northwest to the barren steppes of Central Asia.

With respect to local populations, he took great pains to identify the sources of grievances and find ways to ameliorate them. He could be utterly ruthless in battle, but also appreciated the need for establishing conditions that would allow for a lasting peace. He appears to have truly believed in the mission of making Xinjiang something beyond a military colony. Indeed, some of the half million trees his men planted in the northwest still stand today. His efforts on behalf of refugees bespeak the experiences of his youth and reveal a sensitivity beneath the gruff exterior. At the same time he could also be stubborn and his policy of banning the New Sect and indiscriminately slaughtering all its adherents speaks to his Qing imperialist mindset. The Qing, after all, were still the rightful rulers of China and it was his duty as a loyal official to stamp out challenges to their legitimacy and their orthodoxy.

With respect to the rebellions in question, they were undoubtedly devastating and it is worth briefly recapping a few figures here. The Taiping Rebellion lasted around fifteen years and resulted in an estimated 30–60 million deaths. It spread over most of central and south China, seriously disrupting life at all levels and undermining Qing control. The much lesser known Muslim rebellion of China's northwest and Central Asia encompassed more than one quarter of China's territory and directly affected tens of millions of people. More extreme accounts claim that the population of Gansu declined by over 90% and that the number of Muslims in Shaanxi declined from 700,000–800,000 to less than 60,000.[133] This depopulation later encouraged settlers from other parts of China with the result that these regions have seen their Muslim populations steadily decline in relation to the number of non-Muslims. This immigration also brought customs and technologies from other parts of China, further transforming the region.[134]

Bruce Elleman argues that the defeat of the Nian, Tungan, and Muslim rebels were important for several reasons. First, it spread the use of modern weapons throughout China, most notably into interior regions. Second, more Han Chinese became proficient in using these weapons and once this happened it was much harder for the Manchu Qing rulers to control the weapons themselves or restrict knowledge of their use. So it can be argued that by virtue of empowering their own officials to suppress these regional revolts, the Qing accelerated the growth of regionalism that precipitated their demise a few years later.[135] He adds that the victories in these rebellions by Han Chinese commanders 'proved the Manchu generals were no longer the sole bearers of military prowess in China' and undermined their prestige as the ruling elites.[136]

On the other hand, the Qing most likely would have fallen, or at least been split up, decades earlier had it not been for the efforts of Zuo Zongtang and other officials like him. And he was spot on in recognizing the need for China to become self-sufficient in the areas of weapons manufacturing and ship building. Ironically China has become one of the world's leading ship builders in recent decades, realizing one of Zuo's dreams.

As for Zuo himself, it is worth quoting his biographer W.L. Bales: 'He was born in obscurity of poverty, schooled himself under great difficulties, suffered many adversities, entered an official career when well past middle age, and by his energy and genius rose to the highest posts open to a Chinese during the Manchu period. He was not prepared either by education or by early training for the profession of arms, yet he became the foremost soldier of his generation in China'.[137] Certainly then, his legacy is worthy of more than just a spicy sweet chicken dish, is it not?

Notes

1. While there are dozens of biographies of Zuo in Chinese, there are far fewer in English, as discussed below. For a brief career overview of Zuo in English, see Hummel, ed., *Eminent Chinese of the Qing Period*, 949–54. For his official biography in Chinese, see Zhao, et al., comps., *Qing shigao*pp, 12023–12035. Hereafter *QSG*. For his traditional chronological biography, see Zhengjun, "Zuo Wenxiang gong nianpu [Chronological Biography of Zuo Zongtang]," 539–757,

hereafter *ZWN*. The best recent modern Chinese biography is Dongliang, *Zuo Zongtang.*

2. For example, Zuo's own collected works encompass fifteen weighty volumes in their modern edition. See Zongtang, *Zuo Zongtang quanji* [Complete Works of Zuo Zongtang]. Hereafter *ZZQJ*.

3. Bales, *Tso Tsung-t'ang,* There are other, more specialized studies, most notably those of Lanny Fields and Immanuel Hsu, cited below.

4. See *QSG*, 12,023, and Fields, "*Tso Tsung-t'ang (1812–1885),*" 5–9.

5. See Yang, *Zuo Zongtang,* 5–7; and Fields, *Campaigns,* 53–56.

6. Fields, *Tso Tsung-t'ang and the Muslims,* 3–4.

7. *ECQP*, 949, and Yang, 10–11.

8. For more on the Opium War and its background, see Wakeman, "The Canton Trade and the Opium War," 163–212. Hereafter *CHC* 10. Also see Fay, *The Opium War, 1840–1842.*

9. Chen, *Tso Tsung-t'ang,* 1–2.

10. Yang, *Zuo Zongtang,* 15–16, and *ZZQJ*, 10, 14–16.

11. Yang, *Zuo Zongtang,* 18–19.

12. There are myriad accounts of the Taiping Rebellion in English and Chinese. Useful general narratives include Michael with Chang, *The Taiping Rebellion: A History with Documents*; and Kuhn, "The Taiping Rebellion," in *CHC* 10, 264–317. For biographies of the founder, Hong Xiuquan, who believed himself to be Jesus' younger brother, see Spence, *God's Chinese Son: The Taiping Heavenly Kingdom of Hong Xiuquan* and *ECQP*, 226–32.

13. *QSG*, 12,023.

14. Yang, *Zuo Zongtang,* 21.

15. Qi was famous for a pair of military manuals he published in the late Ming that emphasized drill, training, proper leadership and coordination of small formations. He was credited with eradicating the so-called Japanese pirate (*wokou*) threat in southeast China, and for inventing new techniques for fighting the Mongols, using war wagons and firearms. Zuo, as will be seen below, employed similar tactics himself in the northwest. For studies of Qi, see his biography in Goodrich and Fang, *Dictionary of Ming Biography*, 220–24; Zhongyi, *Qi Jiguang Zhuan*. For Qi's influence on strategists in the Taiping Rebellion, see Sim and Liu, "Zeng Guofan's Applications of Qi Jiguang's Doctrines in Crushing the Taiping Uprising," 93–104; and *CHC* 10, 286–87.

16. See note 13 above.

17. Chen, *Tso Tsung-t'ang*:6–7; and Fields, "Campaigns," 73.

18. See Chen, *Tso Tsung-t'ang: Pioneer Promoter of the Modern Dockyard*, 8–12.

19. Zeng Guofan was also a native of Hunan and one of the ablest of the late Qing officials and military commanders. Along with Zuo he is credited with pioneering the use of smaller, independently raised militia units to quell the massive late Qing uprisings. Many of his friends, relatives, and protégés held prominent positions in the Qing government or military. Personally, he and Zuo were not particularly friendly, a dislike no doubt exacerbated by their respective egos. For a brief biography of Zeng, see *ECQP*, 842–47.

20. *ZNP* 3, 29, and Bales, *Tso Tsung-t'ang: Soldier and Statesman of Old China*, 135–36.

21. *ZNP* 2, 34.

22. Yang, *Zuo Zongtang,* 27.

23. Ibid., 28–29.

24. Ibid., 29.

25. See the table in ibid., 36.
26. See Yang, *Zuo Zongtang*, 33–34, on this incident.
27. Chu was a designation for the region encompassing northern Hunan, southern Henan, Anhui, and Jiangxi provinces, derived from an ancient state that had existed there.
28. *QSG*, 12,024.
29. Ibid., 12,025.
30. Ibid., 12,025; and *CHC* 10, 454.
31. See Bales, *Tso Tsung-t'ang: Soldier and Statesman of Old China*, 139–41.
32. Incidentally, these mercenaries were mostly French. Ibid, 153.
33. Yang, *Zuo Zongtang*, 45.
34. *QSG*, 12,026.
35. Ibid, 12,026. The full record of Zuo's exploits in Zhejiang can be found in Qin and Zhongying, *Ping Zhe jilue* [Record of the Pacification of Zhejiang].
36. Bales, *Tso Tsung-t'ang: Soldier and Statesman of Old China*, 177–78.
37. On these battles, see *QSG*, 12,026–27.
38. Yang, *Zuo Zongtang*, 53–54.
39. *ZNP* 3, 40, and Bales, *Tso Tsung-t'ang: Soldier and Statesman of Old China*, 184.
40. For a translation of one of Zuo's proposals for the naval yard, see Teng and Fairbank, *China's Response to the West*, 79–83.
41. For a biography of Shen, see *ECQP*, 527–29.
42. Again, there are many detailed sources and studies pertaining to the Nian. English, see Teng, *The Nien Army and their Guerrilla Warfare*; and Chiang, *The Nien Rebellion*. In Chinese, see Xueqin, comp., *Qin ding jiaoping Nian (fei) fanglue*[Campaign History of the Pacification of the Nian Bandits]; and Chongqi, ed., *Nian jun ziliao bieji*[Historical Materials on the Nian Army]. For a review of Chinese research on the Nian, see Perry, ed., *Chinese Perspectives on the Nien Rebellion*. The rise of the Nian has been attributed to a combination of geographic conditions, the impact of high rents and other rebellions upon the region, natural disasters and the emergence of secret societies and salt smuggling operations. See Perry, *Chinese Perspectives*, 10–12, and *CHC* 10, 312–15.
43. *QSG*, 12,027.
44. Ibid.
45. Yang, *Zuo Zongtang*, 78–79.
46. *QSG*, 12,028.
47. See Bales, *Tso Tsung-t'ang: Soldier and Statesman of Old China*, 200–202.
48. Ibid., 199–200.
49. Yang, *Zuo Zongtang*, 79; and *ZZQJ* 11, 7–8.
50. *ZZQJ* 11, 13–14.
51. Yang, *Zuo Zongtang*, 79.
52. Ibid., 80–81.
53. *CHC* 10, 312.
54. See Liu, 'The Ch'ing Restoration,' in *CHC* 10, 460. On the late Ming, early Qing situation see Swope, *On the Trail of the Yellow Tiger*.
55. *CHC* 10, 461–63.
56. Perry, *Chinese Perspectives on the Nien Rebellion*, 33.
57. *ZZQJ*, 11, 33.
58. Yet another protégé of Zeng Guofan. Li Hongzhang, was one of the preeminent late Qing officials and diplomats. He would become a great rival of Zuo,

in part because of their differing opinions on how to deal with the threats of the Western imperialists and how to prioritize military spending. For a biography of Li, see *ECQP*, 317–24.

59. See note 49 above.

60. Yang, *Zuo Zongtang*, 81–82.

61. Yang, *Zuo Zongtang*, 83; and *QSG*, 12,028.

62. *QSG*, 12,029.

63. *ZZQJ* 11, 99.

64. Yang, *Zuo Zongtang*, 86.

65. Ibid., 87.

66. The Dowager Empresses were initially the regents for the boy Emperor Tongzhi, but managed affairs for pretty much his entire short reign. Cixi came to be the most powerful, and probably the most reviled, figure in the empire. For her biography, see *ECQP*, 698–704. For a recent revisionist and controversial take on her life and reign, see Chang, *Empress Dowager Cixi*.

67. *ZNP*, 4, 48–49.

68. Ibid., 49–50; and Bales, 211. At this time 3.3 *taels* were equivalent to about 1 pound sterling or $4.86 in US dollars.

69. Fields, "Campaigns," 103–104.

70. Fields, *Tso and the Muslims*, 62.

71. Elleman, *Modern Chinese Warfare, 1795–1989*, 65.

72. Fields, *Tso and the Muslims*, 65. Also see Kim, *Holy War in China*, 159–60 on the various Sufi sects operating in the region.

73. Fields, *Tso and the Muslims*, 67–69.

74. Ibid., 70.

75. For a much more thorough discussion of the background to these uprisings and conditions on the steppe, see Kim, *Holy War in China*, 1–72. On the Qing economic penetration of Central Asia and government policies there, see Millward, *Beyond the Pass, 1759–1864*.

76. On corrupt Qing governance in Central Asia, see Han, *Zuo Wenxiang gong zai xibei*, 24–25. Hereafter *ZWZX*.

77. *ZWZX*, 25.

78. Ibid., 26.

79. On the background in northwest China, also see *ZWZX*, 21–24.

80. Yang, *Zuo Zongtang*, 89.

81. Ibid., 89.

82. See Atwill, *The Chinese Sultanate*, 84–115.

83. These taxes had first been set up in the wake of the Qing's 'unequal treaties' with the Western powers and were designed to soften the blow of reparations by providing new revenue streams for the cash-strapped state at the local level. Of course, the Western powers pressured the Qing to keep import taxes as low as possible and China had effectively lost tariff autonomy so the *lijin* became a regular feature of the tax system. For one of Zuo's requests, see *ZZQJ* 4, 26–28.

84. Fields, *Tso and the Muslims*, 80–81.

85. See Yang, *Zuo Zongtang*, 90, for details of postings and duty assignments.

86. Bales, 234.

87. For one of Zuo's victory reports, see *ZZQJ* 4, 11–14.

88. *QSG*, 12,029.

89. Yang, *Zuo Zongtang*, 91.

90. *QSG*, 12,029.
91. *ZZQJ* 11, 149–50.
92. *QSG*, 12,029. It is unclear whether the reference to ravaging Han women was a charge frequently leveled at Muslims. It generally appears in sources pertaining to Han Chinese rebels as well, indicating the general lack of propriety of such groups.
93. This was apparently engineered by Ma Hualong through his official contacts, possibly even Mutushan. See Bales, 241–42.
94. Yang, *Zuo Zongtang*, 93.
95. For Zuo's criticism of Mutushan's policy of suasion, see *ZZQJ* 4, 191–92.
96. *ZZQJ* 4, 179–84.
97. *QSG*, 12,030; Yang, 94; and *ECQP*, 951.
98. Yang, *Zuo Zongtang*, 96. Some sources put the execution date two weeks earlier. See *ECQP*, 951.
99. Bales, *Tso Tsung-t'ang: Soldier and Statesman of Old China*, 262–65.
100. *ZNP* 5, 34–35.
101. Yang, *Zuo Zongtang*, 96.
102. Bales, *Tso Tsung-t'ang: Soldier and Statesman of Old China*, 273–75.
103. See Chen, *Tso Tsung-t'ang: Pioneer Promoter of the Modern Dockyard*, 52–56.
104. *QSG*, 12,030. The Ili Crisis is treated at length in Hsu, *The Ili Crisis*.
105. *QSG*, 12,030–31.
106. Yang, *Zuo Zongtang*, 100.
107. Ibid; and Bales, 290–92.
108. Yang, *Zuo Zongtang*, 100.
109. *ZZQJ*, 12, 419.
110. Fields, "Campaigns," 126. Another version of this quote has Zuo saying, 'The only distinction is between innocent and rebellious. There is none between Han and Muslims'. See Liu and Smith, "The Military Challenge," 228. Hereafter *CHC* 11.
111. See *ZZQJ* 5, 282; Fields, *Tso and the Muslims*, 84–85; and Yang, 102–104.
112. Fields, *Tso and the Muslims*, 87.
113. Yang, *Zuo Zongtang*, 103.
114. Ibid., 104.
115. Yang, *Zuo Zongtang*, 105–106.
116. On the backdrop to the war and the emergence of Yakub Beg, see Kim, *Holy War in China*, 37–97.
117. *QSG*, 12, 031.
118. Ibid., 12,031.
119. On Zuo's memorial pleading his case, see *QSG*, 12,032; and Fields, *Tso and the Muslims*, 82.
120. A new Qing ruler, Guangxu, nephew of Cixi, had ascended the throne in 1875.
121. Elleman, *Modern Chinese Warfare*, 77; and *CHC* 11, 238–39.
122. *ZZQJ* 6, 265–70.
123. Elleman, *Modern Chinese Warfare*, 76–78. On sizes and location of grain stores, see Yang, *Zuo Zongtang*, 127.
124. *ZZQJ*, 12, 16–17.
125. Yang, *Zuo Zongtang*, 131.
126. *ZZQJ* 12, 117–18.
127. Yang, *Zuo Zongtang*, 167. Bai had previously skirmished with the Chinese further south, receiving limited aid from Yakub Beg.

128. Hodong Kim argues that Yakub Beg had been convinced, probably by the British, that the Qing were amenable to a diplomatic solution so he supposedly instructed his men not to fight vigorously. There was also a rumor circulating that Yakub Beg would be spared if he sent the Chinese outlaws Bai Yanhu and Yu Xiaohu to the Qing. See Kim, 169–72.
129. Yang, *Zuo Zongtang*, 137.
130. Kim, *Holy War in China*, 167.
131. *QSG*, 12,032.
132. See *QSG*, 12, 032, and Kim, *Holy War in China*, 169–70, who argues for the stroke explanation.
133. Chu, *The Moslem Rebellion in Northwest China*, vii.
134. Hou, "Tongzhi Huimin qiyi fou xibei diqu renkou qianyi ji yingxiang," 68–72.
135. See Elleman, 57–58.
136. Elleman, 58.
137. Bales, *Tso Tsung-t'ang*, 5.

Disclosure statement

No potential conflict of interest was reported by the author.

Bibliography

Atwill, David G. *The Chinese Sultanate: Islam, Ethnicity, and the Panthay Rebellion in Southwest China, 1856–1873*. Stanford: Stanford University Press, 2006.
Bales, W.L. *Tso Tsung-t'ang: Soldier and Statesman of Old China*. Shanghai: Kelly & Walsh, 1937.
Chang, Jung. *Empress Dowager Cixi: The Concubine who Launched Modern China*. New York: Alfred A. Knopf, 2013.
Chen, Gideon. *Tso Tsung-t'ang: Pioneer Promoter of the Modern Dockyard and Woolen Mill in China*. New York: Paragon Books, 1968.
Chiang, Siang-tseh. *The Nien Rebellion*. Seattle: University of Washington Press, 1954.
Chu, Wen-djang. *The Moslem Rebellion in Northwest China, 1862–1878*. Paris: Mouton & Co., 1966.
Elleman, Bruce. *Modern Chinese Warfare, 1795–1989*. London: Routledge, 2001.
Fan, Zhongyi. *Qi Jiguang zhuan* [Biography of Qi Jiguang]. Beijing: Zhonghua shuju, 2002.

Fay, Peter Ward. *The Opium War 1840–1842*. Chapel Hill: University of North Carolina Press, 1997.

Fields, Lanny. "Tso Tsung-t'ang (1812–1885) and his Campaigns in Northwestern China, 1868–1880." Ph.D. Diss., Indiana University, 1972.

Fields, Lanny. *Tso Tsung-t'ang and the Muslims: Statecraft in Northwest China, 1868–1880*. Kingston, Ontario: Limestone Press, 1978.

Goodrich, L. Carrington, and Chaoying Fang, eds. *Dictionary of Ming Biography*. Vol. 2. New York: Columbia University Press, 1976.

Hou, Chunyan. "Tongzhi Huimin qiyi fou xibei diqu renkou qianyi ji yingxiang [An Investigation into the Impact upon Population in the Muslim Revolts of the Northwestern Regions during the Tongzhi Reign]." *Shaanxi daxue xuebao* 1997, no. 3 (1997): 68–72.

Hsu, Immanuel C.Y. *The Ili Crisis: A Study in Sino-Russian Diplomacy, 1871–1881*. Oxford: Oxford University Press, 1965.

Hummel, Arthur W., ed. *Eminent Chinese of the Qing Period*. Revised ed. Great Barrington, MA: Berkshire, 2018.

Kim, Hodong. *Holy War in China: The Muslim Rebellion and the State in Chinese Central Asia, 1864-1877*. Stanford: Stanford University Press, 2004.

Kuhn, Philip A. "The Taiping Rebellion." In *The Cambridge History of China Volume 10: Late Ch'ing, 1800-1911, Part One*, edited by John K. Fairbank, 264–317. Cambridge: Cambridge University Press, 1978.

Liu, Kwang-ching, and Richard Smith. "The Military Challenge: The North-west and the Coast." In *The Cambridge History of China Volume 11: Late Ch'ing, 1800-1900, Part 2*, edited by John K. Fairbank and Kwang-ching Liu, 202–273. Cambridge: Cambridge University Press, 1980.

Luo, Zhengjun. "*Zuo Wenxiang gong nianpu* [Chronological Biography of Zuo Zongtang]." In *Xuxiu siku quanshu* [Supplement to the Four Treasuries]. 1200 vols, 539–757. Vol. 557. Shanghai: Shanghai guji chubanshe, 1997.

Michael, Franz, and Chung-li Chang. *The Taiping Rebellion: A History with Documents*. Vol. 3. Seattle: University of Washington Press, 1966.

Millward, James A. *Beyond the Pass: Economy, Ethnicity, and Empire in Qing Central Asia, 1759-1864*. Stanford: Stanford University Press, 1998.

Nie, Chongqi, ed. *Nian jun ziliao bieji* [Historical Materials on the Nian Army]. Shanghai: Shanghai guji chubanshe, 1958.

Perry, Elizabeth, ed. *Chinese Perspectives on the Nien Rebellion*. Armonk, NY: M.E. Sharpe, 1981.

Qin, Hancai. *Zuo Wenxiang gong zai Xibei.i* [Zuo Zongtang in the Northwest]. Shanghai: Shanghai shudian, 1989.

Qin, Xiangye, and Chen Zhongying. "Ping Zhe jilue [Record of the Pacification of Zhejiang]." *Jindai Zhongguo shiliao congkan* [Compilation of Historical Materials on Modern China], Vol. 24 comp. Shen Yunlong, Taibei: Wenhai chubanshe, 1966–1973.

Sim, Y.H.Teddy, and Sandy J.C. Liu. "Zeng Guofan's Applications of Qi Jiguang's Doctrines in Crushing the Taiping Uprising." In *The Maritime Defence of China: Ming General Qi Jiguang and Beyond*, edited by Y.H. Teddy Sim, 93–104. Singapore: Springer, 2017.

Spence, Jonathan. *God's Chinese Son: The Taiping Heavenly Kingdom of Hong Xiuquan*. New York: W.W. Norton, 1996.

Swope, Kenneth M. *On the Trail of the Yellow Tiger: War, Trauma, and Social Dislocation in Southwest China During the Ming-Qing Transition*. Lincoln: University of Nebraska Press, 2018.

Teng, S.Y. *The Nien Army and their Guerrilla Warfare*. Paris: Mouton & Co., 1961.

Teng, Ssu-yu, and John King Fairbank. *China's Response to the West*. Cambridge, MA: Harvard University Press, 1979.

Wakeman, Frederick, Jr. "The Canton Trade and the Opium War." In *The Cambridge History of China Volume 10: Late Ch'ing, 1800–1911, Part One*, edited by John K. Fairbank, 163–212. Cambridge: Cambridge University Press, 1978.

Yang, Dongliang. *Zuo Zongtang* [Biography of Zuo Zongtang]. Beijing: Renmin wenxue chubanshe, 2014.

Zhao Erxun (ed.), et al. comps. *Qing shigao* [Draft History of the Qing]. Vol. 4. Beijing: Zhonghua shuju, 1998.

Zhu, Xueqin. "comp. *Qin ding jiaoping Nian (fei) fanglue* [Campaign History of the Pacification of the Nian Bandits]." In *Zhongguo fanglue congshu* [Collected Campaign Histories of China] 32 vols. Vol. 1. Taibei: Chengwen chubanshe, 1968.

Zuo, Zongtang. *Zuo Zongtang quanji* [Complete Works of Zuo Zongtang], 15. Changsha: Yuelu shushe, 2014.

A predisposition to brutality? German practices against civilians and *francs-tireurs* during the Franco-Prussian war 1870–1871 and their relevance for the German 'military *Sonderweg*' debate

Bastian Matteo Scianna

ABSTRACT
The German Sonderweg thesis has been discarded in most research fields. Yet in regards to the military, things differ: all conflicts before the Second World War are interpreted as prelude to the war of extermination between 1939–1945. This article specifically looks at the Franco-Prussian War 1870–71 and German behaviour vis-à-vis regular combatants, civilians and irregular guerrilla fighters, the so-called francs-tireurs. The author argues that the counter-measures were not exceptional for nineteenth century warfare and also shows how selective reading of the existing secondary literature has distorted our view on the war.

On 1 September 1870, the battle of Sedan decided the first phase of the Franco-Prussian War. The French Army was beaten and Emperor Napoleon III marched into captivity with over 100,000 of his soldiers. Under Prussian guidance, troops from several German states had won a series of victories since the start of the campaign in early August. In Bazeilles, a small town near Sedan, French Marines and National Guards put up a fierce resistance. After being repelled, the Bavarian troops shelled the village with artillery before infantry units resumed the attack. In the heat of battle, some surrendering soldiers were shot out of hand and over 400 houses burned down. Convinced that civilians had illegally taken part in the battle, the Bavarians subsequently captured around one hundred suspects, but released them unharmed the next day.[1] International media reports and public outrage in Paris and Berlin led to mutual accusations of atrocities – fact and fiction were mixed. However, investigations after the war established that 'only' 39 civilians were either killed or wounded during the fighting in Bazeilles, in contrast to 2,600 dead soldiers on each side.[2]

The incident hinted at the many problems that the Germans encountered until the end of the war in May 1871: the continued French resistance impeded any clear separation between civilians and combatants, as calls for a *levée en masse* led to the recruitment of *francs-tireurs* (literally free shooters)[3] to supplement the new armies. Bazeilles also demonstrated the influence of the press and the precarious nature of slaughter narratives about German 'barbarism'. They cast a long shadow and would be evoked in the opening stages of the First World War in order to describe 'German atrocities'.[4]

The history of the German military and the 'totalisation' of war have sparked numerous debates over time. Not least due to the wars in Iraq and Afghanistan, references to historical 'lessons learned' from insurgencies and occupations have flourished over the last decade. But flawed examples and unsustainable *longue durée* arguments can dangerously twist historical facts and denude them of their context.[5] Especially in the German case a toxic mix of an uncritical Anglo-American fascination with Teutonic 'super soldiers' that often overlooked war crimes, the long-overdue debate that debunked the myth of a chivalrous 'clean Wehrmacht',[6] and crimes in the colonies were conflagrated since the early 2000s.[7] Some of these debates have showed a tendency, however, to analyse German military history through the lenses of the Second World War and have distorted our view on the events of 1870–1871.

In particular, the German atrocities in Belgium in 1914 have been interpreted as an almost inescapable escalatory step after the Franco-Prussian War. John Horne and Alan Kramer's influential study on 1914 also devoted special attention to German actions against *francs-tireurs* and civilians in 1870–1871. They argued that the institutionalised memory of the *Franktireurkrieg* influenced the harsh German reprisals against alleged partisans during the advance through Belgium and northern France in 1914, during which German forces killed approximately 6,500 civilians.[8] The initial debate following Horne and Kramer's book led to clarifications in their argument that are often overlooked.[9] Yet, other scholars who built on their research even described the German operations in 1914 as a deliberate terror campaign,[10] without offering context or comparisons.[11] Along similar lines, Isabel Hull saw 1870–1871 as a precedent that changed the German Army's behaviour towards civilians and irregulars.[12] She argued it was the origin of ruthless military culture centred on 'military necessity', which led to the subsequent atrocities in German colonies and during the First World War. Other authors followed this nexus between Imperial Germany, colonial war, and genocide in the Second World War and drew continuities from 'Windhoek to Auschwitz'.[13] However, several scholars have shown that the German military did not learn any operational lessons or import cultures of violence from colonial campaigns.[14]

In contrast to prior associations of German occupational regimes during the First World War as precursors to Nazi practices,[15] recent scholarship has painted a more balanced picture.[16] It is often overlooked that the atrocities in 1914 occurred during the chaotic days of an invasion and not during an occupation. The last years have seen an upsurge of new studies on German atrocities in Belgium in 1914, which were accompanied by intense debates,[17] but also comparisons to other theatres, which have repulsed ideas of a new German military *Sonderweg* in terms of mass crimes.[18] Alexander Watson upheld the notion that reminiscences of 1870–1871 played a role in 1914, but argued that German actions were neither 'unusual nor was their conduct out of place compared to other contemporary armies' norms of violence; if anything, they were milder, and therefore 'attempts by historians to present the atrocities as a prelude or pointer to Nazi genocide and annihilation warfare in eastern Europe three decades later lack credibility'.[19] Also Peter Lieb repulsed the idea of a German military *Sonderweg* in the East between 1914–1919.[20]

The next logical question would be if 1870–1871 could credibly be described as 'prelude or pointer' to 1914, and to what extent it can be classified as the foundation for a 'German way of COIN'.[21] The Franco-Prussian War – and in particular the 'people's war' after September 1870 – has been framed as a conflict 'on the road to total war'.[22] Most scholars agree that elements of restrained *ancien régime* warfare existed alongside more total aspects, such as new technology, a full mobilisation of national economic and human resources, and excesses of violence.[23] Yet, also Hull agreed with Howard that the German Army by and large behaved in a disciplined manner in 1870–1871, without wreaking 'absolute destruction'.[24] Despite the fact that the *francs-tireurs* have received very little serious study, most reference works on the Franco-Prussian War still imply German ruthlessness and the alleged headaches the *francs-tireurs* had caused them.

The doctoral dissertations of Sanford Kanter and Paul Hatley specifically studied the guerrilla warfare in 1870–1871, but are frequently overlooked. Hatley described the *francs-tireurs'* highly diverse character and their low effectiveness in combat, which merely 'prolonged the inevitable' defeat for two or three months.[25] Kanter's work stressed that the resistance movement is largely a myth.[26] Indeed, he claimed that there was neither a people's war, nor widespread destruction by German forces. His findings were backed by similar assessments by Michael Howard and German historians who based their research on primary sources. Frank Kühlich, Frank Becker, and Heidi Mehrkens all remained sceptical as to the actual level of the insurgency and arguments hinting at an escalation of violence.

Still, several authors continue to claim that there was a fierce counterinsurgency, which left a lasting 'influence on the German military until 1914, and to some extent even during the Second World War'.[27] Mark Stoneman did not consider 1870–1871 as all-out slaughter, but indirectly implied that there was *something different* that set a precedent for later.[28] The short essay by Marcus Jones also followed Stoneman's argument, and cited Bazeilles (!) as an example that the *francs-tireurs* prolonged the Franco-Prussian War.[29] David Stone claimed to provide a view from 'inside the German army', but without using German primary sources, rather he perpetuated myths about the influence and signifi-cance of the *francs-tireurs*.[30] Henri Ortholan offered detailed descriptions of the armies at the Loire and in Eastern France, without, however, consulting primary sources.[31] The same holds true for the account by Alain Gouttmann, which focused little on the *francs-tireurs*.[32] Also Colonel Armel Dirou relied largely on French post-war memoirs – with all the inherent problems and biases – to depict the *francs-tireurs'* operations and a German trend towards 'total war' based on a flawed reading of Clausewitz's search for decisive victory.[33] Still, Dirou regarded these irregular units as unorganised and in constant quarrels with regulars and civilians. Thus, he acknowledged that the government had to tame them in order to avert a prolonged civil war.

Indeed, the main lesson the French Army drew from 1870–1871 was to avoid placing their bets on the National Guard, a *levée en masse*, or still less the *francs-tireurs*, let alone praise their efforts.[34] The latter were too closely related to the Commune and could function neither as a role model for the French military (who had made a first step at regaining prestige by defeating the Commune), nor for society at large, given the Third Republic's political infighting, anti-militarism, and civilian-military tensions.[35] Calls for a Republican Army expressed the desire for an organised professional army as a school of the nation[36] ˉ not marauding and hapless *francs-tireurs*.[37] These two concepts must not be confused when analys-ing 1870–1871, or the German debates afterwards. Yet the neglect of French scholarship (and nineteenth-century context) exemplifies the often-narrow focus of scholars who work on the German Army,[38] and it is astonishing how little serious study the Franco-Prussian War has attracted despite its importance.[39]

Therefore, this chapter will examine the nature of German reactions to the French war effort under the Gambetta government after September 1870. First, it will give additional context on nineteenth-century counter-insurgency, before second, assessing the actual scale of the *francs-tireurs*, and, third, the German counter-measures. This will include a brief look at the treatment of regular soldiers and civilians in 1870–1871. This chapter will argue that the scale and intensity of *francs-tireurs* was very limited and no serious threat to overall operations in 1870–1871. Far from celebrating them as heroes, the majority of Frenchmen – who had no interest in irregular resistance – shunned the *francstireurs* whose little-fruitful actions were quickly forgotten after the war. The German counter-measures were in line with nineteenth-century practices and should not, therefore, be seen as starting point of a Germanmilitary *Sonderweg*.

Counter-insurgency in the nineteenth century

During the Napoleonic Wars, the revolutionary rhetoric of a people's war forged a rather positive image of the partisan in Germany, while many insurgencies across Europe were suppressed with varying degrees of brutality.[40] However, irregular combat threatened the established monarchical order and the norms of *ancien régime* warfare.[41] When the anti-Napoleon coalition advanced into France in 1813, they reached out to local notables to establish good relations.[42] Yet tensions escalated in eastern France when half-starving allied soldiers and the population contested for scarce food during winter; but these episodes did not spiral out of control and never had any influence on the campaign.[43] The irregular resistance of the *corps-francs* was negligible and French peasants created a new daily routine under occupation,[44] leading to a 'passive resistance' against their own government.[45] Unsurprisingly, the memory of the occupations in 1814 and 1815–1818 had been 'largely erased from the national cultural memory in favour of more glorious events, except during subsequent conflicts, when they were reconstructed as, first, justification for revenge against France's "hereditary enemy" Germany and, second, evidence of France's capacity for regeneration through defeat'.[46] Thus, it is important to note the longevity of the intrinsically linked political myths of 'brutal occupation' and 'heroic irregular resistance', and their selective use.

In Prussia, the so-called *Befreiungskriege* and their propagandised memory played an essential role in stirring up anti-French sentiments and forming a collective identity, despite the persistent regional differences.[47] Even though the soldiers in 1870–1871 often linked current events to the Napoleonic Wars, this did not mean that they could derive any practical lessons from it[48] – especially as the men were trained to fight a (regular) dynastic war.[49] As conflicts between the great powers were largely evaded after the Congress of Vienna, the monarchs successfully managed to retain the monopoly on violence by taming the partisan and keeping war separated from society. Conscription served as an instrument for improving social cohesion and as a disciplining process for nation-building.[50] Soldiers were intended as tools for upholding the dynastic order and not to waging war *among* the people. Consequently, the Prussian Army was employed – much like other European armies – to suppress internal unrest, and proved itself as backbone of the monarchical order in 1848.[51]

Besides domestic duties, small war operations remained a minor issue in the war against Denmark in 1864 or during the 1866 campaign against Austria-Hungary.[52] However, some senior German officers had gained experience through liaison duties or service under foreign flags before holding important command positions in 1870–1871. August von Werder (1808–1887), the XIV Corps' Commander, lost an arm during his service with the Russians in the Caucasus, and Ludwig von der Tann (1815–1881), commander of the I Bavarian Corps, observed French operations in Algeria in 1843 and headed a free corps in the First Danish War. August Karl von

Goeben (1816–1880), commander of the VIII Corps and later the First Army, even published on his deeds in the First Carlist War and in Morocco. Yet such experiences were not widespread among NCOs or the rank-and-file in 1870. Furthermore, the handful of writers on small wars had very little influence – what mattered was great-power war.[53]

One must not forget that even in countries whose armies were more strongly involved in small wars, such as the British, Americans, French, Polish, Spanish, Russians, Austrians, and Italians, there were no concise counter-insurgency doctrines, and the importance attached to irregular warfare was secondary at best.[54] Insurgencies were brutally suppressed: looting, pillaging, and requisitioning were commonplace in the nineteenth century, as were reprisals against civilians or alleged irregulars.[55] In conclusion, the involvement in counter-insurgency operations constitutes a minor paradigm in Prussian military history and theoretical writings prior to 1870, and reminiscences of the Napoleonic Wars could not substitute practical or theoretical schooling in anti-partisan warfare – when the German armies advanced into France, they were not expecting irregular resistance, and nor were they trained to counter it.[56]

German atrocities 1870–1871?

In 1866, Prussia had won a decisive victory over Austria that changed the balance of power on the continent.[57] Merely four years later, a quarrel over the Spanish throne officially started the Franco-Prussian War. Despite Bismarck's manipulations, the war was initially perceived as a legitimate German response against French aggression.[58] The German armies prevailed in the bloody battles of August 1870, and advanced into France, where the campaign culminated on 1 September with the victory at Sedan.[59] After merely four weeks, the German forces had defeated most of the highly experienced French soldiers and taken close to 300,000 prisoners. The Prussian General Staff had proven more capable at managing the fog of war, manoeuvring and controlling the troops,[60] and the superior German artillery outweighed the strength of the French Chassepot rifle.[61] Given the norm of European war, the powers would now resort to the negotiation table. But in Paris, Léon Gambetta (1838–1882) and General Louis-Jules Trochu (1815–1896) established a new Government of National Defence on 4 September 1870.[62] Gambetta refused to surrender and used inflammatory revolutionary rhetoric to stir up a nationwide *levée en masse*. The Germans were forced to mobilise over one million men in order to control the vast territory, repel possible attempts to relieve a Paris under siege, and fight on the different fronts.[63] In fact, French resistance was centred on the capital, the Loire, and the so-called Army of the East (around Belfort, Besançon, and in the Vosges). The following will first analyse German behaviour *vis-à-vis* French regular soldiers and civilians, second, attempt to assess the scale of the new French forces; and finally, investigate the countermeasures employed to fight irregular resistance where it emerged.

No brutalisation: the treatment of regulars and civilians

The sudden collapse and chaotic retreat of the French Army made a great impression on the populace. Seeing their protector demoralised and beaten, they expected the cruel-natured enemy to turn against them.[64] Many 'slaughter narratives' and false rumours led to collective fears bordering panic.[65] Yet, in many cases the German forces functioned rather as policemen and not in a way the myth of the 'ruthless occupiers' would have us believe. For example, *francs-tireurs* patrolling the villages outside Paris told atrocity stories about *les Prussiens* in order to make the inhabitants abandon their houses, which they then looted.[66] When the civilians encountered Prussians, they returned to their destroyed and plundered houses, trusting them more than the *francs-tireurs*,[67] and peasants even turned in their compatriots to the Germans out of indifference to the struggle or fear of reprisals.[68]

The interactions were largely marked by the peaceful passivity of the population and proper behaviour of German troops.[69] However, tensions with civilians did erupt, particularly when combat operations were conducted in urban centres, which the German soldiers particularly detested.[70] Moreover, when they were unable to track down *francs-tireurs* after ambushes, their anger was at times directed against civilians – especially if they suspected them of being informants or disguised rebels.[71] Sometimes, hostages were taken in order to assure the payment of contributions or to increase the safety of German troops.[72] This was, however, a custom of war[73] and there is no proof that any hostages were harmed during the conflict.[74] Many misunderstandings and merciless actions did occur, but this was still far from outright mutual hatred and deliberate slaughter.[75] By the way of comparison, the much-publicised three-week German bombardment of Paris had resulted in 97 casualties, whereas up to 20,000 Frenchmen fell victim to the outright civil war during the crushing of the Commune.[76] After the armistice in January 1871, almost no incidents occurred across France, which left François Roth concluding that 'the occupation after the Franco-Prussian War was neither a terror regime, nor an arbitrary regime'.[77]

Indeed, it is more than questionable that a 'personal hatred of the French'[78] influenced German behaviour, as the relations with civilians and professional French soldiers were rather "normal".[79] The French (metropolitan) regulars' bravery was respected and they were perceived as equals.[80] There prevailed a soldierly ideal of war as honourable duel, diametrically opposed to any personal hatred. The good treatment of French POWs[81] is another indicator that contradicts the claim of boundless brutalisation. Despite the lacking preparation for such large prisoner numbers and the beginning of harsh winter climate the mortality rate remained low: in Bavarian camps 1,508 out of 39,339 POWs had died by mid-February 1871 (3.8 per cent) and in the North German Confederation 7,230 out of 285,124 (2.5 per cent); low percentages that led to a positive review by a French delegation, which had been allowed to inspect the camp conditions.[82]

Again, only by comparisons we can better understand the German conduct: when the French Army of the East fled with 85,000 men to neutral Switzerland and was interned there, around 1,700 men perished in captivity (2.0 per cent),[83] and the mortality rate in both Confederate and Union camps during the American Civil War was much higher – out of 195,000 imprisoned Union soldiers approximately 30,000 died in captivity (15.4 per cent), and around 26,000 of the 215,000 Confederate prisoners did not survive (12.9 per cent).[84] Thus, when analysed in context, the treatment of French prisoners can hardly account for German brutalisation. *Francs-tireurs* did not always enjoy prisoner of war status,[85] but were often taken captive.[86] After the capitulation of Strasbourg, for example, many *francs-tireurs* were released on their promise not to take up arms again and no one was executed.[87] Most interestingly, the few existing documents indicate that German POWs in the hands of *francs-tireurs* also fared relatively well, which precipitated German comportment *vis-à-vis* captured *francs-tireurs*.[88] This would support the findings on relations between regular soldiers, which were marked by informal agreements that often kept violence as low as possible in a sort of 'live-and-let-live' system.[89]

In conclusion, the German forces did not deliberately or systematically target civilians, who largely remained unharmed passive bystanders; nor did their treatment of French regular soldiers hint at any radicalisation processes. Thus, it is vital to differentiate between the treatment of regular uniformed soldiers, civilians, and *francs-tireurs*, while it is also useful to take a closer look at the German reactions and counter-measures against irregular resistance.

Every Frenchman to the front?

Any analysis has to differentiate between the new regular units, the (mostly uniformed) National Guards and the less structured, less disciplined men in the *Garde Mobile* and the *francs-tireurs*.[90] The lines between these units remained blurry, however, as manpower fluctuated and both sides had an interest in over-stating their numbers: it provided the Germans with an excuse for setbacks and the French government could celebrate an alleged volunteer spirit. Further *franc-tireur* forces were rallied in and around Paris,[91] accompanied by calls for a *guerre à outrance*.[92] In the (less enthusiastic) provinces,[93] the Republican Government of National Defence conscripted around 5,000 new troops per day for their regular formations, which increased the National Guard to 320,000 men, and overall force levels reached 830,000 by February 1871.[94] Yet, this does not mean that they were all combat-ready frontline troops or that they could be employed *en bloc* against the German occupiers.

It is virtually impossible to give precise estimates of the various *francs-tireurs* formations, let alone document their losses. Michael Howard estimated 57,300 *francs-tireurs* under arms (based on French official sources published after 1900) and one thousand German fatalities due to their actions.[95] French reports on killed *Prussiens* vary, but many are too high, for example in the Vosges.[96] Stéphane Audoin-Rouzeau stated that it is impossible to provide exact numbers, and listed sources offering a range between 17,000 and 141,000 *francs-tireurs*.[97] Forrest acknowledged the same problems, and cited over 20,000 volunteers for Paris and an additional 11,502 in the main recruiting areas – for both the *Garde Mobile* and the *francs-tireurs*.[98] Dirou listed 393 battalions with 72,000 *francs-tireurs* (including 3,000 officers and NCOs) that he deemed operative between September 1870 and February 1871 – without acknowledging the highly fluid character of these formations.[99] In fact, only by disregarding traditional patterns of French history could we believe the myth of a rural population eager to enrol for a distant government.[100] Many men joined the irregular forces to dodge conscription, and desertion posed a severe problem – especially after setbacks and the sharp decline of morale in January 1871.[101] Most *francs-tireurs* formations never gained any reasonable combat efficiency or military discipline; they wore no proper uniforms and suffered from poor supply, which led to many quarrels with civilians.[102] The French government recognised these problems: the *francs-tireurs* needed official combatant status and stricter discipline. Severe measures in regards to the latter were taken in October and November 1870, including court-martialling.[103]

Other scholars' estimates on insurgent levels rely mainly on Howard, whereas Sanford Kanter calculated the number of insurgents killed. Based on an unpublished official French inquiry after the war, he placed the total of *francs-tireurs* killed in action at 53.[104] Furthermore, he stated that the overall damage in the whole country was constantly exaggerated, while in reality the occupied regions had barely experienced any disruption.[105] Kühlich also rejected the idea that there had been a full-scale people's war and dismissed as absurd the claims about 1870–1871 representing a new kind of warfare.[106] Wolfgang Etschmann also held that the *francs-tireurs* did not hamper any operations, setting the death toll at several hundred fatalities on both sides.[107] The exact numbers will remain unknown, but they were marginal in comparison to overall French casualties. In total, the French army lost 24,031 soldiers in action (plus an additional 28,896 men who died from combat-inflicted wounds and diseases such as typhus and dysentery), 89,228 wounded and 14,138 missing; while most civilian losses occurred during the sieges of Paris (275), Belfort (262), Mézières (53), and Strasbourg (400).[108] In comparison, the German states had on average around 800,000 men in the field, of whom 28,208 died in combat or from battle-inflicted wounds, almost 15,000 died from illnesses, and 88,488 were wounded. A great majority of casualties (on both sides) occurred in the initial encounters in August and September until the battle of Sedan, i.e. the first phase of the war that preceded the 'guerrilla campaign'.[109] In fact, German fatalities dropped from 12,299 in August (1.6 per cent of combat strength), to 6,788 in September, 4,999 in October, 3,392 in November, and saw a quick rise to 4,476 in December (0.3 per cent of combat strength) and 4,141 in January (as German troop levels also rose to almost one million), before ultimately falling to 3,277 between February and June.[110] Even if one accepts Howard's account of roughly one thousand German fatalities due to *francs-tireurs* activities during the entire war, it can still be hardly seen as a serious threat. In general, the military results of Gambetta's call to arms were a disaster – the troops were poorly equipped and officered, and never stood a chance against professionally trained and combat hardened German soldiers.[111] But how did the Germans react when they encountered irregular resistance?

Sticks and carrots: German counter-measures

The German high command was upset about the continued French war effort, which they perceived as a futile and unnecessary prolongation of hostilities.[112] The soldiers had anticipated (another) brief campaign and believed in fighting a just war against French aggression.[113] Likewise, international opinion found it dishonourable to continue a war with a *levée en masse* that Paris had started and already lost.[114] Given the nature of insurgencies, Röhkramer argued 'it is understandable why the German soldiers were primarily concerned with their own safety. Without playing down individual cases of excessive cruelty, one has to conclude that a national war necessarily includes violence of this kind. The soldiers are less responsible than the circumstances of war'.[115] Thus the soldiers in the field gradually came to accept the harsh measures taken (or claimed to be taken) by some of their comrades,[116] but were wholly inadequately prepared for such a kind of warfare. As Stoneman noted, the Germans had expected a fair and open fight, and not 'an enemy who would not shoot and then run, hide or pretend to be an innocent civilian'.[117]

The Prussian King had compared irregular resistance to banditry – to be punished with ten years' imprisonment, or execution if individuals had harmed German troops.[118] This official proclamation left room for commanders on the ground and we should not take gruesome political rhetoric, for example also by Bismarck, for the same thing as actual procedures in the field.[119] The Germans struggled to differentiate between the diverse units, but were willing to ascribe the *francs-tireurs* combatant status if they were wearing uniforms, carried their weapons openly, and operated under military command – according to the customs of war.[120] Indeed, the legal situation in 1870 was far from clear. The Geneva Convention of 1864 and the Saint Petersburg Declaration of 1868 had only regulated the treatment of the wounded and banned certain projectiles, and harsh reactions to irregulars were commonplace in the nineteenth century.[121] We should also bear in mind that 'what in 1870–1871 were arguably legal reprisals would by the twentieth century be regarded as war crimes'.[122] Still, the German Army distinguished between the war-mongering of the Government of National Defence, the regular soldiers, the mostly-peaceful inhabitants, and the *francs-tireurs*.

The German forces were spread all over France: from Paris to the Vosges. Their main aim was to uphold the siege of Paris and suppress the newly recruited armies that operated mainly in the Loire region and near the Vosges. The Germans did not change their operational procedure considerably and the counter-measures entailed limited active elements, besides more active patrolling of railways and roads. In order to protect their lines of communication and to counter-raid the enemy, they established small 'flying columns', which chiefly resulted in unsuccessful and frustrating endeavours.[123] The patrolling duties put further strain on the infantry, as they often had to shield the precious (and scarce) cavalry units[124] or march along the railway lines.[125] In short, the lack of any coherent counter-insurgency strategy during the war was apparent. But had there been any need to develop one?

We repeatedly come across the number of around 100,000 Germans deployed (of over one million mobilised) in order to protect their lines of communication and rear areas for the 500,000 front line troops.[126] However, it was almost exclusively the *Landwehr* (i.e. second- or third-rate battalions) who secured these areas.[127] Any army needs to safeguard the lines of communication to secure operational freedom and supply. Accordingly, the safeguarding of railways should not prima facie be interpreted as a countermeasure forced upon the Germans by guerrilla activity. Further, the actual raids against railways and bridges posed no serious threat to the outcome of the war. The most successful such raid – against a bridge in Fontenoy-sur-Moselle on 22 January 1871 – destroyed its target, but it occurred very late in the war, and the bridge was fully repaired within two weeks.[128] At the same time, the defence of Dijon, often depicted as a large-scale encounter, 'merely' resulted in 25 Prussian fatalities and had no strategic consequences.[129] Thus, it is questionable whether the actions of the *francs-tireurs* (which are difficult to separate from those of the Army of the East in general) were really 'frequent occurrences' that 'imposed crippling shortages on the Prussians'.[130]

Nevertheless, the German general staff did have difficulties supplying their men on foreign soil. Lothar Sukstorf even went so far as to speak of a 'German crisis' in early December, when due to a lack of supply and reserve forces, marching duties and sickness, German troops' combat-readiness was critically low, with repercussions for discipline and morale.[131] In some cases, German authorities allowed 'wild' food requisitioning, but the situation never spiralled out of control, not even during the harsh winter. Furthermore, Sukstorf did not attribute these difficulties to the *francs-tireurs*.

However, safeguarding supplies and controlling the territory were not the only concerns. All surviving evidence on active engagements by *francs-tireurs* mention either ambushes or fierce resistance in urban centres. At the Loire, the favourable terrain helped the newly levied troops harass the occupying forces.[132] Many reports speak vaguely of ambushes or single shots fired at enemy columns, which were difficult to attribute. In response such incidents, the Germans often marched to the nearest village. On 18 October 1870, German troops reached Châteaudun, a town of 6,000 inhabitants, where approximately 4,000 National Guards and *francs-tireurs*[133] refused to surrender and fortified houses for the defence. They repelled the first wave of attackers, after which German artillery 'prepared' the town for a second advance.[134] The French rallied civilians to join the battle, and the evolving fight 'made Bazeilles pale by comparison'.[135] While many houses burned, the fierce fighting continued throughout the night, with innocent residents also perishing.[136] Nevertheless, there were no reprisals, and after the battle the Germans took 44 *francs-tireurs* prisoner.[137] Other than the number of burned houses and captured French, we have no reliable data on the number of fatally wounded.[138] Châteaudun was immediately used as a symbol for resistance and German brutality.[139] But this case also demonstrates, that even in fierce urban encounters, there were limitations to operational procedures. *Francs-tireurs* were rarely shot outright, despite the high command's gruesome rhetoric. Wartime propaganda and the post-war desirability of painting the picture of a brutal invader should not distort our view of the facts. Examples such as Bazeilles and Châteaudun were the exception: in countless towns, the mayor and other notables managed to preclude a defence.[140] The Bavarians, for example, behaved terribly in Orléans, but nothing comparable happened in Rouen, Nancy, Reims, Dijon or Tours. Roth has argued that the taking of tributary payments was the most common form of reprisals against civilians,[141] while Mehrkens concluded that 'a consistent procedure [vis-à-vis the francs-tireurs] cannot be demonstrated by the source material. The individual commanders decide[d] themselves on the spot, how to react to attacks and how to deal with captured national guards and francs-tireurs'.[142] Even in the rare cases where towns were burned down, the civilians had left already.[143] Thus, even the sporadic acts of irregular resistance did not necessarily lead to bloodshed – situational factors were the main driver for the few excessive applications of force.[144]

Conclusion

The German soldiers did not adopt a 'kill anything that moves' policy in 1870–1871. As argued above, the Germans were placing their bets on deterrence, and resorted to demanding monetary payments. Reprisals against alleged insurgents were 'in comparison to other wars, particularly in the twentieth century, rather mild'.[145] The views of the Germans in 1870–1871 were comparable to other historical examples – not least the US Civil War: the soldiers lacked counter-insurgency experience, and detested the specific nature and methods of guerrilla warfare, rather than the overall French resistance or the French people in general.[146] Their treatment of regular soldiers, POWs and civilians fell within the norms of war. On the other hand, the Germans held little sympathy for the irregular francs-tireurs. Sometimes they shot them on the spot, in other cases they were put on trial and even after that often unharmed. The German forces clearly committed acts of excessive violence that would today be seen as war crimes, i.e. the killing of innocent civilians. However, based on the almost unanimous scholarly verdict, these reactions were far from exceptional or excessive, and declarations that demanded a ruthless dealing with francs-tireurs and enemy resistance in general should always be compared to actual deeds. There was neither a carte blanche nor a 'Commissar Order', and German soldiers could be held responsible for illegitimate acts of violence. Situational factors, such as combat stress, fears of subversion, and threats to supply, were the main reasons for excessive violence and sometimes-harsh symbolic punishments.[147] In this regard, the German occupation was not very different from 1813–1815, where relations had been rather cordial, and deteriorated only where a combined set of situational factors emerged.

The Franktireurkrieg should finally be recognised as what it was: an exaggerated myth. As Alan Forrest stated in accordance with Kanter: it 'was an invention, a convenient alibi for failure, since in many parts of the country the call for a partisan insurrection was met with embarrassing indifference, especially among the peasantry, while agriculture was largely left undisturbed by the Prussian invasion, even in many departments that lay directly in the invader's path'.[148] Rather, the idea of a spontaneous rush to the colours fitted the French revolutionary narrative and the goal of creating an image of German ruthlessness, which could be used as rallying calls in 1914 or after 1940. Yet politicised wartime propaganda – which various scholars have debunked – continues to serve as a point of reference for *longue durée* arguments about the German military.

Pieper maintained that the Germans had 'a predisposition towards unusually brutal counterinsurgency'.[149] However, he also cited Hull's assessment of German East Africa (1905–1907), which is backed up by Bührer's findings, and Lieb's research on Ukraine 1918, as counterexamples to German brutalisation. If we add to this the almost unanimous scholarly verdict that the Germans did not exercise extreme measures in 1870–1871, as well as the more differentiated analysis of German atrocities of 1914, the picture of an exceptional 'German Way of War' before the Second World War must be questioned. Further research has to investigate whether an alleged 'guerrillaphobia'[150] and 'dread of irregular warfare' was inherent or even 'endemic'[151] in the German Army after 1870–1871. After all, which armies do like guerrilla wars? And why should the German military have been daunted by *cauchemars* of a phenomenon that even the French acknowledged to be a failure? Only transnational and contextualised comparisons – based on archival research, or, at least full use of available secondary literature – can provide anything resembling satisfactory answers and enhance our understanding of the German armies between 1870 and 1945.

Notes

1. Mehrkens, *Statuswechsel*, 114.
2. Initial French reports had set the civilian death toll at 2,000 civilians: Showalter, *Wars of German Unification*, 279.
3. The term *franc-tireur* dates back to Napoleonic times and described an early form of light infantry separate from the regular army. In 1870–1871 these were mainly local volunteer units, often not fighting directly under government command. There existed a great variation and confusion in the usage of terms such as guerrilla, partisan, insurgent etc. Even today there are many hybrid forms, with the irregular character of the person in question representing the common denominator, in contrast to a regular uniformed soldier that visibly operates under a state authority and chain of command. Thus here the terms insurgent, guerrilla, partisan, and *franc-tireur* will be used interchangeably.
4. Mehrkens, *Statuswechsel*, 113, 127.
5. Gumz, "Reframing the Historical Problematic of Insurgency," 555–9; Scheipers, "Counterinsurgency"; and Porch, *Counterinsurgency*.
6. On the 'Wehrmacht debate' see Hartmann et al, *Verbrechen der Wehrmacht*.
7. An example for such selective argumentation is Melson, "German Counterinsurgency Revisited"; now extended in Melson, *Kleinkrieg*.
8. Horne and Kramer, *German Atrocities*, 89–174.

9. Anderson, "A German Way of War?"; Horne and Kramer, "German Atrocities in the First World War: A Response"; and Anderson, "How German is it."

10. Lipkes, *Rehearsals*; Zuckerman, *The Rape of Belgium*; and Nelson, "Ordinary Men."

11. Kramer later embedded the German case in a wider European context: see Kramer, *Dynamic of Destruction*; also Alan Kramer, "German War Crimes 1914 and 1941."

12. Hull, *Absolute Destruction*, 117–30. See also Hull, *A Scrap of Paper*; Messerschmidt, "Völkerrecht und 'Kriegsnotwendigkeit'."

13. Zimmerer, *Windhuk nach Auschwitz*; Madley, "From Africa to Auschwitz"; and Baranowski, *Nazi Empire*.

14. Kuß, *Deutsches Militär*; Bührer, *Die kaiserliche Schutztruppe*; and Gerwarth and Malinowski, "Der Holocaust als 'kolonialer Genozid'."

15. Liulevicius, *War Land*.

16. Lieb, "Suppressing insurgencies"; Kauffmann, *Elusive Alliance*; Becker, *Les cicatrices rouges*; and De Schaepdrijver, *Military Occupations*.

17. Münch, *Bürger in Uniform*, 182ff; and Bönker, "A German Way of War." Spraul, *Der Franktireurkrieg* has to be treated with caution; more useful is Keller, *Schuldfragen*. Yet, his findings have been met with fierce criticism; see Kramer and Horne, "Wer schiesst hier aus dem Hinterhalt." A good overview on this debate is provided by Pöhlmann,"Habent sua fata libelli."

18. On the Russian invasion in Galicia, see Prusin, *Nationalizing a Borderland*; on East Prussia; Watson, "Unheard-of Brutality"; and on Serbia, Gumz, *Resurrection and Collapse*, 44ff.

19. Watson, *Ring of Steel*, 132.

20. Lieb, "Der deutsche Krieg im Osten."

21. The literature on the lessons the German Army drew from 1870–1871 is vast. Yet, most authors focused on the regular people's war and not on how the Germans intended to counter guerrillas; an often-overlooked exception is Potempa, "Der kleine Krieg." Potempa argued that the Germans were preoccupied by threats to their lines of communication in future conflicts, not least from irregular forces. For general debates in German military journals, see Pöhlmann, "Das unentdeckte Land," and the classic accounts, Showalter, "From Deterrence"; Förster, "Facing 'People's War"; Foley, *German Strategy*, 14–37; and Echevarria, *After Clausewitz*.

22. Förster and Nagler, *On the Road to Total War*; for general reference, see Wawro, *Franco-Prussian War*; Audoin-Rouzeau, *1870*; Roth, *La guerre*; and Gersdorff and Groote, *Entscheidung 1870*.

23. Mehrkens, *Statuswechsel*, 256; Howard, *Franco-Prussian War*; Becker, *Bilder*; and Kühlich, *Die Deutschen Soldaten*.

24. Hull, *Absolute Destruction*, 119.

25. Hatley, "Prolonging the Inevitable."

26. Kanter, "Defeat 1871," 31ff; condensed in Kanter, "Exposing the Myth."

27. Pieper, "German Approach," 2; He upholds the same argument in Pieper, "From Fighting." See also Dirou, *La guérilla*, 121ff.

28. Stoneman, "The Bavarian Army."

29. Jones, "Fighting," 189.

30. Stone, 'First Reich,' 183ff.

31. Ortholan, *L'Armée de l'Est*; and Ortholan, *L'Armée de la Loire*.

32. Gouttman, *La grande défaite*.

33. Dirou, *La guérilla*, 225ff.
34. Instead, the army attempted to regain lost ground by augmenting the professionalism, discipline, and civic education of the regular forces; see Porch, *The March*, 35–7; Chanet, *Vers l'armée nouvelle*. After all, even the French 'army leaders had no faith in the conscript soldier' and the 'French people were unwilling to accept the rigors of compulsory military service', see Challener, *The French Theory*, 5, 31–2.
35. Varley, *Under the Shadow*, 203, 205.
36. Crépin, *Défendre la France*, 331ff.
37. Forrest, *The Legacy*, 147, 157ff, 171ff.
38. Excellent in depicting the reciprocal relationship is Nolan, *The Inverted Mirror*; see also Schivelbusch, *Culture of Defeat*; and Tison, *Comment sortir*.
39. This was not the case in the late nineteenth century: in 1898, a bibliography on the war listed over 7,000 titles, see Howard, *Franco-Prussian War*, vii.
40. Heuser, "Small Wars"; Rink, "The Partisan's Metamorphosis," 27, 33. The internal suppression of the revolt in the Vendée was arguably the most brutal, see also Esdaile, *Popular Resistance*.
41. Bell, *First Total War*, 8–12, 265, 275–81.
42. Hantraye, *Les Cosaques*, 18–9.
43. Leggiere, *The Fall*, 77.
44. Hantraye, *Les Cosaques*, 27, 55, 274–5.
45. Leggiere, *The Fall*, 72–3, 76–8, cited on 77.
46. Haynes, "Remembering," 540.
47. Jeismann, *Das Vaterland*, 76ff.
48. Mehrkens, *Statuswechsel*, 145ff.
49. Clark, "The Wars," 560–5.
50. Frevert, *A Nation in Barracks*, 2, 70.
51. After the Franco-Prussian War, the French and German approach to internal policing greatly varied and the Prussian Army was far less involved: see Johansen, *Soldiers as Police*. It would be interesting to compare the Prussian case – in 1848, but also later – to other armies. The Italian Army, for example, was deployed to a far greater extent against domestic uprisings in the late nineteenth century, see Gooch, *Italian Army*, 12–3.
52. Hahlweg, *Guerilla*, 66–7.
53. Ibid., 62ff. See also Daase and Davis, *Clausewitz on Small War*. Yet, one has to remember that Clausewitz's star only began to rise during the German Wars of Unification and his writings on small wars were not at the forefront of attraction.
54. Beckett, *Modern Insurgencies*, 31, 35; Heuser, *The Evolution*, 398, 405; Rid, "The Nineteenth Century"; Porch, "Bugeaud, Gallieni, Lyautey"; and Beccaro, "Carlo Bianco and *Guerra per bande*."
55. Nabulsi, *Traditions of War*, 21–2.
56. Dirou overlooks these differences, see Dirou, *La guerrilla*, 155–88.
57. On 1866 and the pre-history to 1870 see, Wetzel, *Duel of Nations*; Showalter, *Wars of German Unification*, 123–200; and Wawro, *Franco-Prussian War*, 20–40.
58. Showalter, *Wars of German Unification*, 216ff.
59. For a detailed operational account, see Wawro, *Franco-Prussian War*, 85–229.
60. Sukstorf, *Problematik der Logistik*, 2.
61. On the French high command's problems, see Porch, *The March*, 45ff.

62. Gambetta's motifs and the Government of National Defence are described vividly in Mayeur, *Léon Gambetta*.
63. For an in depth description, see Wawro, *Franco-Prussian War*, 257–98.
64. Roth, *La guerre*, 67–8.
65. Mehrkens, *Statuswechsel*, 97.
66. Hatley, "Prolonging the Inevitable," 152–6. Many such atrocity stories were quickly corrected.
67. Forbes, *My Experiences*, vol. II, 63–4, 262.
68. Porch, *The March*, 34.
69. Kühlich, *Die Deutschen Soldaten*, 300, 302; and Stoneman, *Bavarian Army*, 283. For example, the cases of sexual violence were very low, as the German armies maintained a strong civilian behavioural code, see Mehrkens, *Statuswechsel*, 173.
70. Kühlich, *Die Deutschen Soldaten*, 353.
71. Rindfleisch, *Feldbriefe*, 28, 109–12. Indeed the French at times used civilians for intelligence, Foudras, *Les Francs-Tireurs*, 67–8.
72. Lassberg, *Mein Kriegstagebuch*, 155. After the Germans were shot at in Châteauneuf, they took the mayor prisoner and threatened to burn the village if more attacks took place, which ended hostilities, Foudras, *Les Francs-Tireurs*, 44.
73. Nabulsi, *Traditions*, 34–6. More problematic was the placing of hostages on trains or using them as shields during reconnaissance missions, where they were exposed to greater dangers: see Cardinal, *Deutsch-Französischer Krieg*, vol. II, 170.
74. Mehrkens, *Statuswechsel*, 188, 195–6.
75. Rohkrämer, "Daily life," 504.
76. Over this timespan the Germans lost several hundreds to counter-battery fire and French mortality from diseases was around 4,000. Also in this case, 'most of the population remained as spectators' during the battle for Paris – before the massacres – the *Versaillais* lost 'merely' 1,500 men, see Tombs, *The War*, 162.
77. Roth, "Occupation et liberation," 316–7.
78. Stoneman, *Bavarian Army*, 291. It would be interesting to investigate the Bavarian example further, as the French had been their traditional ally.
79. Mehrkens, *Statuswechsel*, 138.
80. Michael Howard described such phenomena as a 'professionalization restraint' between regular soldiers, in contrast to the treatment of rebels: see Howard, "Constraints on War."
81. There were rumours about the alleged 'brutal nature' of French colonial soldiers, which led to often-brutal behaviour towards them, Becker, *Bilder*, 153–7.
82. Botzenhart, "French Prisoners," 590. Those who died in captivity included soldiers that had been wounded on the battlefield.
83. Ortholan, *L'Armée de l'Est*, 199–00.
84. On both sides the mortality rate hit double digits, while many myths about deliberate neglect have been debunked: see Gillispie, *Andersonvilles of the North*. There are apparent differences between the conflicts, however, which must not be overlooked, see Degler, "The American Civil War."
85. Botzenhart, "French prisoners," 588.
86. Hatley, "Prolonging the Inevitable," 136.
87. Ibid., 124.
88. Kühlich, *Die Deutschen Soldaten*, 428; and Bizzoni, *Impressioni*, 144.
89. Rohkrämer, "Daily life," 500. Here I am also referring to the findings of Tony Ashworth on similar behaviour in the First World War.

90. Detailed in Mehrkens, *Statuswechsel*, 129–35.

91. Some groups existed already before the campaign, drawing from members of shooting clubs, Howard, *Franco-Prussian War*, 249–50.

92. Forrest, *The Legacy*, 127. In general, the urban population seemed more anti-German than that in the countryside – Roth, *La guerre*, 152 – and war enthusiasm varied greatly between urban and rural centres, Audoin-Rouzeau, *1870*, 216, 239.

93. Mainly in the North, East, and along the northern parts of the Loire.

94. Tombs, *The War against Paris*, 2; Audoin Rouzeau, *1870*, 187.

95. Howard, *Franco-Prussian War*, 252.

96. Wolowski, *Campagne*, 70–1; in comparison to the official statistics provided in Löhlein, *Feldzug*, 319.

97. Audoin-Rouzeau, *1870*, 198.

98. With an additional '1,132 in the Ardennes, for instance, 1,165 in the Gironde, 1,250 in the Indre-et-Loire, 1,823 in the Bouches-du-Rhône, 2,807 in the Seine-inférieure, 3,325 in the Nord', Forrest, *The Legacy*, 126.

99. Dirou, *La guérilla*, 20.

100. See the classic argument on the distance between government and rural population in Weber, *Peasants into Frenchmen*. Yet, Weber's arguments have also been qualified in recent years, see Ford, "Peasants into Frenchmen."

101. Roth, *La guerre*, 344; and Audoin-Rouzeau, *1870*, 253–5.

102. Howard, *Franco-Prussian War*, 252–3.

103. The anti-republican General Charles-Denis Bourbaki (1816–1897) struggled immensely to uphold combat morale, see Meier-Welcker, "Der Kampf mit der Republik," 117. See also Bizzoni, *Impressioni*, 210–13; and Foudras, *Les Francs-Tireurs*, 19, 35–8.

104. Kanter, "Defeat 1871," 184–5.

105. Kanter, "Exposing the Myth," 15.

106. Kühlich, *Die Deutschen Soldaten*, 312, 314.

107. Etschmann, "Guerillas und Franctireurs," 40.

108. Roth, *La guerre*, 507–8. Roth did not provide any sources for his numbers, but his account is the most detailed. He also added that the sieges were particularly bloody for the regular armies as well, particularly due to diseases, see Ibid., 509.

109. Howard, *Franco-Prussian War*, 453; Roth, *La guerre*, 508. In 1866, the rapid-firing Prussians had 'consistently killed, wounded, or captured five Austrian soldiers for every casualty of their own', Wawro, *Franco-Prussian War*, 51, 307; thus these numbers hint at the possibility of the French regular army to inflict severe casualties.

110. Engel, *Die Verluste der deutschen Armeen*, 282.

111. The battle of Beaune-la-Rolande on 28 November 1870 is just one example: 9,000 Germans successfully repulsed repeated attacks by 60,000 Frenchmen, at the cost of – both dead and wounded – 850 casualties to 8,000, see Wawro, *Franco-Prussian War*, 271–4; and Porch, *The March*, 10.

112. Förster, *Moltke*, 227, 230.

113. Kühlich, *Die Deutschen Soldaten*, 130–2.

114. Roth, *La guerre*, 61, 136–7.

115. Rohkrämer, "Daily life," 509.

116. Loch and Vette, *Friedrich Clauson von Kaas*, 176. The Prussian officer Kaas received reports of several hundred executed Frenchmen, but was never able to verify the

stories of the boastful Bavarian cavalry: see Ibid., 136. However, it would sustain Stoneman's argument of a special predisposition of the Bavarian cavalry. Yet it is questionable if this was a cultural phenomenon or the situational aspect of cavalry reconnaissance and a peculiar unit culture. Christoph Hertner (University of Bern) is currently researching this field.

117. Stoneman, *Bavarian Army*, 276.
118. Cardinal, *Deutsch-Französischer Krieg*, vol. V, 72–3.
119. As done, e.g. in Dirou, *La guérilla*, 236ff.
120. Howard, *Franco-Prussian War*, 208, 378–1.
121. Mehrkens, *Statuswechsel*, 25–39; Toppe, *Militär und Kriegsvölkerrecht*, 86; and Best, *Humanity in Warfare*.
122. Tombs, "The Wars against Paris," 562.
123. It seems that the Germans seldom attempted night attacks at the *francs-tireurs* in their strongholds, see Bizzoni, *Impressioni*, 266.
124. Kühlich, *Die Deutschen Soldaten*, 316.
125. The cavalry's dealings with *francs-tireurs* and their own unit culture would deserve an independent study.
126. Wawro, *Franco-Prussian War*, 288–9. The exact displacement of forces and orders for rear duties would also represent a desirable subject of future research.
127. Roth, *La guerre*, 384.
128. Cardinal, *Deutsch-Französischer Krieg*, vol. IV, part 2, 247–99, for the most detailed account. According to Wawro – who cited Horne and Kramer – the Prussians went on a 'killing spree' in the village, 'spearing the inhabitants with their bayonets and heaving them into the flames', see Wawro, *Franco-Prussian War*, 279. Horne and Kramer's source is a French study from 1902; see Horne and Kramer, *German Atrocities*, 142, 489. Other reports spoke of monetary contributions only: see Roth, *La guerre*, 409.
129. Molis, *Les Francs-Tireurs*, 175–9. Giuseppe Garibaldi (1807–1882) had rallied volunteers to fight for the French cause, but relations with General Bourbaki and civilians were not always harmonious, see Ortholan, *L'Armée de l'Est*, 39ff.
130. Wawro, *Franco-Prussian War*, 289; and Ortholan, *L'Armée de l'Est*, 215.
131. Sukstorf, *Problematik der Logistik*, 383, 391, 398–9; and Wawro, *Franco-Prussian War*, 286–7.
132. Showalter, *Wars of German Unification*, 295.
133. The exceptionally capable and well-armed formation under the leadership of Ernest de Lipowski, see Tanera, *An der Loire*, 40.
134. Berlit, *Vor Paris*, 38.
135. Showalter, *Wars of German Unification*, 297. Contemporaries also compared it to Bazeilles, see Tanera, *An der Loire*, 49.
136. In the heat of battle, the irregular units also shot at French marines who lost 38 men in friendly fire incidents, as they were mistaken for Germans due to their blue uniforms: see Wawro, *Franco-Prussian War*, 264–5.
137. Schneider, "Der Krieg in französischer Sicht," 195.
138. Large parts of the town burned down, yet German accounts claiming that wind was the driving factor may have some element of truth, as only the eastern parts of the town burned down; see Hatley, "Prolonging the Inevitable," 173.
139. Even a street in Paris was named after it, see Favre, *The Government*, 216ff.

140. Audoin-Rouzeau, *1870*, 214–5; Hoenig, *Der Volkskrieg*, vol. I, 182ff. Especially the mayors were vital in protecting the needs of the population and assuring a system of collaboration with the German occupiers, see Parisot, "De la négociation."
141. Roth, *La guerre*, 373, 376.
142. Mehrkens, *Statuswechsel*, 141.
143. Kühlich, *Die Deutschen Soldaten*, 319. When a Saxon officer was killed near Beauvais the town had to pay 400,000 francs (around $40,000 today) as compensation, whereas Héricourt was burned down after an ambush, see Wawro, *Franco-Prussian War*, 238.
144. Rohkrämer, "Daily life," 511–3. He also argued that it was far from developing into a 'total war'.
145. Kühlich, *Die Deutschen Soldaten*, 317.
146. Becker, *Bilder*, 225.
147. Hantraye, *Les Cosaques*, 26, 31, 273.
148. Forrest, *The Legacy*, 127.
149. Pieper, "German Approach," 2.
150. Ibid., 9.
151. Chickering, *Imperial Germany*, 84.

Disclosure statement

No potential conflict of interest was reported by the author.

Bibliography

Anderson, Margaret L. "A German Way of War?" *German History* 22, no. 2 (2004): 254–258. doi:10.1191/0266355404gh302xx.
Anderson, Margaret Lavinia. "How German Is It?" *German History* 24, no. 1 (2006): 122–126. doi:10.1191/0266355406gh369xx.
Audoin-Rouzeau, Stéphane. *1870: La France Dans La Guerre*. Paris: Armand Colin, 1989.
Baranowski, Shelley. *Nazi Empire. German Colonialism and Imperialism from Bismarck to Hitler*. Cambridge: Cambridge University Press, 2011.
Beccaro, Andrea. "Carlo Bianco and *Guerra per Bande: An Italian Approach to Irregular Warfare*." *Small Wars & Insurgencies* 27, no. 1 (2016): 154–178. doi:10.1080/09592318.2016.1122924.
Becker, Annette. *Les Cicatrices Rouges, 14–18: France Et Belgique Occupées*. Paris: Fayard, 2010.

Becker, Frank. *Bilder Von Krieg Und Nation: Die Einigungskriege in Der Bürgerlichen Öffentlichkeit Deutschlands, 1864–1913*. Munich: De Gruyter, 2001.

Beckett, Ian. *Modern Insurgencies and Counter-Insurgencies: Guerillas and Their Opponents since 1750*. London: Routledge, 2001.

Bell, David. *The First Total War: Napoleon's Europe and the Birth of Modern Warfare as We Know It*. Boston, MA: Houghton Mifflin, 2007.

Berlit, Bruno, ed. *Vor Paris Und an Der Loire 1870 Und 1871. Feldpostbriefe*. Kassel: Fischer, 1872.

Best, Geoffrey. *Humanity in Warfare: The Modern History of International Law of Armed Conflicts*. New York, NY: Columbia University Press, 1980.

Bizzoni, Achille Antonio. *Impressioni Di Un Volontario All'esercito Dei Vosgi*. Milan: Sonzogno, 1874.

Bönker, Dirk. "A German Way of War? Narratives of German Militarism and Maritime Warfare in World War I." In *Imperial Germany Revisited. Continuing Debates and New Perspectives*, edited by Sven Oliver Müller and Cornelius Torp, 227–238. Oxford: Berghahn, 2011.

Botzenhart, Manfred. "French Prisoners of War in Germany, 1870–71." In *On the Road to Total War: The American Civil War and the German Wars of Unification, 1861–1871*, edited by Stig Förster and Jörg Nagler, 587–595. Cambridge: Cambridge University Press, 1997.

Bührer, Tanja. *Die Kaiserliche Schutztruppe Für Ostafrika. Koloniale Sicherheitspolitik Und Transkulturelle Kriegführung 1885–1918*. Munich: Oldenbourg, 2011.

Cardinal von Widdern, Georg. *Deutsch-Französischer Krieg 1870–71. Der Krieg an Den Rückwärtigen Verbindungen Der Deutschen Heere*. Vol. 6. Berlin: Eisenschmidt, 1893–1899.

Challener, Richard D. *The French Theory of the Nation in Arms 1866–1939*. New York, NY: Columbia University Press, 1955.

Chanet, Jean-François. *Vers L'armée Nouvelle. République Conservatrice Et Réforme Militaire, 1871–1879*. Rennes: Presses Universitaires de Rennes, 2006.

Chickering, Roger. *Imperial Germany and the Great War, 1914–1918*. Cambridge: Cambridge University Press, 2014.

Clark, Christopher. "The Wars of Liberation in Prussian Memory: Reflections on the Memorialization of War in Early Nineteenth-Century Germany." *The Journal of Modern History* 68, no. 3 (1996): 550–576. doi:10.1086/245342.

Crépin, Annie. *Défendre La France. Les Français, La Guerre E Le Service Militaire, De La Guerre De Sept Ans À Verdun*. Rennes: Presses Universitaires de Rennes, 2005.

Daase, Christopher, and James W. Davis, eds. *Clausewitz on Small War*. Oxford: Oxford University Press, 2015.

De Schaepdrijver, Sophie, ed. *Military Occupations in First World War Europe*. New York, NY: Routledge, 2015.

Degler, Carl N. "The American Civil War and the German Wars of Unification: The Problem of Comparison." In *On the Road to Total War: The American Civil War and the German Wars of Unification, 1861–1871*, edited by Stig Förster and Jörg Nagler, 53–71. Cambridge: Cambridge University Press, 1997.

Dietrich von, Lassberg. *Mein Kriegstagebuch Aus Dem Deutsch-französischen Kriege 1870/71*. Munich: Oldenbourg, 1906.

Dirou, Armel. *La Guérilla En 1870. Résistance Et Terreur*. Paris: Bernard Giovanangeli, 2014.

Echevarria, Antulio J. *After Clausewitz. German Military Thinkers before the Great War*. Lawrence, KS: University Press of Kansas, 2000.

Engel, Ernst. *Die Verluste Der Deutschen Armeen an Offizieren Und Mannschaften Im Kriege Gegen Frankreich 1870 Und 1871*. Berlin: Verlag des Königlichen Statistischen Bureaus, 1872.

Esdaile, Charles, ed. *Popular Resistance in the French Wars: Patriots, Partisans and Land Pirates*. New York, NY: Palgrave Macmillan, 2005.

Etschmann, Wolfgang. "Guerillas Und Franctireurs, 1866 Und 1870." In *Freund Oder Feind?: Kombattanten, Nichtkombattanten Und Zivilisten in Krieg Und Bürgerkrieg Seit Dem 18. Jahrhundert*, edited by Erwin Schmidl, 31–44. Frankfurt: Lang, 1995.

Favre, Jules. *The Government of the National Defence: From the 30th of June to the 31st of October 1870*. London: King, 1873.

Foley, Robert. *German Strategy and the Path to Verdun: Erich Von Falkenhayn and the Development of Attrition, 1870–1916*. Cambridge: Cambridge University Press, 2005.

Forbes, Archibald. *My Experiences of the War between France and Germany*. Vol. 2. London: Hurst and Blackett, 1871.

Ford, Caroline. "Peasants Into Frenchmen Thirty Years After." *French Politics, Culture & Society* 27, no. 2 (2009): 84–93. doi:10.3167/fpcs.2009.270205.

Forrest, Alan. *The Legacy of the French Revolutionary Wars. The Nation-In-Arms in French Republican Memory*. Cambridge: Cambridge University Press, 2009.

Förster, Stig. "Facing 'people's War': Moltke the Elder and Germany's Military Options after 1871." *Journal of Strategic Studies* 10, no. 2 (1987): 209–230. doi:10.1080/01402398708437297.

Förster, Stig, ed. *Moltke: Von Kabinettskrieg Zum Volkskrieg: Eine Werkauswahl*. Bonn: Bouvier, 1992.

Foudras, Théodorit. *Les Francs-Tireurs De La Sarthe, Journal D'un Commandant*. Chalon-sur-Saone: Mulcey, 1872.

Frevert, Ute. *A Nation in Barracks: Modern Germany, Military Conscription and Civil Society*. Oxford: Berg, 2004.

Gerwarth, Robert, and Stephan Malinowski. "Der Holocaust Als 'kolonialer Genozid'? Europäische Kolonialgewalt Und Nationalsozialistischer Vernichtungskrieg." *Geschichte Und Gesellschaft* 33 (2007): 439–466. doi:10.13109/gege.2007.33.3.439.

Gillispie, James M. *Andersonvilles of the North. The Myths and Realities of Northern Treatment of Civil War Confederate Prisoners*. Denton, TX: University of North Texas Press, 2008.

Gooch, John. *The Italian Army and the First World War*. Cambridge: Cambridge University Press, 2014.

Gouttman, Alain. *La Grande Défaite 1870–1871*. Paris: Perrin, 2015.

Gumz, Jonathan. "Reframing the Historical Problematic of Insurgency: How the Professional Military Literature Created a New History and Missed the Past." *Journal of Strategic Studies* 32, no. 4 (2009): 553–588. doi:10.1080/01402390902986972.

Gumz, Jonathan. *The Resurrection and Collapse of Empire in Habsburg Serbia, 1914–1918*. Cambridge: Cambridge University Press, 2009.

Hahlweg, Werner. *Guerilla. Krieg Ohne Fronten*. Stuttgart: Kohlhammer, 1968.

Hantraye, Jacques. *Les Cosaques Aux Champs-Elysées: L'occupation De La France Après La Chute De Napoléon*. Paris: Belin, 2005.

Hartmann, Christian, Johannes Hürter, and Ulrike Jureit, eds. *Verbrechen Der Wehrmacht. Bilanz Einer Debatte*. Munich: Beck, 2005.

Hatley, Paul. "Prolonging the Inevitable: The Franc-Tireur and the German Army in the Franco-German War of 1870–1871." PhD diss., Kansas State University, 1997.

Haynes, Christine. "Remembering and Forgetting the First Modern Occupations of France." *Journal of Modern History* 88, no. 3 (2016): 535–571. doi:10.1086/687527.

Heuser, Beatrice. "Small Wars in the Age of Clausewitz: The Watershed between Partisan War and People's War." *Journal of Strategic Studies* 33, no. 1 (2010): 139–162. doi:10.1080/01402391003603623.

Heuser, Beatrice. *The Evolution of Strategy: Thinking War from Antiquity to the Present.* Cambridge: Cambridge University Press, 2010.

Hoenig, Fritz. *Der Volkskrieg an Der Loire Im Herbst 1870.* Vol. 6. Berlin: Mittler, 1893–1897.

Horne, John, and Alan Kramer. *German Atrocities 1914. A History of Denial.* New Haven, CT: Yale University Press, 2002.

Horne, John, and Alan Kramer. "German Atrocities in the First World War: A Response." *German History* 24, no. 1 (2006): 118–121. doi:10.1191/0266355406gh367xx.

Howard, Michael. *The Franco-Prussian War.* London: Routledge, 1991.

Howard, Michael. "Constraints on War." In *Laws of War. Constraints on Warfare in the Western World*, edited by Michael Howard, George J. Andreopoulos, and Mark R. Shulman, 1–11. New Haven, CT: Yale University Press, 1994.

Hull, Isabel. *Absolute Destruction: Military Culture and the Practices of War in Imperial Germany.* Ithaca, NY: Cornell University Press, 2005.

Hull, Isabel. *A Scrap of Paper. Breaking and Making International Law during the Great War.* Ithaca, NY: Cornell University Press, 2014.

Jeismann, Michael. *Das Vaterland Der Feinde: Studien Zum Nationalen Feindbegriff Und Selbstverständnis in Deutschland Und Frankreich 1792–1918.* Stuttgart: DVA, 1992.

Johansen, Anja. *Soldiers as Police. The French and Prussian Armies and the Policing of Popular Protest, 1889–1914.* Aldershot: Ashgate, 2005.

Jones, Marcus. "Fighting 'this Nation of Liars to the Very End': The German Army in the Franco-Prussian War, 1870–1871." In *Hybrid Warfare: Fighting Complex Opponents from the Ancient World to the Present*, edited by Williamson Murray and Peter R. Mansoor, 171–198. Cambridge: Cambridge University Press, 2012.

Kanter, Sanford. "Defeat 1871: A Study in 'after-war'." PhD diss., University of California, Los Angeles, 1972.

Kanter, Sanford. "Exposing the Myth of the Franco-Prussian War." *War & Society* 4, no. 1 (1986): 13–30. doi:10.1179/106980486790303871.

Kauffmann, Jesse. *Elusive Alliance: The German Occupation of Poland in World War I.* Harvard: Harvard University Press, 2015.

Keller, Ulrich. *Schuldfragen. Belgischer Untergrundkrieg Und Deutsche Vergeltung Im August 1914.* Paderborn: Schöningh, 2017.

Kramer, Alan. *Dynamic of Destruction: Culture and Mass Killing in the First World War.* Oxford: Oxford University Press, 2007.

Kramer, Alan. "German War Crimes 1914 and 1941. The Question of Continuity." In *Imperial Germany Revisited. Continuing Debates and New Perspectives*, edited by Sven Oliver Müller and Cornelius Torp, 239–250. Oxford: Berghahn, 2011.

Kramer, Alan, and John Horne. 2012. "Wer Schiesst Hier Aus Dem Hinterhalt?" *Frankfurter Allgemeine Zeitung*, March 1, 2018. https://www.faz.net/aktuell/feuille ton/massaker-in-belgien-im-ersten-weltkrieg-15472194.html

Kühlich, Frank. *Die Deutschen Soldaten Im Krieg Von 1870/71: Eine Darstellung Der Situation Und Der Erfahrungen Der Deutschen Soldaten Im Deutsch- Französischen Krieg.* Frankfurt: Lang, 1995.

Kuß, Susanne. *Deutsches Militär Auf Kolonialen Kriegsschauplätzen. Eskalation Von Gewalt Zu Beginn Des 20. Jahrhunderts.* Berlin: Links, 2010.

Leggiere, Michael V. *The Fall of Napoleon. Volume I. The Allied Invasion of France, 1813–1814.* Cambridge: Cambridge University Press, 2007.

Lieb, Peter. "Suppressing Insurgencies in Comparison: The Germans in the Ukraine, 1918, and the British in Mesopotamia, 1920." *Small Wars & Insurgencies* 23, no. 4–5 (2012): 627–647. doi:10.1080/09592318.2012.709765.

Lieb, Peter. "Der Deutsche Krieg Im Osten Von 1914–1919. Ein Vorläufer Des Vernichtungskriegs?" *Vierteljahreshefte Für Zeitgeschichte* 65, no. 4 (2017): 465–506. doi:10.1515/vfzg-2017-0029.

Lipkes, Jeff. *Rehearsals. The German Army in Belgium, August 1914.* Leuven: Leuven University Press, 2007.

Liulevicius, Vejas. *War Land on the Eastern Front: Culture, National Identity, and German Occupation in World War I.* Cambridge: Cambridge University Press, 2001.

Loch, Thorsten, and Markus Vette, eds. *Friedrich Clauson Von Kaas. "potsdam Ist Geschlagen". Briefe Aus Dem Deutsch-Französischen Krieg 1870/71.* Freiburg: Rombach, 2016.

Löhlein, Ludwig. *Feldzug 1870–71. Die Operationen Des Korps Des Generals Von Werder.* Berlin: Mittler, 1874.

Madley, Benjamin. "From Africa to Auschwitz: How German South West Africa Included Ideas and Methods Adopted and Developed by the Nazis in Eastern Europe." *European History Quarterly* 35, no. 3 (2005): 429–464. doi:10.1177/0265691405054218.

Mayeur, Jean-Marie. *Léon Gambetta. La Patrie E La République.* Paris: Fayard, 2008.

Mehrkens, Heidi. *Statuswechsel: Kriegserfahrung Und Nationale Wahrnehmung Im Deutsch-Französischen Krieg 1870/71.* Essen: Klartext, 2008.

Meier-Welcker, Hans. "Der Kampf Mit Der Republik." In *Entscheidung 1870. Der Deutsch-Französische Krieg,* edited by Ursula von Gersdorff and Wolfgang von Groote, 105–164. Stuttgart: DVA, 1970.

Melson, Charles D. "German Counterinsurgency Revisited." *Journal of Military and Strategic Studies* 14, no. 1 (2011): 1–33.

Melson, Charles D. *Kleinkrieg: The German Experience with Guerrilla Warfare, from Clausewitz to Hitler.* Philadelphia, PA: Casemate, 2016.

Messerschmidt, Manfered. "Völkerrecht Und "kriegsnotwendigkeit" in Der Deutschen Militärischen Tradition Seit Den Einigungskriegen." *German Studies Review* 6, no. 2 (1983): 237–269. doi:10.2307/1428529.

Molis, Robert. *Les Francs-Tireurs Et Les Garibaldi. Soldats De La République 1870–1871 En Bourgogne.* Paris: Tirésias, 1995.

Münch, Philipp. *Bürger in Uniform. Kriegserfahrungen Von Hamburger Turnern 1914 Bis 1918.* Freiburg: Rombach, 2009.

Nabulsi, Karma. *Traditions of War: Occupation, Resistance, and the Law.* Oxford: Oxford University Press, 1999.

Nelson, Robert L. "Ordinary Men' in the First World War? German Soldiers as Victims and Participants." *Journal of Contemporary History* 39, no. 3 (2004): 425–435. doi:10.1177/0022009404044448.

Nolan, Michael E. *The Inverted Mirror: Mythologizing the Enemy in France and Germany, 1898–1914.* New York, NY: Berghahn, 2005.

Ortholan, Henri. *L'Armée De La Loire, 1870–1871.* Paris: Bernard Giovanangeli, 2005.

Ortholan, Henri. *L'Armée De l'Est 1870–1871.* Paris: Bernard Giovanangeli, 2009.

Parisot, Guillaume. "De La Négociation Comme Instrument D'occupation Pacifiée Et D'exploitation Économique Efficace Pendant La Guerre De 1870–1871." In *Le Temps Des Hommes Doubles: Les Arrangements Face À L'occupation, De La Révolution Française À La Guerre De 1870*, edited by Jean-François Chanet, Annie Crépin, and Christian Windler, 279–302. Rennes: Presses Universitaires de Rennes, 2013.

Pieper, Henning. "The German Approach to Counterinsurgency in the Second World War." *The International History Review* 37, no. 3 (2014): 1–12.

Pieper, Henning. "From Fighting 'francs-tireurs' to Genocide: German Counterinsurgency in the Second World War." In *Insurgencies and Counterinsurgencies. National Styles and Strategic Cultures*, edited by Beatrice Heuser and Eitan Shamir, 149–167. Cambridge: Cambridge University Press, 2016.

Pöhlmann, Markus. "Das Unentdeckte Land. Kriegsbild Und Zukunftskrieg in Deutschen Militärzeitschriften." In *Vor Dem Sprung Ins Dunkle. Die Militärische Debatte Über Den Krieg Der Zukunft 1880–1914*, edited by Stig Förster, 21–131. Paderborn: Schöningh, 2016.

Pöhlmann, Markus. "Habent Sua Fata Libelli. Zur Auseinandersetzung Um Das Buch 'german Atrocities 1914.'" Portal Arbeitskreis Militärgeschichte. Accessed November 20, 2018. http://portal-militaergeschichte.de/http%3A//portal-militaergeschichte.de/poehlmann_habent

Porch, Douglas. *The March to the Marne. The French Army 1871–1914*. Cambridge: Cambridge University Press, 1981.

Porch, Douglas. "Bugeaud, Gallieni, Lyautey: The Development of French Colonial Warfare." In *Makers of Modern Strategy: From Machiavelli to the Nuclear Age*, edited by Peter Paret, 376–407. Oxford: Oxford University Press, 1986.

Porch, Douglas. *Counterinsurgency. Exposing the Myths of the New Way of War*. Cambridge: Cambridge University Press, 2013.

Potempa, Harald. "Jahrbuch Innere Führung 2010. Die Grenzen Des Militärischen." In *Der Kleine Krieg in Der Deutschen Militäpublizistik: Das Militär- Wochenblatt 1871–1900*, edited by R. Hammerich Helmut, Uwe Hartmann, and Claus von Rosen, 134–151. Berlin: Miles, 2010.

Prusin, Alexander V. *Nationalizing a Borderland: War, Ethnicity, and Anti-Jewish Violence in East Galicia, 1914–1920*. Tuscaloosa, AL: University of Alabama Press, 2005.

Rid, Thomas. "The Nineteenth Century Origins of Counter-insurgency Doctrine." *Journal of Strategic Studies* 33, no. 5 (2010): 727–758. doi:10.1080/01402390.2010.498259.

Rindfleisch, Georg Heinrich. *Feldbriefe 1870–71*. Göttingen: V&R, 1891.

Rink, Martin. "The Partisan's Metamorphosis: From Freelance Military Entrepreneur to German Freedom Fighter, 1740 to 1815." *War in History* 17, no. 1 (2010): 6–36. doi:10.1177/0968344509348291.

Rohkrämer, Thomas. "Daily Life at the Front and the Concept of Total War." In *On the Road to Total War: The American Civil War and the German Wars of Unification, 1861–1871*, edited by Stig Förster and Jörg Nagler, 497–518. Cambridge: Cambridge University Press, 1997.

Roth, François. "Occupation Et Libération Des Départments Lorrains." In *La Guerre De 1870/71 Et Ses Conséquences: Actes Du XXᵉ Colloque Historique Franco- Allemand Organisé À Paris 1984–1985*, edited by Philippe Levillain and Rainer Riemenschneider, 313–317. Bonn: Bouvier, 1990.

Roth, François. *La Guerre De 1870*. Paris: Fayard, 1990.

Scheipers, Sibylle. "Counterinsurgency or Irregular Warfare? Historiography and the Study of 'small Wars'." *Small Wars & Insurgencies* 25, no. 5–6 (2014): 879–899. doi:10.1080/09592318.2014.945281.

Schivelbusch, Wolfgang. *The Culture of Defeat. On National Trauma, Mourning, and Recovery.* London: Granta, 2003.

Schneider, Fernand Thiébaut. "Der Krieg in Französischer Sicht." In *Entscheidung 1870. Der Deutsch-Französische Krieg*, edited by Ursula von Gersdorff and Wolfgang von Groote, 165–203. Stuttgart: DVA, 1970.

Showalter, Dennis. "From Deterrence to Doomsday Machine: The German Way of War, 1890–1914." *Journal of Military History* 64, no. 3 (2000): 679–710. doi:10.2307/120865.

Showalter, Dennis. *The Wars of German Unification.* London: Bloomsbury, 2004.

Spraul, Gunter. *Der Franktireurkrieg 1914.* Berlin: Frank & Timme, 2016.

Stone, David. J. A. *'First Reich'. Inside the German Army during the War with France 1870–71.* London: Brassey's, 2002.

Stoneman, Mark. "The Bavarian Army and French Civilians in the War of 1870– 1871: A Cultural Interpretation." *War in History* 8, no. 3 (2001): 271–293. doi:10.1177/096834450100800302.

Sukstorf, Lothar. *Die Problematik Der Logistik Im Deutschen Heer Während Des Deutsch-Französischen Krieges, 1870/71.* Frankfurt: Lang, 1994.

Tanera, Carl. *An Der Loire Und Sarthe.* Munich: Beck, 1906.

Tison, Stéphane. *Comment Sortir De La Guerre? Deuil, Mémoire Et Traumatisme (1870–1940).* Rennes: Presses Universitaires de Rennes, 2011.

Tombs, Robert. *The War against Paris 1871.* Cambridge: Cambridge University Press, 1981.

Tombs, Robert. "The Wars against Paris." In *On the Road to Total War: The American Civil War and the German Wars of Unification, 1861–1871*, edited by Stig Förster and Jörg Nagler, 541–564. Cambridge: Cambridge University Press, 1997.

Toppe, Andreas. *Militär Und Kriegsvölkerrecht: Rechtsnorm, Fachdiskurs Und Kriegspraxis in Deutschland 1899–1940.* Munich: Oldenbourg, 2008.

Varley, Karine. *Under the Shadow of Defeat. The War of 1870–71 in French Memory.* New York, NY: Palgrave Macmillan, 2008.

Watson, Alexander. "'Unheard-of Brutality' Russian Atrocities against Civilians in East Prussia, 1914–1915." *Journal of Modern History* 86, no. 4 (2014): 780–825. doi:10.1086/678919.

Wawro, Geofffrey. *The Franco-Prussian War: The German Conquest of France in 1870–1871.* Cambridge: Cambridge University Press, 2003.

Weber, Eugen. *Peasants into Frenchmen. The Modernization of Rural France, 1870– 1914.* Stanford, CA: Stanford University Press, 1976.

Wetzel, David. *A Duel of Nations. Germany, France, and the Diplomacy of the War of 1870–1871.* Madison, WI: University of Wisconsin Press, 2012.

Wolowski, Ladislas. *Campagne De 1870–71. Corps Franc Des Vosges (armée De L'est). Souvenirs.* Paris: Chamuel, 1871.

Zimmerer, Jürgen. *Von Windhuk Nach Auschwitz. Beiträge Zum Verhältnis Von Kolonialismus Und Holocaust.* Münster: LIT, 2007.

Zuckerman, Larry. *The Rape of Belgium: The Untold Story of World War I.* New York, NY: New York University Press, 2004.

The campaign of the lost footsteps: the pacification of Burma, 1885-95

Ian F. W. Beckett

ABSTRACT

What Rudyard Kipling called the 'campaign of lost footsteps' was the longest campaign fought by the Victorian army. The conquest of Upper Burma, an area of 140,000 sq. miles with a population of four million, took only three weeks in November 1885 and was accomplished with minimum cost. However, the removal and deportation of the Burmese King and dismantling of all trad-itional authority led to growing resistance to British rule leading to an increasingly difficult guerrilla war. Though the Burmese guerrillas were charac-terised by the British as mere bandits or *dacoits*, many were former soldiers along with Buddhist monks. The extremely difficult nature of campaigning in the terrain and climate of Burma was not sufficiently appreciated by the War Office, who viewed the conflict as a 'subaltern's war' and 'police' work. Intended regime change was also not accompanied by any consideration of the likely implications. Prolonged insurgency necessitated deploying a force far larger than originally intended; though order was finally secured by 1895, the campaign proved destructive of Burmese society while British recruitment of hill tribes into the police and armed forces sowed the seeds for future divisions.

On 15 March 1889, 24-year-old Rudyard Kipling, a journalist with the Allahabad-based newspaper, *The Pioneer*, made a two-day visit to Burma. He was vividly impressed by the colours of Rangoon, a young Burmese girl he saw at Moulmein, and the sound of windblown bells. A year later 'The Road to Mandalay' appeared in *Barrack Room Ballads*.[1] Mandalay, which Kipling entirely imagined, lay 635 miles to the north of Rangoon up the Irrawaddy, the real road to Mandalay. British troops had used the river in completing the conquest of Burma in 1885 but fighting continued. There was a darker side to Kipling's reflections. A school friend, Lieutenant Robert Dury, had been killed at Minhla in November 1885. References to Dury's death appeared in a number of Kipling's works and he wrote of the Irrawaddy in May 1889,[2]

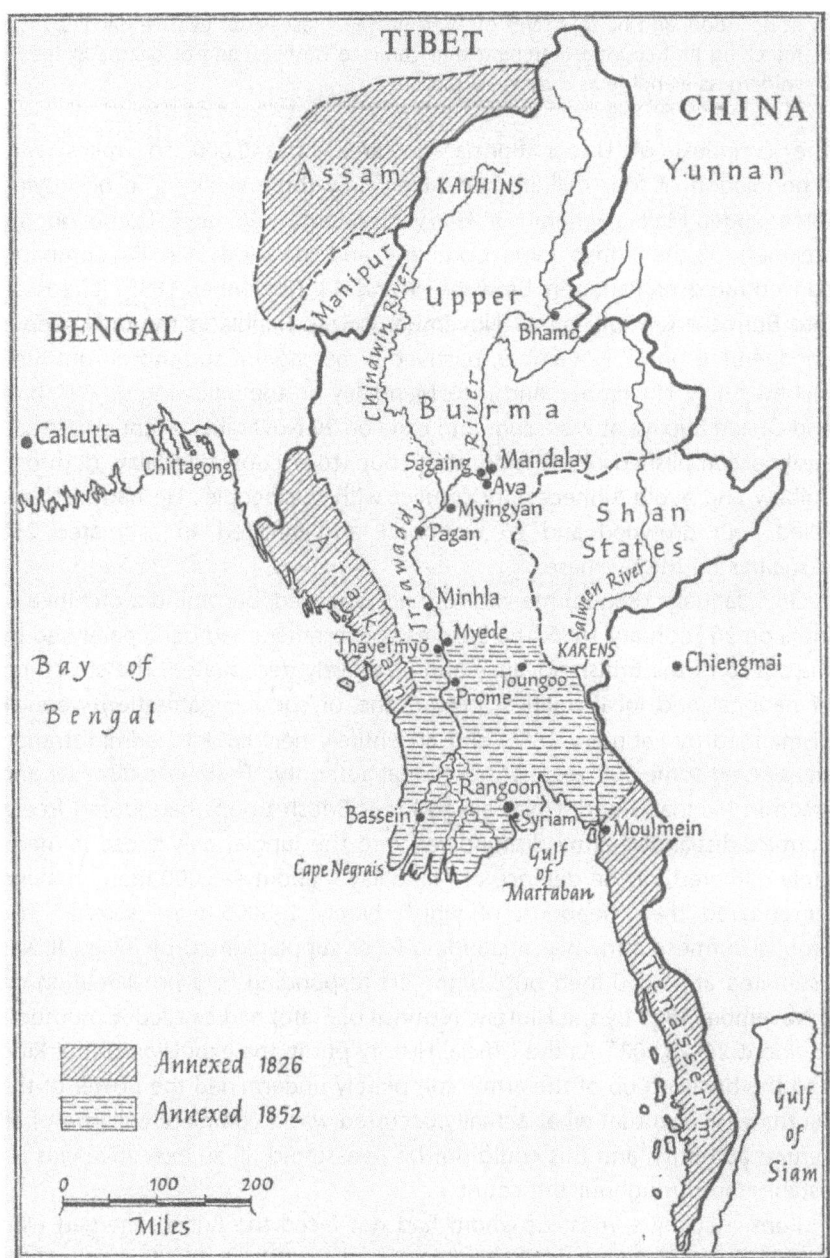

Figure 1. Map comes from Stewart (1972), frontispiece.

I reflected that I was looking upon the River of the Lost Footsteps – the road that so many many men of my acquaintance had travelled, never to return, within the past three years .... They had gone up the river and they had died. At my elbow stood one of the workers in New Burma, going to report himself at Rangoon, and he told tales of interminable chases after evasive dacoits, of marching and counter-marching that came to nothing, and of deaths in the wilderness as noble as they were sad.

The conquest of Upper Burma, an area of 140,000 sq. miles with a population of four million, had taken only three weeks. The occupying force under Major General Sir Harry Prendergast VC and, borne on the steamers of the British India Company and Irrawaddy Flotilla Company, concentrated at Rangoon between 5 and 11 November 1885. It crossed into Burmese territory on 11 November, seized Minhla in the only serious engagement on 17 November, received a message of surrender from King Thibaw on 27 November, and took Mandalay on the following day. Thibaw and Queen Supayalat were sent into exile on 29 November. Prendergast had easily accomplished his mission of a 'coup' to occupy Mandalay, dethrone Thibaw and avoid 'unnecessary conflict with the people'. He had lost four killed, four drowned and 26 wounded, and inflicted an estimated 250 casualties on the Burmese.[3]

On 1 January 1886 Burma was formally annexed, becoming a province of India on 26 February 1886. The Burmese government had been paralysed by the speed of the British advance and effectively decapitated. The structures of national and local political power, and of social organisation were all dismantled by February 1886, and an entirely new colonial administration supplanted some 300 years of traditional authority.[4] There was disorder and arson in Mandalay and it was alleged that British troops had looted freely. Burma's disbanded army disappeared into the jungle, only those immediately gathered for the defence of Mandalay – about 4–5,000 men – having surrendered their weapons, of which barely 2–3000 were seized.[5] The regular Burmese army was a standing force supplemented by levies. It was estimated at 15,650 men but volunteers responding to a proclamation on 7 November 1885 by the Hluttaw (Council of State) had swelled its numbers to about 24–28,000.[6] As the Official History put it, the expulsion of the King and the breaking up of the army 'completely undermined the power of the Hlutdaw [sic]; so that what actually occurred was a complete collapse of all central authority, and this could not be re-asserted till an executive was re-established throughout the country'.[7]

Former soldiers, most of whom had not faced the British, merged with existing bandits known as dacoits to wage guerrilla struggle. Burmese kings often had little control over distant officials and attempted rule over ethnic minorities such as the Chins, Karens, Kachins, Mons, and Shans frequently led to rebellion. At the time of the British invasion, much of the Burmese army was engaged in fighting the Shans, dacoits and assorted dissidents with real royal authority not extending much beyond Mandalay. These groups were no more willing to accept British administration than

Burmese, adding to the complexity of what became the longest campaign fought by the Victorian army. What one political officer characterised as a war of 'little affairs' took ten years.[8] It embraced not only the Third Anglo-Burmese War (1885–86) but also major expeditions against the Chins, Kachins, Lushais and Shans between 1889 and 1895.[9]

First, the article will describe the background to the conquest and pacification before turning, secondly, to the nature of Burmese opposition to occupation. Thirdly, it will deal with the problems of campaigning in Burma before, fourthly, turning to the pacification campaign. Fifth, the paper will deal with the campaigns against the hill tribes before reaching a conclusion.

I

Dealing first with Burma prior to 1885, the ruling dynasty at the court of Ava was established in the mid-eighteenth century. It was an absolute monarchy with a rigid hierarchical system. Succession was supposed to pass from brother to brother before it passed from father to son, but polygamy meant there were multiple heirs. Fathers tried to nominate sons rather than brothers as heirs, and sons plotted to succeed fathers. Invariably, when one ruler was replaced, rival claimants were murdered. So far as the British were concerned, the instability of Burma posed a threat to the north eastern frontier of India but there were also equally significant economic considerations. Burmese incursions into the small independent states of Manipur and Assam and into the British-protected state of Cachar resulted in the First Anglo-Burmese War (1824–26). In February 1826 Burma ceded Arakan and Tenasserim – amounting to two fifths of the kingdom – while renouncing interest in Assam, Cachar and Manipur; agreeing to a commercial treaty; and paying an indemnity. The commercial treaty yielded little of value since the Burmese declined to trade in either rice, a staple food for its people, or in silver. Teak became the basis of trade. General corruption led to increasing attempts at extortion on the part of Burmese officials towards British merchants seeking to penetrate further into the interior as teak forests in Lower Burma became exhausted. Following various incidents, a British blockade of Rangoon resulted in the Second Anglo-Burmese War (1852–53). The Burmese ceded Lower Burma and hundreds of square miles of valuable teak forest in June 1853.[10] The loss of Lower Burma denied the kingdom access to rice surpluses and increased migration rates to the south.[11]

Modernisation was attempted by King Mindon (r. 1853–78) but under the pliable and indolent Thibaw, who succeeded to the throne in 1878, declining revenues, increasing economic difficulties, and poor harvests exacerbated the breakdown of royal administration. The British perception was that the real power behind the throne was Thibaw's capricious wife and half-sister, Supayalat, who was held responsible for the slaughter of 31 of Thibaw's male siblings. A trading dispute gave the British the excuse to act. In reality, they were alarmed by a Franco-Burmese treaty in January 1885. From the new capital of Mandalay, Thibaw imposed a £2.3 million fine on the Bombay-Burma Trading Corporation for allegedly exporting more teak than they had bought. Commercial interests in Rangoon were calling for annexation, a policy favoured by the Secretary of State for India, Lord Randolph Churchill, who wanted a quick victory in time for the forthcoming general election. Voting began before occupation was completed and the result was Gladstone's short-lived third administration. An ultimatum was despatched on 22 October 1885 demanding Thibaw accept British control over Burma's foreign policy; appoint a British envoy with direct access to the King; agree on arbitration of the fine; and facilitate British trade with China through Burma. Thibaw rejected the ultimatum on 5 November 1885 and occupation followed. British officials with experience of Burma opposed annexation on the grounds of the likely military and financial costs. The alternative of a protectorate and indirect rule foundered on the lack of any acceptable successor to Thibaw, and the perceived incapacity of the Hluttaw or any local administration to function effectively as a guarantor of British interests given the rapid breakdown of any local authority and of any pretence of law and order following occupation.[12]

II

Turning to Burmese opposition to British rule, the initial occupation had been easily achieved. Its speed prevented Burmese plans to block the Irrawaddy and to complete the Italian designed defences of Minhla, Ava and Sagaing. But, far from welcoming the British as expected, the Burmese resented the removal of a King who if weak was still semi-divine in their eyes. Opposition was also fuelled by unwittingly symbolic evidence of disrespectful British attitudes to traditional authority such as the use of parts of the royal palace as an Anglican chapel, a British club and the Chief Commissioner's office; and rumours of wanton destruction of property, ill-treatment of monks and women, and summary executions.[13]

The increasing spread of resistance has been characterised as deriving from a combination of Burmese patriotism, millenarianism and continuing banditry.[14] A number of royal or claimed royal princes had evaded capture and deportation. Some such as the Myinzaing Prince, who died of fever in August 1886, proved a magnet for initial resistance. So did monks such as U Ottama, who was only captured in July 1889. Monks (pongyi) lost status and influence through the removal of the monarchy and the Hluttaw's proclamation in November 1885 invoked the supposed threat to the Buddhist faith, portraying the British as heretics. While Buddhism extolled non-violence, the upheaval following occupation and the apparent threat to faith justified it. Some historians see Buddhist participation, like that of princes, as attempts to restore the traditional order that had all but disintegrated during Thibaw's reign. Certainly, Buddhism was indistinguishable from the concept of Burmese kingship and national identity.[15] Although they had also lost status and authority, many of the nobility and gentry did not contest British rule. They were supplanted, however, by Burmese with local influence such as Boh Swe, hereditary *thugyi* (headman) of Mindat. Boh Swe was eventually killed in October 1887. There were then also the pure dacoits such as Bo Hla U, an active bandit on the Lower Chindwin since 1883, who simply turned into a resistance leader: he was killed in an internecine quarrel in April 1887.[16]

Revolt spread to Lower Burma in 1886, the Chief Commissioner Sir Charles Bernard recognising in March that 'these outbreaks are not dacoity but incipient rebellion of a part of the population evoked by affairs in Upper Burma'.[17] Since Burmese police, officials and headmen were implicated, it reinforced the decision to bypass existing hierarchies and centralise administration under British officials.[18] Some collaborators were found although the British also tried none too successfully to use the Buddhist Primate or Thathanabaing to counter the influence of monks within the opposition movement.[19]

The sporadic nature of resistance in Lower Burma raises another aspect that has been noted with particular reference to the area of the Toungoo to Mandalay railway, which the British began to construct in September 1886 and completed in March 1889. It is argued that the Burmese regular army was effectively 'armed rural folk' and that the developing conflict reflected traditional Burmese rural warfare culture. Resistance was not so much to the British – and thereby indicative of national or proto-national loyalties- as to any centralising authority. Accordingly, alongside royal princes, monks or bandit leaders, village communities themselves could take up arms with internal conflict between villages where some backed the British. In this respect, there was a separate civil war running alongside the overall pacification campaign.[20] Moreover, it was one in which the traditional taking of heads was common. The practice spread to British and especially Indian troops as suggested by Kipling's poem, 'The Grave of the Hundred Head':[21]

> Then a silence came to the river,
> A hush fell over the shore,
> And Bohs that were brave departed.
> For the Burmans said
> That a white man's head
> Must be paid for with heads five-score

Taking of heads by troops was prohibited in theory in 1888.

It was convenient for the British authorities to refer to all opponents as dacoits since this suggested that they were the enemy of the British and the Burmese population alike.[22] Thus, the CinC in India, Lieutenant General Sir Frederick Roberts, believed that dacoity was 'not the work of "patriots" but of restless spirits who have lived by the trade all their life'.[23] It also fitted into a narrative tradition with respect to British expectations of Burma and its society since the first Anglo-Burmese War.[24] Whatever the source of leadership or motivation, dacoit bands led by those characterised as *bohs* could range from a handful to as many as 4,000 men. Invariably the British distinguished between *bohs* and their followers, suggesting the significance of local leadership. Of 71 of the most prominent *bohs* between 1885 and 1894, 32 were killed or executed, three died and four surrendered.[25] It is impossible to gauge the numerical strength of opposition to the British or the losses inflicted on the Burmese.

Intimidation was applied by the dacoits to the local population in order to ensure their assistance although, in many cases, banditry was so long established that coercion was not required.[26] Troops disarmed villagers, which left them more at the mercy of dacoits.[27] Grattan Geary of the *Bombay Gazette*, a critical observer, recorded of the plight of villagers: 'They will be shot as dacoits if they have arms; if they have none, they will be robbed and possibly murdered by the dacoits.' The choice, he claimed, was 'dacoiting or being dacoited'.[28] It was exceptionally difficult to gain any information and equally hard to distinguish between dacoits and peaceful villagers.[29] Given the intimidation and the lack of intelligence, it was concluded that the only course was to make villagers 'fear us more than the bandits'.[30]

III

Turning to campaigning, Burmese forces were poorly armed with mostly flintlock muskets, bows and spears. In defence, however, they could be difficult to dislodge from solid teak loop-holed stockades although they rarely stood for long in such structures before melting into the jungle. They were at their most effective as guerrillas. While British resources and firepower would ultimately prevail, this did not make campaigning in Burma any easier. A constant was the problem presented by climate and terrain. Field Marshal Viscount Slim was later to write of Burma, 'It could fairly be described as some of the world's worst country, breeding the world's worst diseases, and having for half the year at least the world's worst climate.'[31]

The hot season lasted from February to May, with daytime temperatures easily reaching 105 degrees Fahrenheit and humidity increasing through April and May. Some relief came with the onset of the South West Monsoon in June but, equally, torrential rain until October meant impassable waterways and unhealthy conditions in which mosquitoes flourished. Rainfall of

15" in 24 hours was not unusual in the Arakan with 142" of the annual 170" falling in just three months. In the hills bordering India annual rainfall averaged 500". Only in the cool season from November to January were campaigning conditions at all pleasant. The terrain was equally difficult. The mountains of the Arakan Yomas ran north to south the entire length of the country merging in the north with the jungle clad and deceptively named Lushai Hills rising up to 12,000 feet. The Chin Hills rose up to 10,000 feet. In the case of the Chin-Lushai Hills, one British officer recorded, 'Approached through malarial valleys and *terai* which decimates the troops with sickness before they reach the inhabited heights, mountains and over paths so bad that sometimes it is only with infinite difficulty that five miles a day are accomplished.'[32] Also running broadly north to south in deep valleys through the mountainous forests and jungles were the great rivers of the Irrawaddy, the Sittang, the Chindwin and the Salween, all of which impeded west to east communications. The main north to south artery was the Irrawaddy, navigable for 800 miles as far north as Bhamo.

The jungle was not uniform. Bamboo trees and waist high elephant grass characterised the Arakan but, between the Irrawaddy and the Chindwin, it gave way to teak forest. North of Mandalay it was mostly impenetrable jungle. Brigadier General George White, commanding in Mandalay wrote on 3 January 1886 that the country was 'the most difficult I ever saw or thought of to plan operations in – everywhere water & jungle, both of which defy calculation as to time of march. The natives know nothing of distance, & the country is unsurveyed'.[33] Troops relied on tracks and paths that had been well trodden over the years but this could make them vulnerable to ambush. Tracks were also difficult to negotiate in heavy 'ammunition' boots.[34] Dacoits likened British columns to buffalo moving through reeds which closed behind them. It was rare for the frequently fruitless marches to be less than seven hours.[35] It could be a frightening environment of constant animal and insect noise, especially so at night.

In consistently high temperatures and humidity, disease flourished. In the monsoon season there was malaria, prickly heat and ringworm, and in the dry season jungle ticks caused scrub typhus. The Burmese equivalents of bluebottles were a constant, as were leeches. Malaria, dysentery and cholera posed the greatest dangers. Cholera broke out even before the fall of Mandalay.[36] Between November 1885 and October 1886, for example, there were 91 British and Indian fatalities from action but 930 deaths from disease with a further 2,032 invalided to India.[37]

In the course of 1889 the military hospital at Shwebo in Upper Burma admitted 643 patients, the majority suffering from diarrhoea or ague. That at Toungoo in Lower Burma had 538 admissions, mostly cases of malaria, but also including 41 admissions as a result of 'local violence'. Thayetmyo, on the frontier between Upper and Lower Burma, had 548 admissions, mostly from dysentery.[38] In August 1886 one important post at Napé at the foot of the Arakan Yomas, had to be withdrawn through persistent sickness there: over 500 men were invalided during the rainy season.[39] At any one time 20 per cent of the field force was incapacitated in 1885–86.[40]

IV

In terms of the pacification campaign, the force initially provided Prendergast totalled 11,844 men of whom 3,029 were British. Two brigades were drawn from the Madras Army and one from Bengal.[41] The total cost was a modest 20 lakhs, one lakh being the equivalent of 100,000 rupees.[42] No land transport was provided beyond some 2,000 Punjabi coolies, many of whom proved too old to carry the burdens required of them as well as succumbing to disease.[43] Transport was supplemented therefore in December 1885 by 100 royal elephants and 300 ponies from the Manipur Cavalry but half the elephants were without mahouts and untrained, and half the ponies were unserviceable.[44]

George White complained that the force had been equipped 'as a fish to fight a dog'.[45] British control extended no further than the range of the expeditionary force's rifles with insufficient troops to occupy the country as a whole. Initially, there were only five civil officers available.[46] Flying columns – usually no more than 200 men – were organised but often arrived too late to prevent dacoits raiding villages or to catch them. White commented on 17 July 1886,[47]

> The villagers, having cause to recognize that we are too far off to protect them, lose confidence in our power and throw in their lot with the insurgents. They make terms with the leaders and baffle pursuit of those leaders by round-about guidance or systematic silence. In a country, itself one vast military obstacle, the seizure of the leaders of the rebellion, though of paramount importance, thus becomes a source of greatest difficulty.

White concluded that the most effective response would be 'close occupation of the disturbed districts by military posts'.[48] That required additional troops to counter what appeared a many-headed hydra.[49]

Initially, ten posts were established along the Irrawaddy. This was soon extended to 25 as it was recognised that these were necessary to maintain military ascendancy, provide centres for well affected Burmese, protect lines of communication, and support the civil power. Posts were extended to cover the routes from Toungoo to Mandalay and from Toungoo to Thayetmyo, which together with those along the Irrawaddy enclosed central Burma in a rough triangle.[50] Posts usually consisted of a bamboo stockade and parapet, fronted by a ditch filled with thorns or prickly pear, and a liberal use of bamboo spikes beyond. Paths were cleared between posts to facilitate movement and communication. By the summer of 1886 there were 43 posts in Upper Burma and 47 in Lower Burma concentrated on the Sittang and the Irrawaddy. By March 1887 there were 141 posts in Upper Burma. In addition, Irrawaddy Flotilla Company steamers routinely carried military guards or police to prevent dacoits crossing the rivers, as well as controlling piracy.[51]

There were over 100 engagements between April and July 1886 alone with 29 in August and 27 in September.[52] A typical action was that on 4 July 1886 when Jemadar Imam Khan of the 26th Madras Infantry with 30 men, escorting a convoy, was attacked by 200 dacoits. They were driven off without loss, the dacoits losing an estimated 16 dead.[53] One district commissioner, A. R. Colquhorn, wrote a paper in October 1886 on the pacification of the Vendée during the French Revolutionary Wars. The evolving British method certainly bore some resemblance with progressive occupation of the country-side with lines of garrisoned posts with constant patrols operating from them. Flying columns would operate beyond the line of posts with the civil administrators and police tackling the areas inside the lines.[54]

By 31 December 1886 the force had been increased to 31,571 men in Upper Burma and 4,176 in Lower Burma, the cost now amounting to 121½ lakhs.[55] Upper Burma had six military districts, six constituting brigade commands and two independent commands. The field force was now reckoned to yield a ratio of one man for every 3½ sq. miles of country compared to the original one man for every seven sq. miles.[56]

The pattern of campaigning hardly varied in terms of the desultory nature of engagements. As was noted by the Viceroy, Lord Dufferin, of a minor engagement at Chenbyut in October 1887, the deaths of two officers gave the affair more importance than it deserved.[57] One official account of the cold weather operations of 1887–88 recorded,[58]

> The story of the year is a record of endless marches by day and night, through dense jungle where the path could hardly be traced, along paths so thick in mud that the soles of men's boots were torn off as they marched, over sandy tracts devoid of water, over hills where there were no paths at all. Rarely was there the chance of an engagement to cheer the troops; stockades were found empty, villages deserted, camps evacuated, and yet everywhere there was the probability of a sudden ambush from every clump of trees or line of rocks, or at any turn of the road.

In 1888, there were five engagements in March, four in April, three each in May and June, two each in July and August, one in September, two in October, and three in November, resulting in five British and Indian dead. Winter season operations saw 51 engagements between December 1888 and February 1889, with 32 British and Indian deaths.[59]

The pacification campaign proved controversial from the start, George White writing to Roberts in June 1886 that it would be best to keep events out of the press for the 'operations we are constantly undertaking which are so necessary would not long attract the tide of public sympathy towards us involving as they do heavy loss nearly always inflicted on the enemy with, I am happy to say, very slight loss on ourselves'.[60] White professed 'the strongest repugnance to wholesale butchery'.[61]

There had been unwelcome publicity given to the campaign by the correspondent of The Times, Edward Moylan. Moylan was deported from Mandalay in December 1885 for evading censorship regulations and reporting on the initial disorder. Moylan then used his considerable political contacts to be restored to his position, writing increasingly critical reports on the slowness of the pacification campaign and what he regarded as Prendergast's incompetence. Moylan was a paid representative of the Bombay-Burma Trading Corporation, Prendergast installing a different Burmese chief minister to the one the company preferred.[62] Moylan then happened on a story in January 1886 that, during a series of 22 exemplary executions of dacoits at Mandalay, the Provost Marshal, Colonel W. W. Hooper, had not only questioned men about to be executed by firing squad to try and extract intelligence but also, as an amateur photographer, delayed the order to fire so as to capture the death at the exact moment he exposed his negative.

Hooper had photographed executions when not actually on duty but had not delayed the order to fire to set up the camera.[63] Inevitably questions were raised in Parliament by Liberals and Irish Nationalists. Grattan Geary, who was critical of Hooper, believed that undue severity only made dacoits liable to fight more desperately than might otherwise have been the case.[64] The executions, however, were believed to be a necessary response to the rising guerrilla activity. Colonel Edward Sladen, Prendergast's chief political officer, reported that 'the severest measures were imperatively called for as a terror to evil-doers and the public safety'.[65] Nonetheless, Dufferin immediately forbade further executions for fear that they 'will strengthen the hands of those who may be disposed to criticise our conquest of the country'.[66] Sir Charles Brownlow reported that Lord Randolph Churchill was 'hysterical'. Such was public opinion – 'equally jumpy in its applause & its censure' – that Prendergast had gone to 'number one' of generals to be removed.[67] He was relieved of the command in March 1886 and Hooper reprimanded and sent back to India with no chance of further promotion. The first Chief Commissioner, Bernard, was equally a victim of Moylan's continuing criticism in January 1887.

Dufferin defended Burma policies in October 1886, arguing that a now critical British public had expected too much: 'They thought that the bloodless campaign of last November, and the capture of Theebaw and of his capital, had finished off the business; and they were delighted with the idea of having acquired a new Province in so inexpensive a manner. The

Government of India, on the contrary, never indulged in these sanguine anticipations ... '[68] Rather similarly, Sir Charles Brownlow, the Assistant Military Secretary for India at the War Office, noted, 'The senseless British public went into the same hysterical rapture about the occupation of Mandalay as it did about Tel-el-Kebir. Its want of judgement and foresight, as proved by subsequent events in both instances is not very creditable to it and hence its impatience and disappointment.'[69] Equally, the Secretary of State for War, W. H. Smith, suggested the public was 'very impatient knowing as little of the difficulties in the way of operations in Burmah as they would of operations in the moon'.[70]

Seen as a safe pair of hands, Lieutenant General Sir Herbert Macpherson, CinC of the Madras Army, succeeded Prendergast in September 1886. White was considered too junior for the command. When Macpherson died of fever within days of arrival in October, Roberts took personal command of operations. Roberts set priorities and issued detailed guidance to column commanders. Columns must be provisioned for ten days with grain for horses and ponies to be obtained on the march. Supply depots had to be established and particular care taken of horses, ponies and pack animals. Columns must be carefully co-ordinated and local guides obtained. Roberts stressed that 'the chief object of traversing the country with columns is to cultivate friendly relations with the inhabitants, and at the same time to put before them evidences of our power, thus gaining their good-will and their confidence'. The 'broadest possible margin' had to be drawn between dacoits, and villagers coerced into assisting them. The inhabitants, therefore, had to be treated properly. Military operations, however, had to inflict 'the heaviest loss possible' and column commanders were permitted to inflict punishments if civil officers were not present. Villages should be disarmed but a certain number could be returned on licence to 'responsible villagers'. Roberts also suggested means of combatting ambushes with designated flanking parties told off to sweep into the jungle when a column was fired upon.[71]

Since Roberts believed troops 'should make their presence felt everywhere', he set an ideal column at 50–100 British infantry, 100–200 Indian infantry, 30 mounted infantry, and two mountain guns. Charles Callwell, the doyen of British small wars theory, felt larger columns might perhaps have had more impact.[72] It might also be noted that Callwell maintained that while villages which harboured dacoits were destroyed, cattle carried off and crops impounded, 'great care had to be exercised not to punish villages which were merely victims of dacoity'. Callwell claimed that, even where fines were levied to induce villagers to disarm, the population was friendly since operations were not directed against the people as a whole.[73]

The instructions displayed considerably more recognition of 'hearts and minds' than Roberts's conduct of operations in the Second Afghan War

(1878–80), suggesting perhaps that he had learned lessons from the widespread criticism his methods had then attracted.[74] In February 1887 Roberts handed over command to White. White had been commanding Upper Burma since April 1886 although the Duke of Cambridge had refused repeatedly to grant White rank as temporary major general in view of his lack of seniority on the list of colonels.[75] Another repercussion of the ongoing controversies was that the Queen's son, the Duke of Connaught had been under consideration for the Madras command. He was ruled out on political grounds given the Madras army was nominally responsible for Burma. Accordingly, Lieutenant General Sir Charles Arbuthnot, who had only just arrived to command in Bombay, was transferred to Madras with Connaught given the less politically sensitive Bombay command.[76]

Given the perceived unreliability of the Burmese, the new civil and military police intended to augment troop strength was raised mainly in India: military police were organised into battalions and bore some resemblance to the armed police in the Madras presidency.[77] The first police levy in February 1886 was for 1,122 men from Indian regiments to be supplemented by an additional 2,240 men recruited from northern India. Two more levies were authorised in March 1886 for the protection of the railway to be constructed between Toungoo and Mandalay, and for the Mogaung area. They were again drawn from northern India but also from Gurkhas. Further police were raised in March 1887.[78] Pay was not as attractive as for the army and there was a tendency to rely on lower caste recruitment. Wastage was also high, not least from disease. Generally, the notion of 'martial races' pervaded the process: Sikhs and Gurkhas were preferred. The hill tribes, notably the Karens, were seen as sufficiently 'martial' and more resistant to disease, with increased recruitment from 1892.[79] Subsequently, Chins, Kachins and Karens were also preferred for the Burma Rifles.[80] It might be noted that there was criticism from a 'martial races' perspective of the Madras army being entrusted with the bulk of operations following the initial occupation,[81] with attempts by the Bengal army to assume control.[82]

As in the case of the new civilian administration, police officers were found from military officers or civilian officials from other provinces of India.[83] In the Lower Chindwin District, Captain Frederick Raikes, the assistant commissioner, found it difficult in 1886 to man all 16 police posts suggested, as well as keeping the recommended 100 men in reserve at district headquarters and a further 70 as a mounted force since this assumed all the police were effective. Of the 781 police, 370 were untrained recruits leaving only 241 trained men for the 16 posts once the reserve and mounted men were deducted. In addition, weapons issued were those captured from dacoits rather than modern firearms and the two inspectors sent could not speak any Burmese. Accordingly, Raikes had to request troops for he could not otherwise guarantee law and order or revenue collection. Detachments from the 18th Bengal Infantry were provided for nine posts and Raikes was able to employ 'loyalists' at six others. Even then a Chin raid on one village in January 1887 resulted in the decapitation of 12 villagers including seven women and three children before troops could reach it. Raikes despaired 'how much a state of things can exist even in the most uncivilised country'.[84]

By March 1887 there were 9,000 police in Lower Burma and over 17,000 military and civil police in Upper Burma under the command of Colonel Edward Stedman as Inspector-General. Burma was far more heavily policed in terms of the ratio of police to population than any other part of British India.[85] By the end of 1888, military police numbers in Upper Burma were 19,177, manning 192 separate posts.[86] In 1888 the minimum number of police in any post was raised from 25 to 40 with the minimum strength of any patrol fixed at ten men. That year military police casualties were 46 dead and 76 wounded compared to an estimated 312 dacoits killed and 721 captured.[87] Civil and military police throughout Burma reached the strength of around 32,000 in 1888, a total not fully reduced until the 1920s.[88]

Troops were steadily reduced, priority being given to those units that had been engaged in the initial occupation.[89] The Burma Field Force was reclassified as a garrison in April 1888. In May 1888 Upper Burma was reorganised into three brigades and five separate commands. In April 1889 a Burma District Command was established with the three districts of Mandalay, Myingyan and Rangoon. The regular garrison was now reduced to 16,080 officers and men and included three newly raised battalions from Burma comprising the Kubo Valley Police Battalion, the Chin Levy and the Shan States Levy re-designated as the 1st, 2nd and 3rd Burma Infantry. The majority were Indians and Gurkhas but Karens, Shans and Kachins could be enlisted.[90] At the beginning of 1887 the police manned 56 posts compared to 142 military posts. By 1889 only 41 were held by troops but 192 by police.[91]

Cavalry and mounted infantry made a particular contribution to pacification. A small force of 98 mounted infantry riding Burmese ponies was improvised by Edward Browne from the Rangoon Volunteer Rifles, Lower Burma police, and his own regiment – the Royal Scots Fusiliers – to give the British more mobility during the invasion.[92] When commanding Upper Burma, White requested three cavalry regiments for the intended 1886–87 winter operations. Mounted men not only offered greater mobility and some ability to outflank opponents but the Burmese were not used to horses and were terrified of them although, invariably, mounted *bohs* were the first to run and could not generally be caught. At peak, four Indian cavalry regiments were used but it was difficult to keep horses alive, 666 out of 2,092 being lost between October 1886 and October 1887 to diseases such as relapsing fever (*surra*), lumbar paralysis (*kumri*), and anthrax.[93] In addition, the mounted infantry was built up to a force of 825 divided into companies of 75 men, each attached to district headquarters and armed with carbines and artillery sword bayonets. Few had any experience of mounted infantry work and Indian soldiers had little knowledge of horse-mastership and took far longer to train to ride. There were few copies of the British regulations available in India so Lieutenant Colonel William Penn Symons, whose services White had specifically requested, wrote his own instructions.[94] Roberts considered Symons to have contributed more to pacification than anyone other than White.[95] Such was the perceived value of mounted infantry that it was decided to retain it as an element within the regular garrison in 1894. In April 1887, 1,600 men had been so employed but it would now be established at 215 British and 300 Indian soldiers with increased pay to persuade infantrymen to volunteer for detached service.[96] Artillery was less successful as mules could not readily keep up with infantry and there were rarely defined targets. Gardner machine guns were often inaccurate due to the jolting of tripods and frequently broke down but they had a good morale effect when employed.[97]

Effective measures introduced alongside purely military operations included rewards offered for information; pardons for surrenders; and the relocation of the rural population to more easily defended locations, which disrupted crop cultivation but cut the dacoits off from their base of support. Many villages were burned. Large scale disarmament was instituted by Bernard's successor as Chief Commissioner, Sir Charles Crosthwaite, in 1887. General pardons were often made conditional on the surrender of guns. Disarmament became especially effective when coupled with the Upper Burma Village Regulation. The latter was drafted by Crosthwaite in February 1887 and implemented from October 1887. He regarded it as 'the most effective weapon in our battery for the restoration of peace and order'.[98] Traditional *thugyi* were replaced by appointed headmen who were made responsible for law and order, combing the function of *thugyi* and policeman. Thus, villagers had to respond to the headman's gong and perform any service required on pain of 24 hours in the stocks. All Burmese except monks had to offer the *shikho* salutes to British officers and officials previously advanced only to elders, monks and statues of the Buddha.[99] The Lower Burma Village Act was implemented in 1889, meaning that Lower Burma was subjected to the same regulations as Upper Burma.

Despite the contrary evidence of some officials, Crosthwaite made assumptions as to the significance of the *thugyi* in Burmese village society based on Indian experience and previous attempts to establish a village police in Lower Burma. Essentially, the Regulation and its various additional amendments imposed a uniform system on Burma although this was not always carried out consistently at local level.[100] It has also been argued that the British sense of pacification meant sufficient order to collect revenue: a wider psychological pacification was never realised.[101]

From the point of view of pacification, the significance of the Village Regulation was that district commissioners were able to order the fencing of villages, the mobilisation of villagers, collective fines, and the removal of those suspected of supporting or sympathising with dacoits.[102] The latter led to large-scale removals while collective fines proved especially effective, as in the case of breaking up the support base for U Ottama around Minbu in November 1889. One notorious dacoit, Bo Ya Nyun, surrendered in May 1890 after rewards and pardons for defectors had robbed him of support.[103] Strict licensing of guns was introduced in May 1888 with requirements for license holders to act as special constables and loss of guns resulting in confiscation of all within a village.

Railway construction between Toungoo and Mandalay assisted movement of troops but also served to provide wage income for up to 24,000 men, offering further reason for supporting the British administration, albeit that it was actually forced labour.[104] Indeed, it was suggested that the railway was 'one of the most pacifying influences in the eastern districts'.[105] In the case of the Kanhow tribe among the Chins, it was suggested that those chiefs taken captive in the 1890–91 winter campaign and sent to Rangoon 'were much impressed by the wonders of civilisation and the evidence they saw of the power of the white man'. Supposedly they vowed 'never again to withstand British arms'.[106]

On occasions, operations were hampered by civil-military friction. Roberts was determined to get William Lockhart to Burma although there were more senior officers available and Lockhart received command of the 3rd Brigade in the Burma Field Force.[107] Lockhart's sector was the Ningyan and Yementhin districts, some eight punitive expeditions being launched by him between October and November 1886. There was some dispute on brigade boundaries between Lockhart and Brigadier General Charles East.[108] Lockhart also proved impatient of the restrictions placed on him by Bernard. White cautioned Lockhart that Bernard was obliged to carry out the orders of the Government of India and that Lockhart should take more account both of British public opinion critical of burning villages, and also of the need to protect the villagers against the dacoits who were intimidating them. Lockhart deprecated what he saw as Bernard's 'tenderness for the people'. In October 1886 White reminded Lockhart that Bernard was acting under government orders and allowances had to be made for villagers whom the British had not been able to protect from dacoit intimidation. Nor did Lockhart comprehend the need to be mindful of British public opinion.[109] Nonetheless, White recognised of Lockhart's methods that districts could not be 'pacified with rosewater'.[110]

During the winter campaign of 1889–90, there was disagreement between Major J. E. Blundell commanding the Tonbon Expeditionary Force and his political officer, G. Shaw. Blundell was criticised generally for his slow progress but the point of contention was the destruction of the village of Manpan, which political officers considered friendly. From Madras, Arbuthnot wrote pointedly that punishments required instant decisions and martial law should be declared whenever the state of a district justified military action.[111]

By 1888 steady progress was being made in Upper Burma, with few remaining larger dacoit bands or prominent leaders, more willingness on the part of the population to assist in the maintenance of law and order, and an apparent recognition that their interests were best served by doing so.[112] Resistance in Lower Burma was never as organised, mostly consisting of small and isolated bands. From an estimated 2,183 dacoits operating in Lower Burma in 1888, the number had declined to an estimated 181 by 1890.[113] The last major expedition in Burma proper was the Wuntho expedition in 1891 on the edge of the Shan States. Attention, therefore, turned to the Shan States and the Kachin and Chin Hills, areas previously in rebellion against the Burmese. The problem was different in that rather than disparate dacoit bands, the hill tribes were well organised.

V

The Shan States comprised over 40,000 sq. miles with a population of 1.2 million. The Limbin Prince, who had planned a rising against Thibaw turned his so-called Limbin Confederacy against those Shan chiefs prepared to co-operate with the British. Initial British columns were committed to support the anti-Limbin chiefs in November 1886 and January 1887, and the Limbin Prince surrendered in May 1887. Resistance continued, however, until January 1889.

George Scott, who became Superintendent of the Northern Shan States, recorded that Shan villages were invariably well fortified with earth breastworks, triple stockades and spiked approaches. Troops found it difficult to move during the rainy season, with the Shans able to return to 'old haunts' as soon as troops departed. Flying columns were more effective in the cold weather season when Gurkhas proved able to outmanoeuvre the Shans even in dense jungle-clad hills although this rarely resulted in much contact.[114] Under the Shan States Act of November 1888, the hereditary *sawbwa* were retained with reduced powers since they appeared more recognisably as the kind of local chiefs with whom the British could work compared to *thugyi*.[115] The Wa were one Shan tribe that never acknowledge British suzerainty but, as they did not generally raid into other areas, they were left alone.[116] Generally, the British approach to all the hill tribes was to defeat resistance, impose an indemnity, insist on a continuation of the annual tributes previously paid to the Burmese King, and to establish a garrison with a resident political officer to represent British power.[117]

In the Kachin Hills, opposition led by U Po Saw resulted in four separate operations being mounted in 1888–89. The Kachins tended to fight from stockades rather than taking to guerrilla tactics but the British still resorted to the destruction of villages, crops and livestock. Between January and May 1889, for example, the Mogaung Column commanded by Captain H. O'Donnell, the Staff Officer for the Burma District, destroyed 24 villages with a total of 355 houses, all household goods, 419,000 lbs of grain, 59 buffaloes, and also fowls and pigs 'without number'.[118] The four expeditions together destroyed 46 villages with 639 houses. The force suffered one dead and 23 wounded. It captured 30 *dahs* (Burmese knives), seven guns, a number of spears and a symbolic golden umbrella but no enemy bodies were found although it was assumed from blood stains that the Kachins had sustained losses.

Operating against the Kachins in April 1891, Lieutenant J. K. Watson of the 4th King's Royal Rifle Corps noted:[119]

> The wily Kachin places himself behind this and directly the head of the column appears round the corner they fire one salvo from their primitive guns and make a bolt – to catch them is almost an impossibility; you can do nothing in the way of flanking parties, the jungle's far too thick, and a file of men 30 or 40 yards ahead of the column is your advance guard. If they see anything suspicious ahead they should get off the path and into the jungle at once and hollow back to the column behind. If the road is stockaded you may be able to work a small party round thro' the jungle unless it's too thick, in which case the obstacle must be rushed. Well, going along with the advanced file is sufficiently exciting.

The Kachins were largely pacified by December 1895 although a police expedition was mounted in March 1898 to expel a force of Chinese-Shans from the Bhamo district.

Like the Kachins, the Chins favoured fighting from stockades but soon adapted to guerrilla tactics. The British initially attempted negotiation but dacoits driven out of Upper Burma encouraged Chin raids down from the hills. The Chin Field Force was organised in response in December 1888. White, who accompanied the expedition, reported the Siyin Chins as the 'most difficult enemy to see or hit I ever fought'.[120] Fort White was established at Tokhlaing. In November 1889 the Chin-Lushai Expedition was mounted to prevent further Chin raids into Burma and to open up a good road between Chittagong and Burma. Two columns totalling 3,608 fighting men and directed by Penn Symons were deployed with an additional one advancing from Chittagong in India under Colonel Vincent Tregear, the earlier Lushai campaign of 1871–72 having been mounted from India.[121] Great difficulties were encountered traversing the Chin Hills and malaria accounted for

207 deaths among troops and followers of the two Burma columns with a further 2,122 invalided: only nine troops or followers were killed.[122]

In the on-going pacification effort among the Chins, British columns could rarely carry more than 10 days' food supplies with them carried by coolies, making it impossible to do much damage to the Chins or their crops. Ambushes were best met by rushes since it was found that the Chins would not stand to fight. They were especially skilful at night and it became practice to ring posts with tins on wires as a warning.[123]

Lieutenant Watson, serving against the Chins in January 1892, recorded:[124]

> At first we tried the 'zubberdushti' (high-handed) method with them; that was when General Faunce was in command; villages were burnt, prisoners taken, and so on. It didn't answer. Then, under Symonds [sic], the reverse was done. The Chins were made much of, etc., etc. That didn't seem to do either, and from what I can gather the policy now seems to be to steer a medium course, – treat them kindly, pay good prices for all local provisions, coolies, etc, but to be down severely on any infringements of our authority. Thus in former years all Chins were allowed to carry arms. Now any native (except red-turbanned ones) seen with a gun has it taken from him. It's just as well too, for there used constantly to be cases of two or three armed Chins meeting a column on a march; they would be allowed to pass, and then calmly secrete themselves in the jungle and have a pot shot at the tail end of the rear guard.

In 1892 larger columns broke up Chin combinations then dispersed into garrisons to harass further the Chins, destroy food supplies and impose disarmament. Herbert Housman, younger brother of the poet, A E. Housman, serving as a NCO with 4th Kings Royal Rifle Corps, witnessed Gurkhas tackling one village in April 1892:[125]

> It was more of a massacre than a fight. A shell was dropped into the village, causing the utmost alarm & surprise, & instead of keeping behind their stockade the Chins rushed out, whether in terror or to attack I don't know. They were shot down like dogs & were soon flying in all directions. Some 20 or 30 being killed, & I saw many running away with shattered arms & damaged legs. I am glad we took no part in this. The Gurkhas are very cool & shoot well, but are terribly cruel.

Two days later, another group of 300 Chins was dispersed by volley fire with 43 killed outright, Housman writing 'the devil-may-care feeling came, & kill as many as you can was the one aim & object I had'.[126]

The Chins proved the most durable opponents of British occupation, the last significant leader – a Siyin – surrendering in May 1894. Over 7,000 weapons were taken from the Chins between 1893 and 1896.[127] The implementation of the Chin Hills Regulations in 1896, taking the Chin Hills into Burma as a scheduled district, then led to further disorder until 1900. By that year, however, the British military garrison in Burma was down to just 10,324 men.[128]

VI

As Kipling suggested, the 'Burmese business was a subaltern's war'.[129] Sir Charles Nairne, while CinC in Bombay, later suggested that subalterns who did well in Burma made their reputation for the future.[130] Among those who served in Burma noted for later distinction were Henry Rawlinson, Henry Wilson, Richard Haking and Ralph Clements, whose career was cut short by fatal appendicitis in 1909. Among brigade commanders, in addition to White, good reputations were made by William Lockhart, William Penn Symons, Cecil East, Robert Low, Edward Stedman and William Gatacre. As White wrote on one occasion, Burma 'somehow finds out weak points very rapidly'.[131]

Yet, while reputations were made in the eyes of the India military authorities, the pacification of Burma was not viewed in the same way at the War Office. Seven separate bars were awarded for the India General Service Medal relating to operations in Burma between 1885 and 1893.[132] Pleas for a separate medal, however, were rejected. Cambridge did not believe that Burma ranked as a campaign. Both Roberts and White felt therefore that there had been too little reward for Burma, while some of those serving in 'police work' in Burma believed that issuing only a clasp was an economy measure.[133] The Military Secretary, Sir George Harman, suggested there had been too little action and the Prince of Wales, too, was led to believe there had been few casualties.[134] Roberts was obliged to reduce the list of those being put forward for honours.[135] He felt keenly the difference in the number of awards for recent actions in Egypt, writing 'so long as Egyptian heroes can get decorations for a picnic lasting as many weeks you must not be surprised at my pressing the claims of those who have done what seems to me, more valuable work'.[136] Eventually, three VCs were won during the pacification, all by medical officers.[137] Subsequently, as operations in Burma became a long and exacting pacification campaign, few brigade commanders or staff officers were willing to undertake a full five year term when regiments were being rotated after three years.[138]

Burma never proved to be of real commercial value to the British Empire beyond the Burmah Oil Company that began operations in 1886. In the end, too, Burma remained under British control for only 62 years after 1885, and that was disrupted by the serious Tharawaddy Revolt between 1930 and 1932, and by Japanese occupation between 1942 and 1945. The pacification of Burma, however, has contemporary resonance. Intended regime change was not accompanied by any consideration of the likely implications. The initial force deployed was not sufficient to ensure proper security in the aftermath of occupation. Prolonged insurgency necessitated deploying a force far larger than originally intended or anticipated. Evolving military and civil measures eventually brought order but proved destructive of Burmese society. British preference for the recruitment of hill tribes into police and armed forces equally sowed seeds for future divisions. The pacification of Burma was not just a campaign of lost footsteps but is one of forgotten footsteps.

Notes

1. Webb, "Kipling and Burma", 10–19.
2. Kipling, *From Sea to Sea and Other Sketches*.
3. Intelligence Branch, Army Headquarters, *Frontier and Overseas Expeditions from India*. 144, 146.
4. Thant, *The Making of Modern Burma*, 5.
5. Browne, *The Coming of the Great Queen*, 171–2.
6. Myint-U, *Making of Modern Burma*, 170; and Ni Mi Myint, *Burma's Struggle against British Imperialism, 1885–95*, 34.
7. British Library, Asia, Pacific and Africa Collection (hereafter APAC), IOR/L/MIL/17/19/31/1, H. E. Stanton, *The Third Burmese War, 1885, 1886 and 1887*, 52.
8. Hall, *The Soul of a People*, 66.
9. For an overview, see Roy, *The Army in British India*, 58–76.
10. For a popular, albeit outdated, survey of the military aspects of the first two wars, see Bruce, *The Burma Wars, 1824–1886*.
11. Myint-U, *Making of Modern Burma*, 106–7.
12. Stewart, *The Pagoda War*, 110–7; and Myint-U, *Making of Modern Burma*, 194–8.
13. Myint-U, *Making of Modern Burma*, 199–200.
14. Ibid, 202–7.
15. Wingfield, "Buddhism and Insurrection in Burma," 345–67.
16. Myint, *Burma's Struggle*, 65–6.
17. Ibid, 78.
18. Myint-U, *Making of Modern Burma*, 210–8; and Crosthwaite, *The Pacification of Burma*, 54.
19. Myint, *Burma's Struggle*, 95–6.
20. Charney, "Armed Rural Folk Elements of Pre-colonial Warfare in the Artistic Representations," 155–81.
21. First published in *The Week's News*, 7 January 1888, and then in *Departmental Ditties and Other Verses*, 120–4.
22. Browne, *Coming of Great Queen*, 277.
23. National Army Museum (hereafter NAM), Roberts Mss, 7101-23-100-1, Roberts to Churchill, 28 February 1886.
24. Keck, "Involuntary Sightseeing: Soldiers as Travel Writers," 389–407.
25. Myint, *Burma's Struggle*, 218–21.
26. Hall, *Soul of People*, 61–4; and Crosthwaite, *Pacification*, 103–4.
27. "Mandalay in 1885–88," 47–76, at 50.
28. Geary, *Burma, After the Conquest*, 45–6, 74.
29. Browne, *Coming of Great Queen*, 270–5.
30. Crosthwaite, *Pacification*, 103–4.
31. Slim, *Defeat into Victory*, 169.
32. Newland, *The Image of War*, 6–7.
33. Durand, *The Life of Field*, I, 326.
34. APAC, IOR/L/MIL/7/9182, Report on Operations, 31 March 1888 to 6 July 1889.
35. Stanton, *Third Burmese War*, 72–4.

36. Stewart, *Pagoda War*, 90.
37. APAC, IOR/L/MIL/17/19/31/3, and Bodé, *The Third Burmese War*, 37.
38. The National Archives (hereafter TNA), WO 334/95, Sick Reports.
39. *Frontiers and Overseas Expeditions*, 220–1.
40. *Frontiers and Overseas Expeditions*, 228.
41. Stanton, *Third Burmese War*, 32.
42. APAC, IOR/L/MIL/7/9181.
43. Browne, *Coming of Great Queen*, 192.
44. *Frontier and Overseas Expeditions*, 168.
45. Durand, *Life of White*, I, 326.
46. *Frontier and Overseas Expeditions*, 164–5, 178.
47. Nisbet, *Burma Under British Rule – And Before*, I, 111.
48. Crosthwaite, *Pacification*, 15.
49. APAC, White Mss, Eur Mss, F108/3, White to Chesney, 14 August 1886.
50. *Frontier and Overseas Expeditions*, 193.
51. Nisbet, *Burma Under British Rule*, I, 112–4, 126.
52. *Frontier and Overseas Expeditions*, 197.
53. Bodé, *Third Burmese War*, 16.
54. Ibid, 49–53.
55. APAC, IOR/L/MIL/7/9181, Report on Operations for the Suppression of Brigandage.
56. *Frontier and Overseas Expeditions*, 241.
57. APAC, IOR/L/MIL/7/9/180, Viceroy to India Office, 19 October 1887.
58. *Frontier and Overseas Expeditions*, 289.
59. Newnham Davis, *Burmese War*, Appendix I, iii-xvii.
60. APAC, White Mss, Eur Mss F108/3, White to Roberts, 19 June 1886.
61. NAM, Roberts Mss, 7101-23-90, White to Roberts, 9 July 1887.
62. Vibart, *The Life of General Sir Harry Prendergast VC: The Happy Warrior*, 278–80.
63. Stewart, *Pagoda War*, 118–31, 140–1, 164–70.
64. Geary, *Burma*, 234, 241–5, 267–8.
65. Stewart, *Pagoda War*, 130.
66. Ibid, 127–8.
67. NAM, Roberts Mss, 7101-23-12, Brownlow to Roberts, 28 January 1886.
68. Stewart, *Pagoda War*, 173.
69. NAM, Roberts Mss, 7101-23-12, Brownlow to Roberts, 22 August 1886.
70. TNA, Smith Mss, WO 110/5, Smith to Roberts, 5 November 1886.
71. Robson, *Roberts in India, 1876–93* 357–62.
72. Callwell, *Small Wars*, 141.
73. Ibid, 147–8.
74. Johnson, "General Roberts, the Occupation of Kabul, and the Problems of Transition," 300–22.
75. See, for example, Royal Archives (hereafter RA), VIC/ADDE/1/11444, Cambridge to Roberts, 12 March 1886; VIC/ADDE/1/11467 Cambridge to Roberts, 16 Apl. 1886; and NAM, Roberts Mss, 7101-23-100-1, Roberts to Cambridge, 14 Apl. 1886.
76. TNA, Smith Mss, WO 110/5, Cambridge to Roberts, 3 November 1886; Smith to the Queen, 3 November 1886; and NAM, Roberts Mss, 7101-23-2, Arbuthnot to Roberts, 20 November 1886.
77. Hingkanonta, "The Police in Colonial Burma," 12.
78. APAC, IOR/L/MIL/17/19/31/5, and Stanton, *History of the Third Burmese War*, 136–7.

79. Hingkanonta, "Police in Colonial Burma," 46–88. From a large literature on 'martial races' theory, see Streets, *Martial Races*, 93–101, 132–42.
80. Taylor, "Colonial Forces in British Burma," 195–209.
81. Omissi, *The Sepoy and the Raj*, 13–16.
82. RA, VIC/ADDE/1/11627, Cambridge to Arbuthnot, 5 November 1886; VIC/ADDE/1/12677, Cambridge to Dormer, 20 November 1890.
83. White, *A Civil Servant in Burma*, 157–58.
84. APAC, Raikes Mss, Eur Mss B3291, Raikes Diary, Entries for 28 June 2013 Sept., and 29 October 1886; and 7 and 28 January 1887.
85. Myint, *Burma's Struggle*, 82–5.
86. APAC, IOR/L/MIL/17/19/31/7, and N. Newnham Davis, *History of the Third Burmese War, 1888–89*, 77–8.
87. Crosthwaite, *Pacification*, 97–8.
88. Hingkanonta, "Police in Colonial Burma," 181.
89. APAC, IOR/L/MIL/7/9180, Viceroy to India Office, 14 January 1887.
90. APAC, IOR/L/MIL/17/19/1/8; and Parsons, *History of the Third Burmese War, 1889–90*, 3.
91. Nisbet, *Burma Under British Rule*, I, 141.
92. Browne, *Coming of Great Queen*, 132–3; and Winrow, *The British Army Regular Mounted Infantry, 1880–1913* does not cover Burma.
93. Stanton, *Third Burmese War*, 211–7; and "Notes on Cavalry employed in Upper Burma," 29–38.
94. Durand, *Life of White*, I, 347.
95. NAM, Roberts Mss, 7101-23-99, Roberts to Lansdowne, 8 July 1890.
96. APAC, IOR/L/MIL/7/9199, India Council to India Office, 29 May 1894.
97. Stanton, *Third Burmese War*, 218–9.
98. Crosthwaite, *Pacification*, 81–2, 106.
99. Charney, *A History of Modern Burma*, 7.
100. Iwaki, "The Village System and Burmese Society," 113–43.
101. Aung-Thwin, "The British "Pacification" of Burma," 245–62.
102. Crosthwaite, *Pacification*, 105.
103. Myint, *Burma's Struggle*, 62, 101.
104. Myint, *Burma's Struggle*, 101–2; APAC, Sladen Mss, Eur Mss E290/52, Roberts to Sladen, 3 December 1887; and Winston, *Four Years in Upper Burma*, 69.
105. White, *Civil Servant in Burma*, 168.
106. APAC, IOR/L/MIL/17/19/31/9; and Parsons and Dunn, *History of the Third Burmese War, 1890–91*, 19.
107. NAM, Roberts Mss, 7101-23-100-2, Roberts to Harman, 26 July 1889.
108. APAC, White Mss, Eur Mss F108/3, White to Roberts, 11 May 1887.
109. Durand, *Life of White*, I, 352–4.
110. APAC, White Mss, Eur Mss F108/3, White to Chesney, 12 May 1887.
111. APAC, IOR/L/MOL/7/9189, Arbuthnot to AG, 30 May 1890.
112. APAC, Thirkell White Mss, Eur Mss E254/10(b), Note on State of Upper Burma, 14 July 1888.
113. White, *Civil Servant*, 221–2.
114. APAC, Scott Mss, Eur Mss F278/51, Diary of Expedition to the Shan States.
115. Myint-U, *Making of Modern Burma*, 216–7.
116. *Frontier and Overseas Expeditions*, 426.
117. Aung-Thwin, "British Pacification of Burma," 252.
118. APAC, IOR/L/MIL/7/9182, Report on Mogaung Field Force, 19 Apl. 1889.

119. Pearn (ed.), *Military Operations in Burma, 1890–92: Letters from Lt. J. K. Watson, KRRC*, 10.
120. Crosthwaite, *Pacification*, 303–4.
121. *Frontier and Overseas Expeditions*, 330–2.
122. Ibid, 335.
123. Ibid, 320.
124. Pearn, *Military Operations in Burma*, 35.
125. Bourne (ed.), *Soldier I Wish You Well*, 49.
126. Ibid, 51.
127. *Frontier and Overseas Expeditions*, 355.
128. Nisbet, *Burma Under British Rule*, I, 149.
129. Kipling, "A Conference of the Powers," *Many Inventions*, 33.
130. Jones, "The War of Lost Footsteps," 36–40, at 40.
131. NAM, Roberts Mss, 7101-23-90, White to Roberts, 3 December 1888.
132. Burma 1885–7, Burma 1887–89, Chin-Lushae 1889–90, Burma 1889–92, Lushae 1889–92, Chin Hills 1892–93, Kachin Hills, 1892–93.
133. NAM, Cowell Mss, 2009-02-110-417, Roberts to Cowell, 5 August 1887; APAC, White Mss, Eur Mss F108/3, White to Roberts, 28 May 1887; Howard, *Reminiscences, 1848–90*, 299. and See also APAC, Lyttelton. letter book, Eur Mss F102/43, Lyttelton to Godley, 18 July 1887.
134. NAM, Roberts Mss, 7101-23-59, Pole-Carew to Roberts, 15 and 28 July 1887; 7101-23-100-1, Roberts to Pole-Carew, 5 August 1887.
135. NAM, Roberts Mss, 7101-23-98, Roberts to Dufferin, 30 July 1886.
136. NAM, Roberts Mss, 7101-23-103, Roberts to Brownlow, 16 January 1887.
137. John Crimmin, 1 January 1889; Ferdinand Le Quesne, 4 May 1889; Owen Lloyd, 6 January 1893.
138. NAM, Roberts Mss, 7101-23-2, Arbuthnot to Roberts, 26 July 1890.

Disclosure statement

No potential conflict of interest was reported by the author.

Bibliography

Primary sources

British Library, Asia, Pacific and Africa Collection: India Office Records, Lyttelton Mss, Raikes Mss, Scott Mss, Sladen Mss, Thirkell White Mss, White Mss.
National Army Museum: Cowell Mss, Roberts Mss.
Royal Archives: Cambridge Mss.
The National Archives: W. H. Smith Mss, War Office Mss.

Secondary sources

Ali, M. S., 'The Beginnings of British Rule in Upper Burma: The Study of British Policy and Burmese Resistance, 1885-90', Unpub. Ph.D., London, 1976. doi:10.1084/jem.143.4.741

Aung-Thwin, Michael. "The British "pacification" of Burma: Order without Meaning." *Journal of Southeast Asian Studies* 16 (1985): 245–262. doi:10.1017/S0022463400008432.

Bodé, L. W. *The Third Burmese War, 1885, 1886 and 1887: History of the War from the Annexation of the Country to the Commencement of the Winter Campaign of 1886-87.* Calcutta: Superintendent of Government Printing, 1888.

Bourne, Jeremy, ed. *Soldier I Wish You Well: The Military Poems of A. E. Housman and the Letters from Burma of G. H. Housman.* Bromsgrove: Housman Society, 2001.

Browne, Major Edmond. *The Coming of the Great Queen: A Narrative of the Acquisition of Burma.* London: Harrison & Son, 1888.

Bruce, George. *The Burma Wars, 1824-1886.* London: Hart-Davis MacGibbon, 1973.

Callwell, Charles. *Small Wars: Their Principles and Practice.* 3rd ed. London: HMSO, 1906.

Charney, Michael. *A History of Modern Burma.* Cambridge: Cambridge University Press, 2009.

Charney, Michael. "Armed Rural Folk Elements of Pre-colonial Warfare in the Artistic Representations and Written Accounts of the Pacification Campaign (1886–1889) in Burma." In *Warring Societies of Pre-colonial Southeast Asia: Local Cultures of Conflict within a Regional Context,* edited by Michael Charney and Kathryn Wellen, 155–181. Copenhagen: NIAS Press, 2018.

Crosthwaite, Sir Charles. *The Pacification of Burma.* London: Edward Arnold, 1912.

Durand, Sir Mortimer. *The Life of Field Marshal Sir George White VC.* Vol. 2. Edinburgh: William Blackwood & Sons, 1915.

Geary, Grattan. *Burma, after the Conquest.* London: Sampson Low, Marston, Searle & Rivington, 1886.

Hall, Harold Fielding. *The Soul of a People.* London: Macmillan, 1906.

Hingkanonta, Lalita, 'The Police in Colonial Burma', Unpub. Ph.D., London, 2013.

Howard, Major General Sir Francis. *Reminiscences, 1848-90.* London: John Murray, 1924.

Intelligence Branch, Army Headquarters. *Frontier and Overseas Expeditions from India: Volume 5 - Burma.* Simla: Government Monotype Press, 1907.

Iwaki, Takahiro. "The Village System and Burmese Society: Problems Involved in the Enforcement Process of the Upper Burma Village Regulation of 1887." *Journal of Burma Studies* 19 (2015): 113–143. doi:10.1353/jbs.2015.0008.

Johnson, Rob. "General Roberts, the Occupation of Kabul, and the Problems of Transition, 1879-80." *War in History* 20 (2013): 300–322. doi:10.1177/0968344513483227.

Jones, Martin. "The War of Lost Footsteps: A Re-assessment of the Third Burmese War, 1885-96." *Bulletin of the Military Historical Society* 40 (1989): 36–40.

Keck, Stephen. "Involuntary Sightseeing: Soldiers as Travel Writers and the Construction of Colonial Burma." *Victorian Literature and Culture* 43 (2015): 389–407. doi:10.1017/S1060150314000618.

Kipling, Rudyard. *From Sea to Sea and Other Sketches: Letters of Travel.* Vol. 1st. London: Macmillan, 1900.

"Mandalay in 1885-88: The Letters of James Alfred Colbeck." *SOAS Bulletin of Burma Research* 2 (2004): 47–76.

Myint, Ni Mi. *Burma's Struggle against British Imperialism, 1885-95*. Rangoon: Universities Press, 1983.

Myint-U, Thant. *The Making of Modern Burma*. Cambridge: Cambridge University Press, 2001.

Newland, A. G. E. *The Image of War or Service in the Chin Hills*. Calcutta: Thacker-Spink, 1894.

Newnham Davis, N. *History of the Third Burmese War, 1888-89: The Winter Campaign of 1888-89 and Subsequent Operations up to December 3rd 1889*. Simla: Government Central Printing Office, 1892.

Nisbet, John. *Burma Under British Rule - And Before*. Vol. 2. Westminster: Archibald Constable, 1901.

"Notes on Cavalry Employed in Upper Burma from October 1886 to October 1887." *SOAS Bulletin of Burma Research* 2 (2004): 29–38.

Omissi, David. *The Sepoy and the Raj: The Indian Army, 1860-1940*. Basingstoke: Macmillan, 1994.

Parsons, J. H. *History of the Third Burmese War, 1889-90: The Winter Campaign of 1889-90*. Simla: Government Central Printing Office, 1893.

Parsons, J. H., and E. W. Dunn. *History of the Third Burmese War, 1890-91: The Winter Campaign of 1890-91*. Simla: Government Central Printing Office, 1893.

Pearn, B. R., ed. *Military Operations in Burma, 1890-92: Letters from Lt. J. K. Watson, KRRC*. Cornell University South East Asian Program Data Paper No. 64. Ithaca, NY, 1967.

Robson, Brian, ed. *Roberts in India: The Military Papers of Field Marshal Lord Roberts, 1876-93*. Stroud: Alan Sutton for the Army Records Society, 1993.

Roy, Kaushik. *The Army in British India: From Colonial Warfare to Total War, 1857-1947*. London: Bloomsbury, 2013.

Slim, Field, and Marshal Viscount. *Defeat into Victory*. London: Cassell, 1956.

Stanton, H. E. *The Third Burmese War, 1885, 1886 and 1887: History of the War Prior to the Annexation of the Country*. Calcutta: Superintendent of Government Printing, 1887.

Stanton, H. E. *History of the Third Burmese War, 1885, 1886 and 1887: The Winter Campaign of 1886-87, and Subsequent Operations up to March 31st 1888*. Calcutta: Superintendent of Government Printing, 1889.

Stewart, A. T. Q. *The Pagoda War: Lord Dufferin and the Fall of the Kingdom of Ava, 1885-86*. London: Faber & Faber, 1972.

Streets, Heather. *Martial Races: The Military, Race and Masculinity in British Imperial Culture, 1857-1914*. Manchester: Manchester University Press, 2004.

Taylor, Robert H. "Colonial Forces in British Burma: A National Army Postponed." In *Colonial Armies in Southeast Asia*, edited by Tobias Rettig and Karl Hack, 195–209. London: Routledge, 2009.

Vibart, Colonel Henry. *The Life of General Sir Harry Prendergast VC: The Happy Warrior*. London: Eveleigh Nash, 1914.

Webb, G. W. "Kipling and Burma." *Kipling Journal* 301 (2002): 10–19. 25-32, and 302 (2002)

White, Thirkell, and Sir Herbert. *A Civil Servant in Burma*. London: Edward Arnold, 1913.

Wingfield, Jordan. "Buddhism and Insurrection in Burma, 1886-90." *Journal of the Royal Asiatic Society* 20 (2010): 345–367.

Winrow, Andrew. *The British Army Regular Mounted Infantry, 1880-1913*. London: Routledge, 2016.

Winston, W. R. *Four Years in Upper Burma*. London: C. H. Kelly, 1892.

The *Force Publique's* campaigns in the Congo-Arab War, 1892-1894

Mario Draper

ABSTRACT

Between 1892 and 1894 the *Force Publique* of King Leopold II's Congo Free State engaged in a series of little-known counter-insurgency operations against ivory and slave traders from Zanzibar, commonly referred to as Arabs. Without a particularly strong tradition of imperial service, this chapter argues that the pre-dominantly Belgian officer corps borrowed and adapted methods used by more experienced colonial forces in the 19th Century. Whether taken from existing literature or learned through experience, it reveals that the *Force Publique's* counter-insurgency methods reflected many of the more recognisable aspects of traditional French and British approaches. It suggests that, despite the unique nature of each colonial campaign, basic principles could be adapted by whomsoever to overcome the military and political challenges of colonial conquest. The *Force Publique's* campaigns in the Congo-Arab War, therefore, provide further evidence as to how some base theories could be universally applied.

When the *Société d'Études Coloniales de Bruxelles* published *L'art militaire au Congo* under the direction of Colonel Donny of the Belgian Army in 1897, few could blame late-nineteenth century students of colonial warfare for barely taking note.[1] Its largely forgotten pages offered little in the way of revolutionary approaches to the conduct of small wars. Instead, it focused on the tactical and operational narrative of a small, locally-raised force under the stewardship of a handful of white officers bent on denying the Swahili-speaking Muslim warlords, commonly referred to as Arabs by contemporary Europeans, access to Congolese ivory and slaves. Compared to its contemporaneous publications such as Charles E. Callwell's 1896 book, *Small Wars: their principles and practice*, or Hubert Lyautey's 1900 article, 'Du rôle colonial de l'armée', Donny's exaltation of methods used in the Congo Free State (CFS) paled into insignificance. After all, what lessons could a solitary campaign for economic supremacy, conducted by an inexperienced force, in a largely unknown area of Africa, offer readers that were not already being extrapolated from the myriad campaigns fought by the armies of more established Empires?[2] Yet, *L'art militaire au Congo* was not a useless piece of self-indulgence. It offered, and continues to offer, a window into the approach taken by the predominantly Belgian-officered *Force Publique* to overcome the specific challenges posed by the Congo-Arab War of 1892–1894. Whether taken from existing literature or learned through experience, it reveals that the *Force Publique*'s counter-insurgency methods reflected many of the more recognisable aspects of traditional French and British approaches. It suggests that, despite the unique nature of each colonial campaign, basic principles could be adapted by whomsoever to overcome the military and political challenges of colonial conquest. The *Force Publique*'s campaigns in the Congo-Arab War, therefore, provide further evidence as to how some base theories could be universally applied.

Recent scholarship on modern counter-insurgency has tended to contextualise itself within the historical precedents of the nineteenth century.[3] However, the suggestion that an uninterrupted lineage of doctrinal development can be traced from the late eighteenth century is inherently problematic. Irregular warfare is anything but formulaic. The diversity of colonial operations did not permit for a set of tactical principles to be distilled for consistent application in all scenarios, as was the fashion for conventional warfare in Europe.[4] As Douglas Porch notes, 'Callwell can hardly claim to be the Clausewitz of colonial warfare, but that is precisely the point. From its earliest days, small wars were embraced as a refutation of modern, intellectual, more strategically sophisticated analytical and technological approaches to warfare.'[5] Instead, flexibility in the formulation of attainable objectives, both militarily and politically, came to determine the degree of European success in the colonies. If this meant 'going native' and assimilating oneself in the cultural and military practices of one's enemy, as the French were more wont to do than the British at times, then so be it.[6] These were the kinds of transferable principles that might be adapted to accommodate the vagaries of different colonial campaigns. Certainly, this is where it is possible to trace a degree of universality between the Congo-Arab War and the better-documented French and British experiences. For, as Belgium's pre-eminent scholar on the subject has noted, the *Force Publique*'s success under the command of Francis Dhanis, owed as much to his subordination of military operations to political objectives as it did to his willingness to disregard conventional tactical practices.[7]

Like so many colonial officers before him, and many to come thereafter, Dhanis' approach to colonial campaigning was improvised. It relied on personal qualities of energy, imagination, and skill in both military and political spheres. The Arab campaign was less a pure act of war than a means of taking possession and organising territories, which required officers to transcend the somewhat limited framework of conventional military training received at the *École Militaire*.[8] This was no different to elsewhere. Colonial warfare was anathema to most European armies, whose institutional conservatism and obsession with 'real war' precluded the former from being taught at military academies. Despite Callwell's attempts from the mid-1880s to synthesise colonial experiences beyond the British case in a bid to contribute to an international scholarly debate, the most common means of information transmission was through informal channels; soldier-to-soldier interactions, private study, and personal experience.[9] This, it has been argued, established the basis for 'national traditions' to develop in counter-insurgency methods.[10]

Thus, the French, from Louis-Gabriel Suchet and Thomas-Robert Bugeaud in the first half of the nineteenth century, through to Joseph-Simon Galliéni and his protégé, Lyautey, at its conclusion, slowly developed a population-centric approach to assimilate political and military objectives that would come to form the recognisable concept of the *tache d'huile*, or oil stain.[11] Meanwhile, British concern with costs, manpower, and enemy morale quite often led to a preference for seizing the initiative, which, as the twentieth century dawned, frequently manifested itself in 'butcher and bolt' operations.[12] Although Belgium did not have a strong imperial tradition to draw upon itself, the experiences of its officers attached to the French in Algeria in the 1840s and Mexico in the 1860s, provided a point of reference for aspiring colonial officers.[13] Be it recognising the importance of conciliatory policies in the aftermath of Bugeaud's brutal *razzias*, or the emphasis placed on mobility and tactical flexibility of his, and later Bazaine's, mobile columns in fighting an elusive enemy, clear transferable principles of counter-insurgency operations were in circulation for prospective *Force Publique* officers to digest.[14]

Whether adapted from the nascent pan-European discourse, personal reflection on historical precedents, or an organic arrival at similar conclusions, the *Force Publique* applied a combination of methods in its struggle to pacify and control the Congo region. The primacy of political objectives, the importance of population-centric approaches, the emphasis on morale, and the adoption of local fighting methods suggest a degree of uniformity in the approach to colonial warfare. Therefore, even if the officers during the Congo-Arab War did not act from a position of prior knowledge, there is a strong indication that these commonalities in nineteenth century counter-insurgency methods emerged organically across different armies at around the same time. Without being limited to a 'British way' or a 'French way', there is good reason to believe that an adaptive synergy of core principles was developed by like-minded officers when faced with the unique challenges posed by colonial campaigning. In this sense, the experiences of Dhanis, documented – but, subsequently forgotten – in Donny's work, are as important in understanding the essence of nineteenth century counter-insurgency as those of more famous colonial soldiers and campaigns.

Circumstance clearly played a significant role in determining the CFS's approach in the pacification of the Congo. For this was no ordinary colonial war of conquest in which the invading army sought to force a resistant local population into submission. There was some of that, to be sure, but the primary enemies were themselves interlopers into the host territory. The Arabs had begun to penetrate into the eastern part of the Congo from Zanzibar in search of ivory and slaves in the late seventeenth century. Amidst the internecine struggles of the Congo region, which were aggra-vated by the influx of firearms, these Arab warlords drew strength from continuous expansion. With greater access to animal, human, and the monetary rewards to be harvested therein, came greater desire to push for more.[15] Decades' worth of raids and wars within the old Yeke Kingdom led to 'the spread of violence and overall militarization of society' by the time CFS forces became embroiled in the struggle for economic monopoli-sation from the 1880s.[16] As such, the embryonic *Force Publique* faced quite a task to pacify and expand its own economic reach against a well-armed, well-financed, and shrewd opponent.

From the outset, King Leopold II's imperial venture into the uncharted territory spanning central Africa equally elicited the use of brute force against the autochthone populations of the region. Henry Morton Stanley's numerous expeditions, during which he established trading posts and defensible bases along the Congo River and its tributaries, were frequently bloody affairs. Later, with the creation of the *Force Publique* in 1885, violence in the economic exploitation of the Congo region became systematised and endemic – albeit well-hidden behind a humanitarian veneer.[17] A report by the Interim General Administrator of the Department of the Interior to King Leopold II exemplified this, when it stated that the repression of the slave trade had always been at the heart of the enterprise. It was the Arab '*razzias*' – an interesting use of the term – which compelled Commanders to extend their influence so as to offer local populations protection against these slave runs.[18] However, it was soon evident that white officers were using this liberty to compete directly with the Arab warlords for economic hegemony.[19] Far from shielding indigenous peoples from the cruelties of foreign invaders, the CFS' own intervention exacerbated the use of violence in the region. The likes of Clément Brasseur's 'reign of terror', for example, earned him the local nickname of "*Nkulukulu*" after a bird 'whose inner wings are bloody red".[20] Over time, through the introduction of the *système domanial*, the brutalisation of the Congo became aligned with a system of values that prioritised profit over all else. As Guy Vanthemsche has sug-gested, the CFS could not rely on the metropolis to meet the military and administrative costs of Empire (as was the case elsewhere), resulting in 'a particularly severe and exploitative regime'.[21]

Nevertheless, CFS agents and officers spent much of their time following up hard pacification with something more akin to hearts and minds. Before, during, and after the Congo-Arab War, great lengths were taken to keep certain indigenous chiefs on side. Good relations not only provided the CFS with greater ease of movement through the vast territory, but also afforded it access to auxiliary forces. By simple means of establishing bases for trading and agricultural purposes, expelling the Arab slave raiders, and supplying weapons and gunpowder to support weaker kingdoms in settling old scores, CFS officers and officials were able to mask some of their own brutality.[22] In Dhanis' case, his description as 'a charming man, if a little strange at times, an old African, but of a rare intelligence, speaking all the languages of the country like they were his own,' served him immeasurably well during the campaign.[23] Even before serious hostilities commenced, he managed to subdue the entire Kwango province by means of 'diplomacy', supported by just 80 soldiers and 150 porters.[24] The results separated the local population from the enemy by creating a base of natural support, which then provided keen and willing sources of intelligence to facilitate future military operations.

In this regard, similarities can be drawn with both British and French experiences. Divide and rule was not a new idea, as the establishment of Britain's Indian Army and the general appropriation of social hierarchies broadly testified.[25] However, the manner in which the French began to apply this in Algeria during the 1840s demonstrated first-hand to Belgian observers the importance of marrying long-term political objectives to a military campaign. As Porch has noted:

> Whatever the military arguments in favour of the *razzia*, its long-term effects were baleful. Discipline was difficult to maintain when soldiers were allowed to burn, pillage, and rape. Soon attitudes hardened, sensibilities were anesthetized and any political or military goals beyond utter devastation were lost in an orgy of brutality and excess.[26]

Indigenous populations who were unwilling to 'accept the yoke of conquest' after witnessing the devastation wrought by a flying column could expect 'a war of extermination'.[27] However, for those willing to submit, Bugeaud was prepared to accompany his military successes with a durable political plan to pacify the region. His ideas of Franco-Arab assimilation led to reconciliation between conqueror and conquered by selecting the most influential and willing tribal leaders to govern under the supervision of French officials in the newly created *bureaux arabes*.[28] This did not put an end to the use of maximum force by any means, but the ultimate aim of French imperialism to culturally integrate indigenous peoples demonstrated a willingness to move past the purely military facets of conquest.[29] This was later refined and adapted by the likes of Faidherbe and Galliéni – to relatively good effect – before culminating in Lyautey's policy of 'peaceful penetration'.[30]

Although CFS officers instinctively erred towards the offensive, dealings with the Arab warlords often required them to heed Callwell's' words that a 'delay in entering upon hostilities will but slightly prejudice the chance of ultimate success, [but] any hesitation when operations have commenced is to be deprecated'.[31] In the absence of sufficient infrastructure to adequately take the war to the Arab slavers, it was recommended to pursue an inter-mediary policy of appeasement while CFS strength was built up. To this end, conciliatory methods, such as the nomination of Hamad bin Muhammad bin Juma bin Rajab el Murjebi, otherwise known as Tippu-Tip, to the Governorship at Stanley Falls on 24 February 1887 were taken. This, tem-porarily at least, curbed the frequency of the *razzias* by encouraging the Arabs to sell their ivory to the CFS instead of using large slave caravans to transport their goods to Zanzibar.[32] However, it did not put an end to human trafficking. *Force Publique* officers routinely employed Arab slave-traders as recruiting agents, whose sole access to physically-able young men was through slave raids.[33] Sefu, one of Tippu-Tip's sons, was one such agent paid 10 francs per month with an additional bonus of 100 francs for each able recruit furnished.[34] In this way, as in others, the CFS actually encour-aged the continuation of the slave trade they purported to be fighting against.

Notwithstanding, relations between the CFS and the Arab warlords remained on a knife-edge. With both parties evidently competing for influ-ence and access to untapped riches, there was good reason to believe that something was likely to give. Enforcing the structures of the State on the Arabs and reminding them of their duty to uphold law and order within their jurisdictions was a delicate matter in itself.[35] The flagrant strengthen-ing of the *Force Publique*, the fortification of outposts, and the expansion of infrastructure only exacerbated existing issues. Still, the instructions issued by the Governor General, Camille Jannsen, to Lieutenant Nicolas Isidore Tobback on 30 April 1890, revealed that the Arabs were not believed to have offensive intentions in the near future.[36] As such, CFS officials contin-ued to promote caution in their dealings, for it would be 'unpardonable to [attack] without having every chance of winning'.[37] It was simply unthink-able, given the state of the Congo's existing communications, the as-yet incomplete recruitment of the *Force Publique*, and the lack of control over certain strategic points, to engage and be beaten in a protracted war.[38] This was not to say that further preparations could not be made. For instance, Jannsen advocated fortifying the key defensive positions under CFS control, whilst simultaneously seeking opportunities to secure others in order to demonstrate to the Great Powers 'our rights to our conquests through our ability to defend them'.[39] As long as sufficient explanations and assurances were given to the Arabs to justify any action taken; for instance, declaring that those attacked were taken to be irregulars operating against the express wishes of the legitimate chief with whom the CFS wished to remain on good terms, the Governor General felt certain trouble could be avoided.[40]

Local engagements between the CFS and Arab forces in 1890 and 1891 were not unheard of. In fact, they were to play a significant role in the deterioration in CFS-Arab relations by 1892. The defeat of the former slave, turned vassal of Tippu-Tip, Gongo Lutété on the Lomami; engagements at Ibembo and Majorapa; and Captain Guillaume Van Kerkhoven's expedition in the upper-Ouellé, highlight the frequency with which the two competing empires came to blows.[41] These actions, which often resulted in significant bloodshed were made worse by the unsolicited requisition of ivory from Arab possession. Although supposedly acting within the agreed territorial limits of CFS/Arab influence, the combination of military operations, venturesome commercial activities, and the introduction of a tax on ivory, gave cause for rival Arab warlords in Maniema to set aside their differences. Financially threatened, Mounié Moharra, Sefu's uncle and one-time opponent of Tippu-Tip, declared that 'the white man was too evil to live with'.[42] Rejecting the authority of King Leopold II, the Arab chiefs united in open rebellion.[43] When, in May 1892, the commercial expedition led by the British ivory trader, Arthur Hodister, was massacred at Riba-Riba by order of Sefu, the bleak prospect of a full-scale insurgency became apparent.

The opposing forces differed in numbers, quality, and equipment. It has been estimated that just over 10,000 men were utilised by the CFS during the campaign. Opposing them, the Arab warlords had access to some 100,000 men, although their inability to concentrate their manpower squandered a key advantage.[44] This disparity was further compensated for in training, weapons, and command. Although the embryonic *Force Publique* was not, as yet, at the height of its preparedness, the establishment of a military base at Boma in 1886 had already turned out over 2,000 trained regulars by the end of 1888.[45] By 1891, the *Force Publique* numbered 3,186.[46] These locally-raised troops, recruited on a seven year engagement, gradually began to replace the more expensive, short-service coastal volunteers (predominantly, but not exclusively, Haoussas and Zanzibaris) from neighbouring imperial possessions.[47] Like the British and French before them, the CFS preferred to recruit across a number of ethnicities; though this created problems in itself. Separate quarters, the imposition of French as a language of command, as well as racial stereotyping shaped officers' opinions of the regulars under their command. Tobback, for example, described Haoussas and Bangalas as deferential and compliant, but completely 'ignorant of their profession as soldiers.'[48] However, Lieutenant Emile Lémery's recollection that, 'In war, they [Haoussas] are real lions; they throw themselves at the enemy with a cry, feverishly savage, brandishing their gun in one hand and a terrible machete in the other, of which they make good use,' suggests that indigenous troops could sometimes be forgiven for a lack of European professionalism.[49]

Nevertheless, armed with the latest Albini rifles, the *Force Publique's* regular units consistently out-soldiered and out-gunned the enemy.[50] Although possessing somewhere in the region of 30,000 guns, many of which were percussion pieces, the Arab forces tended to operate independently of one another and lacked the organisation and fire-discipline to make their numeric advantage count.[51] A clear indication of this can be found in, the then, Lieutenant Francis Dhanis' early 1892 engagements against the former slave, turned warlord, Gongo Lutété. Acting independently of the events brewing in Maniema, this vassal of Tippu-Tip had, once again, taken to the field in an attempt to cross the Sankuru River on a slave run. Although outnumbered, Dhanis was able to count on the superior training and equipment of the *Force Publique* to redress the balance of forces. The steadfastness and firepower of Dhanis' regular troops reaped its rewards as Gongo Lutété was defeated four times in three weeks: on 24 April at Mona-Kialo; on 5 and 9 May at Batubenge; and again on 12 May at Kisima-Souri.[52] Comprehending the shift in momentum, Gongo Lutété submitted to Dhanis. His capital of N'Gandu was turned over to CFS control while he, and his entire retinue, offered their services to the *Force Publique*. Ever the pragmatist, Dhanis exhibited shrewd political acumen in accepting the offer, for, not only did it furnish him with thousands of auxiliary troops for the upcoming campaign he planned to wage in Katanga, but it also pacified the region between the Sankuru and Lomami Rivers.

Auxiliaries were an important addition to the CFS' order of battle. Despite their shaky performance at Batubenge on 9 May, during which approximately 500 of them had fled, sheer numbers, combined with local fighting techniques, offered Dhanis tactical and operational flexibility. In addition to Gongo Lutété's warriors, the CFS was able to call upon thousands of irregular soldiers at short notice from pacified regions.[53] Placed under the command of promoted black NCOs, these more mobile units often acted independently from the main force, carrying out raids and *razzias* as they harassed the ephemeral enemy. In short, auxiliaries were used to carry out much of the 'dirty work' during the campaign.[54] It echoed the practice adopted in many nineteenth century colonial campaigns and counter-insurgencies, in which it was 'the disciplined army that [was] obliged to conform its methods to those of [its] adversaries'.[55] Bugeaud's light columns in Algeria were famously reorganised to become 'even more Arab than the Arabs', while mobile forces were raised or converted to gather intelligence and take the fight to irregulars in France's Mexico campaign and countless British expeditions.[56] Understanding the enemy's characteristics and displaying flexibility in meeting them was the surest way of establishing achievable operational objectives to counter the strategic advantage the enemy regularly held.

Invariably, the objective was always to bring the enemy to battle in order to secure a swift end to otherwise costly campaigns. This was imperative for, as pre-1914 German counter-insurgency showed, the absence of 'identifiable strategic targets [could combine] with decentralised command and control to cause operational solutions to expand to fill a vacuum of civilian oversight and vague war aims'.[57] For Donny this meant the destruction of the enemy's men and resources through offensive military action – the complete removal of the Zanzibari Arabs from influence in the Congo region.[58] However, as Callwell had previously noted, 'the disinclination shown by undisciplined warriors to commit themselves to a general engagement' forced European officers to find alternative ways of bringing about a decisive solution.[59] Vigorous operations to secure vital river crossings, capitals, as well as human and material resources were often required to bring an enemy to battle. Thus, General Sir Garnet Wolseley wrote that the objective should be 'the capture of whatever they prize most, and the destruction or deprivation of which will probably bring the war most rapidly to a conclusion.'[60] It also explained why the French became so preoccupied with wars among the people, for as General Pierre le Compte de Castellane put it,

> In Europe, once [you are] master of two or three large cities, the entire country is yours. But in Africa, how do you act against a population whose only link with the land is the pegs of their tents? [...] The only way is to take the grain which feeds them, the flocks which clothe them. For this reason, we make war on silos, war on cattle, [we make] the *razzia*.[61]

In the case of the CFS' campaign in the Congo, the targeting of ivory stocks served the dual purpose of crippling the Arabs' fragile economy and forcing them into giving battle. It also had the subsidiary effect of enriching the State's, and officers' personal, coffers.

Dhanis' actions were clearly inspired by the quest for glory and personal advancement. Seizing the initiative and winning a campaign that would deliver a pacified and organised territory, ripe for economic exploitation, would almost certainly be rewarded by a grateful Sovereign. As a Second Lieutenant in the 8th Line Infantry Regiment, who had passed out 16th in his cohort at the *École Militaire*, colonial aggrandisement equated to opportunity.[62] After all, even Belgian officers who had achieved far less abroad in the past, be it in French service or that of the CFS, had obtained recognition for their efforts as the few men who returned to the Belgian army with any campaign experience.[63] This also proved to be the case for a number of Dhanis' contemporaries, whose search for a release from the 'trammels of European convention' was only matched by the ambition to better otherwise stagnant military careers.[64] To this end, valorous exploits in the name of 'the nation' or the 'noble cause' were frequently recorded with a home audience in mind.[65]

In a sense, this was not a new phenomenon. French and British colonial officers frequently acted hastily and beyond the bounds of respectability in order to establish a reputation back home. Often restrained by governmental policy or popular scrutiny, it was not uncommon to find ambitious men taking matters into their own hands in an attempt to fashion a *fait accomplis*. Whether it was Bugeaud's effective, but inhumane, methods of the '*razzia*', or Marchand's epic, but desperately illogical, march across Africa to Fashoda, the pressure to obtain results that would be lauded and accepted by the *métropole*, was an influential factor in the conduct of war.[66] That Sefu afforded Dhanis with such an opportunity by redirecting his gaze from Katanga to the Maniema and beyond, was viewed as heaven-sent fortune. For with news of the Hodister massacre and the assassination of the Kasongo Residents, Lieutenant Josheph Lippens and Second Lieutenant Henri De Bruyne in November 1892, came a pretext to wage an all-out war on the Arab slavers that cautious CFS administrators had heretofore sought to avoid.

The campaign that ensued required Dhanis to swiftly redirect his forces from Katanga to Maniema. Spread out across a vast distance, initial operations were as much about pushing Sefu back beyond the Lomami as they were about concentrate men and resources.[67] Battles and skirmishes were recorded on almost a daily occurrence in November and December 1892 as Dhanis, supported by Gongo Lutété marched on Nyangwé and Kasongo. Defeating the forces of Mounié Moharra in early January 1893, during which the Arab warlord was killed, CFS troops arrived at the Lualaba River opposite Nyangwé by the end of the month. After an encounter battle that pushed the Arabs beyond the river, Dhanis entered an abandoned Nyangwé on 4 March. By 22 April, Kasongo had also fallen, forcing Sefu to retreat with the remainder of his forces to join with another Arab warlord, Rumaliza, 70 miles to the Southeast at Kabambare. Simultaneously, Captain Louis-Napoléon Chaltin was ordered by the State Inspector, Fivé, to leave his camp at Basoko in the North and join forces with Dhanis at Nyangwé. Steaming down river with 300 men, Chaltin arrived at a flaming Riba-Riba on 30 April 1893 and learned of Mserera's, the former Arab governor, retreat to Stanley Falls.[68] Pushing on, Chaltin reached Stanley Falls just in time to relive the besieged Tobback, whose relationship with Rashid, Tippu-Tip's nephew and successor, had deteriorated markedly since the Hodister massacre. Together, the CFS forces pushed the Arabs out of the region from where they made for Kirundu and then, also, to Kabambare.

These early engagements were characterised by Dhanis' eagerness to organise and direct his disparate forces. Great energy was exerted in unifying forces ahead of engagements, but the distances involved and the fragility of communications meant that separation in the field, 'ever [...] a fruitful source of disaster' according to Callwell, was a necessary evil.[69]

Relying on good intelligence networks, CFS forces could operate independently from one another and of their bases. Provided mobile columns could
supply themselves on the move and were kept abreast of enemy movements, the *Force Publique* could seize the operational initiative with a degree
of confidence. This allowed Michaux, with just a quarter of Dhanis' regulars,
to score a victory on the Lomami River in November 1892 while Dhanis,
supported by Gongo Lutété, dealt with the threat posed by Mounié-Pambé,
Mounié Moharra's son, on the Lualaba in December.[70] Separate engagements in early January 1893 fought by detachments attempting to make
contact with Dhanis' main force, as well as Chaltin's redirected campaign to
relieve Stanley Falls in April, demonstrates the degree to which local initiative was allowed for when dealing with multiple threats.[71] Mobility and
reactiveness were key in this regard. Although concerted efforts were
made at various points to reorganise, regroup, and reinforce, the separation
of forces was countenanced in order to retain the initiative.

Bugeaud had operated under similar principles in Algeria when utilising a
combination of intelligence and mobility to strike deep into areas that
would keep his enemy off-balance.[72] In spite of the humiliating reverses at
Isandlwhana (1879) and Majuba Hill (1881), Callwell could also see the
benefits, under certain circumstances, to trust in the separation of forces
to secure an operational advantage against a disorganised enemy. He wrote,

> If each part of a divided army is in itself a match for whatever force the enemy
> may bring against it, defeat in detail is not to be feared. When dispersion is not
> prejudicial to security it has much to recommend it. The mobility of an army is
> in inverse proportion to its size. Movement in several columns therefore
> facilitates operations. The same forces moreover are at work in preventing
> the massing of the hostile legions against one fraction of the divided host as
> tend to safeguard its communications against organized attack. [...] A strong
> argument in favour of invasion on several lines is [...] in the moral effect
> produced on the enemy by the occupation of wide stretches of territory, and
> in the influence that the appearance of hostile bodies on all sides must exert
> on a people who know not how to turn the situation to account.[73]

The CFS' advances on Nyangwé, Kasongo, Riba-Riba, and Stanley Falls in
short order clearly paid homage to this principle.

In other ways, Dhanis' prioritisation of psychological and moral factors in
tactical considerations also spoke to a patchwork of nineteenth century
counter-insurgency ideals. Fortitude, enthusiasm, and perseverance were
more important than line, square, or shock.[74] *Élan* was often prized above
all else. Lémery's recollections detail the importance of the white officer to
lead the attack, from which point, 'Nothing can stop the force of the *élan*;
either everyone dies or ends up victorious. It is the savage and spontaneous
attack, which throughout this campaign, has been our strength, as we were
always lacking in numbers'.[75] Certainly, there is evidence to suggest that

Dhanis and other *Force Publique* officers deliberately withheld fire in engagements where the possibility of hand-to-hand combat presented itself. While this may well have been part of a broader attempt to conserve precious ammunition – particularly when faced with small-scale skirmishes fought in open order – it soon became evident that the bayonet had an important psychological role to play in the Congo-Arab War. Arab warriors, it was felt, feared cold steel and fled at the sight of it, leaving many suspicious officers as whole-hearted converts to its place on the African battlefield.[76] This was reflected in Donny's tactical summary of operations too, though he noted that it ought to be exclusively used in bush areas, at night, and when forming square.[77]

This is not to say that firepower was not equally prized. As previously stated, the benefits of a well-drilled force laying down a significant weight of shot could also compensate for the deficit in numbers. After all, as Callwell noted, 'Confronted with the rifle and the field piece, assegai and jezail are robbed of their terrors. Individual daring and fanaticism are no match for discipline and mutual reliance.'[78] This spoke to Bugeaud's theories on fire-power in pitched battles as much as it did to Donny's, whose views echoed those of his predecessors and contemporaries in striking fashion. Operating on the offensive but assuming the tactical defensive would allow for the weight of fire to increase the chances of success. It was his opinion that,

> colonial campaigns will be shortened, therefore also rendered more economical, if the troops are comprised of a strong nucleus of veritable soldiers, calm, proven marksmen, well supervised, trained to be stoic rather than to flee, armed with sophisticated weapons and carrying a large quantity of ammunition.[79]

However, colonial warfare rarely offered up many opportunities for decisive fire-action. When they did, as in the case of Omdurman in 1898, the superiority of weaponry and training proved itself to devastating effect.[80] In more open skirmishes, where factors such as terrain, surprise, or manoeuvre might alter the effect of the rifle, it was often deemed prudent to withhold fire and pursue alternative methods to break the will of the enemy. In this sense, mobility and *élan* proved to be as important.

By mid-1893, and the second phase of the campaign, the *Force Publique* began to encounter different tactical challenges. Positional warfare is not something that ordinarily resonates with theories of counter-insurgency or small wars, but, in the case of the Congo-Arab war, it came to define the conduct of operations. Rumaliza, alongside the remnants of other beaten Arab forces, remained the solitary threat to the CFS. When intelligence was received that he was marching on Kasongo with 10,000 men, Dhanis moved immediately to meet him. Crossing the Luama River (a tributary of the Lualaba) the opposing forces met for the first time on 15 October 1893.

However, this was no ordinary encounter. Rather than facing the prospect of an ambush, skirmishing, or a traditional open-order firefight across bush, village, or woodland, the *Force Publique* found Rumaliza's forces well-ensconced in defensive *bomas* (defensive works constructed from hardened clay), from which they proved difficult to dislodge. Dhanis' forces had previously laid siege to towns such as Nyangwe and Kasongo, but specifically designed defensive earthworks demanded tactical adjustments to force a military decision.

'The art of field fortification as understood by antagonists such as we have to deal with in Asia and Africa, and as applied against them,' Callwell wrote, 'is interesting, for it illustrates the advantages derived from the most simple defence works in such wars.'[81] This proved to be the case as Dhanis' force of 400 men failed to break through following an audacious flanking attack. The *bomas* were so well fortified that, even at a distance of 100 yards, the light artillery on hand struggled to make much of an impression.[82] Lacking in men and materiel, Dhanis was forced to await reinforcements as he tried to starve out his opponents. In the meantime, his forces were subjected to constant harassment by the enemy, whose sorties in October caused many casualties, including the death of Lieutenant Pierre Ponthier. By mid-November, the combination of battle casualties and food supplies began to bite. As both sides took stock of events, Rumaliza evaded surveillance and retreated during the night of 15/16 November. Captain de Wouters was placed in charge of the pursuing light column, while Dhanis returned to Kasongo to prepare a better equipped expeditionary force.[83] Contact was made at Ogella on 19 November. The ensuing firefight proved to be a reversal for the *Force Publique* during which Lieutenant de Heusch was also killed. Were it not for the death of Sefu among the multitude of Arab casualties, Dhanis might well have considered the whole sorry episode a complete failure.[84]

As it was, Dhanis led his new expedition of 15 Europeans, 700 regulars, and approximately 2,000 auxiliaries, back towards Rumaliza's newly established defensive *bomas* covering Kabambare.[85] Still short of adequate supplies to force the issue, Dhanis was obliged to lay siege and await further reinforcements from Captain Hubert Joseph Lothaire, who would later be infamously embroiled in the Charlie Stokes Affair in 1895.[86] The junction was made in mid-December 1893, adding a further 300 men to Dhanis' strength.[87] Importantly, Dhanis also brought with him heavier artillery pieces, which had been used to great effect by Chaltin against similar opposition during his campaign to relieve Stanley Falls.[88]

Since Bugeaud's arrival in Algeria, the quest for mobility in combatting irregular forces condemned the role of artillery to that of a secondary importance. Mobile columns relied on speed and logistical self-sustainment. Artillery was considered cumbersome and burdensome to colonial operations.[89] Light field and mountain pieces had found specific roles, but its general experience had shown 'an average expenditure of shell far below what is usual in Continental campaigns.'[90] For the *Force Publique*, however, the moral and operational advantages of deploying artillery was somewhat out of step with contemporaneous trends of thought. Twelve 75mm Krupp mountain pieces were shipped to the Congo in 1890 to be formed into a fully-fledged battery under the command of Michaux. An unknown number of lighter, 37mm Hotchkiss and a further 23, 47mm Nordenfeld guns, built by Cockerill of Seraing, were added to the CFS' arsenal in 1892.[91] The latter were specifically designed for Africa with removable parts allowing for ease of transportation. Apart from the gun itself, which required two men to carry it, its other components and ammunition never exceeded 35 kilograms in weight and could be carried by a single porter. Capable of being assembled in five minutes, these guns were seen as ideal for the Congo. Artillery, in general, was viewed by Donny as having a tremendous moral effect on the enemy and was, as such, indispensable to overcoming specific obstacles during the campaign.[92]

On the banks of the Lulundi River, Dhanis encountered Rumaliza's *bomas* echeloned across the road to Kabambare. On 14 January 1894, an assault on the *bomas* dislodged the Arabs from their defensive works. The artillery was at the heart of the victory. Having isolated the *bomas* from one another ahead of a general assault on the Rumaliza's grand *boma*, an errant shell from a 75mm Krupp gun landed inside the central structure. Setting it ablaze, the Arab forces panicked and fled. Rumaliza escaped with a handful of his followers. The other *bomas* capitulated and their occupants were taken prisoner. Rumaliza fled first to Kabambare before being pursued into German East Africa by Lothaire at the head of a 400-strong column. All that remained was to make contact at Lake Tanganyika with the Anti-Slavery Society forces led by Captain Alphonse Jacques (later raised to Baron Jacques of Diksmuide for his services in the Great War), who had been engaged in operations against one of Rumaliza's lieutenants in the region.[93] This done, by March 1894 the CFS could claim its territory largely free of Arab influence.

At the cost of just 16 Belgian officers and non-commissioned officers (six through sickness), the destruction of the ill-co-ordinated Swahili Empire was the making of those that survived.[94] Promotions followed, and in the case of Dhanis, the title of Baron was bestowed, demonstrating the degree to which imperial service, as in Britain and France, could confer status on those whose careers might otherwise have lacked notoriety. Nevertheless, Dhanis' name does not sit within the pantheon of counter-insurgency practitioners or theorists for one simple reason. His methods did not diverge significantly from the basic principles of those whose actions served to carve out larger empires for France and Britain.

Yet, this is precisely the reason for which his actions, and the Congo-Arab War more generally, ought to be examined. A series of commonalities underpin the experience of most nineteenth century counter-insurgency that cannot be ignored. The subordination of political to military objectives, the assimilation of local culture and practices into pacification methods, the seizure of the operational initiative to counter the strategic deficit, and the importance of a well-drilled cadre to execute flexible tactics in order to bring the weight of technological superiority to bear, are all recognisable facets that transcend national traditions. Whether consciously adopted or developed independently, these basic principles provided many a colonial officer with answers to the vagaries and unpredictabilities that small wars produced. No two campaigns were ever the same and precise rules are anathema to a successful counter-insurgency. However, action taken with the bounds of basic, adaptable principles allowed for the likes of Dhanis to cobble together a hybrid system that led his meagre forces to victory in the Congo-Arab war against a numerous, well-equipped, and highly capable opponent.

Notes

1. Donny, *L'art militaire au Congo*.
2. Callwell, *Small Wars*; Lyautey, "Du rôle colonial de l'armée." To understand the frequency of colonial campaigning in the late nineteenth century, see Beckett, *Modern Insurgencies and Counter-Insurgencies*, 31–2.
3. Rid, "The Nineteenth Century Origins of Counterinsurgency Doctrine"; French, *The British Way in Counter-Insurgency*; and Porch, *Counterinsurgency*.
4. Porch, "Bugeaud, Galliéni, Lyautey," 377.
5. Porch, *Counterinsurgency*, 50.
6. See note 4 above.
7. Marechal, *De 'Arabische' Campagne*, 239–41.
8. Ibid., 241; and Gann and Duigan, *The Rulers of Belgian Africa*, 53.
9. Captain Callwell, "Lessons to be Learnt from the Campaigns."
10. Beckett, *Modern Insurgencies and Counter-Insurgencies*, 24–5, 35.
11. Ibid., 27–9, 40; and Porch, *Counterinsurgency*, 52–4.
12. Ibid., Beckett, 42.
13. For more on Belgian colonial campaigning, see Leconte, *Les Tentatives d'Expansion Coloniale sous le Regne de Léopold 1er*. For Mexico, a detailed account of Belgian involvement in Mexico, see Albert Duchesne, *L'Expédition des volontaires belges au Mexique 1864–1867: au service de Maximilien et de Charlotte.* .
14. Porch, *The French Foreign Legion*, 145–8.
15. Gann and Duigan, *The Rulers of Belgian Africa*, 55–6.
16. Macola, *The Gun in Central Africa*, 84.
17. Guy Vanthemsche, 19–22.
18. Royal Archives, Brussels [RA], Cabinet Léopold II [CLII], Expansion, 144/39, Report to the King on the political and military measures taken to bring

about the repression of the slave trade in the territories of the State. Undated [Likely, 1889].

19. War Heritage Institute, Brussels [WHI], Archives of Belgian Military Abroad [BMA], 43/51 XV/16, Théophile Wahis to Camille Janssen, 31 August 1890.
20. Macola, *The Gun in Central Africa*, 107.
21. Vanthemsche, *Belgium and the Congo*, 22–3.
22. RA, CLII, Expansion, 144/39, Report to the King. Undated [Likely, 1889]; WHI, BMA, 43/55 XV/263/18, Henri Doquier to Merette & Constant Desmet, 3 August 1892; and Macola, *The Gun in Central Africa*, 93.
23. Marechal, *De 'Arabische' Campagne*, 240; WHI, BMA, 43/55 XV/314, Émile Lémery Papers, clipping from *Le Soir*, 5 August 1955. This was part of a series of articles printed between 4–6 August 1955 based on rediscovered correspondence by Lémery's nephew. And *Le Baron Dhanis au Kwango et pendant la campagne arabe*. 21.
24. Pakenham, *The Scramble for Africa*, 438.
25. Beckett, *Modern Insurgencies and Counterinsurgencies*, 34. For a general appreciation of British recruitment policy in India, see Omissi, *The Sepoy and the Raj. The Indian Army 1860–1940*; Roy, *War and Society in Colonial India, 1807–1945*; and *The Army in British India: From Colonial Warfare to Total War, 1857–1947*; and Johnson, "General Roberts, the Occupation of Kabul," 304, 311–2.
26. Porch, "Bugeaud, Galliéni, Lyautey," 380–1.
27. Vandervort, *Wars of Imperial Conquest in Africa*, 68.
28. Rid, "Origins of Counterinsurgency Doctrine," 739–40; and Sullivan, *Thomas-Robert Bugeaud*, 99–100.
29. Beckett, *Modern Insurgencies and Counterinsurgencies*, 41.
30. Porch, "Bugeaud, Galliéni, Lyautey" 388; and *Counterinsurgency*, 51–4.
31. Callwell, *Lessons to be Learnt*, 363.
32. RA, CLII, Expansion, 144/39, Report to the King. Undated [Likely, 1889].
33. WHI, BMA, 43/51 XV/17, Edmond van Eetvelde to Camille Janssen, 15 September 1890.
34. WHI, BMA, 43/51 XV/13, Théophile Wahis to Camille Janssen, 30 July 1890.
35. WHI, BMA, 43/51 XV/16, Théophile Wahis to Camille Jannsen, 31 August 1890.
36. WHI, BMA, 43/51 XV/12, Copy of Instructions from the Governor General to Nicolas Isidore Tobback, 30 April 1890.
37. Ibid.
38. Ibid.
39. Ibid.
40. Ibid.
41. *La Belgique Militaire*, No. 1052, 31 May 1891, 657–8; and Lord Wah, "Le Congo: le retour de Dhanis – La Guerre Arabe," 29.
42. Boulger, *The Congo State*, 161–2; and WHI, BMA, 43/51 XV/36, Lieutenant Sibieux to Nicolas Isidore Tobback, 14 March 1892.
43. Packenham, *The Scramble for Africa*, 436–7.
44. Marechal, *De 'Arabische' Campagne*, 233–4.
45. Deuxième Section de l'État-Major de la Force Publique, *La Force Publique de sa Naissance à 1914*, 36–9.
46. Gann and Duigan, *The Rulers of Belgian Africa*, 79.
47. Marechal, *De 'Arabische' Campagne*, 233–4; Vanderstraeten, "Léopold II et la Force Publique," 12; and De Boeck, *BAONI*, 34–35.

48. WHI, BMA, 43/51 XV/35, Nicolas Isidore Tobback to Camille Jannsen, 27 February 1892.
49. WHI, BMA, 43/55 XV/314, Émile Lémery Papers, clipping from Le Soir, 5 August 1955.
50. Vanderstraeten, "Léopold II et la Force Publique," 15.
51. Marechal, De 'Arabische' Campagne, 233–4.
52. Le Soir, 21 August 1892; and Le Baron Dhanis au Kwango, 18–19.
53. RA, CLII, Expansion, 144/39, Report to the King. Undated [Likely, 1889]; and Marechal, De 'Arabische' Campagne, 233.
54. Marechal, De 'Arabische' Campagne, 234.
55. Callwell, Lessons to be Learnt, 362.
56. Porch, Counterinsurgency, 20; The French Foreign Legion, 147–9; and Spiers, The Late Victorian Army, 295, 317–8.
57. Hull, Absolute Destruction, 134–5.
58. Donny, L'art militaire au Congo, 17.
59. Callwell, Lessons to be Learnt, 364.
60. Beckett, Modern Insurgencies and Counter-Insurgencies, 33.
61. Vandervort, Wars of Imperial Conquest, 68.
62. Musée Royal d'Afrique Centrale [MRAC], Francis Dhanis Papers, HA.01.0003/40, Letter of Promotion to Second Lieutenant, 12 May 1884.
63. Draper, The Belgian Army and Society from Independence to the Great War, 51, 74–6.
64. Gann and Duignan, The Rulers of Belgian Africa, 61–4.
65. Marechal, De 'Arabische' Campagne, 240–1. An interesting example of this was Lieutenant Henri Doquier, who hoped that his service in the Congo-Arab War and his decision to risk the perils of another expedition in 1896, would ensure his promotion to Captain, which was 'in his sights'. See WHI, BMA, 43/55 XV/263/26, Henri Doquier to Merette & Constant Desmet, 27 November 1893; 43/55 XV 263/1–33, Henri Doquier to Merette & Constant Desmet, 9 June 1896; and 43/55 XV 263/33, Henri Doquier to Constant Desmet, 2 November 1896.
66. Porch, "Bugeaud, Galliéni, Lyautey," 380–2, 385–6, 401.
67. MRAC, Francis Dhanis Papers, HA.01.0003/140, Francis Dhanis to Alexandre Delcommune, 28 December 1892.
68. Hinde, The Fall of the Congo Arabs, 215; and Vanderstraeten, "Léopold II et la Force Publique," 21.
69. Callwell, Lessons to be Learnt, 367.
70. MRAC, Francis Dhanis Papers, HA.01.0003/140, Francis Dhanis to Alexandre Delcommune, 28 December 1892; Wah, "Le Congo: le retour de Dhanis," 32; and Le Baron Dhanis au Kwango, 19.
71. Wah, "Le Congo: le retour de Dhanis," 33; and Le Baron Dhanis au Kwango, 20.
72. Porch, "Bugeaud, Galliéni, Lyautey," 378–9.
73. See note 69 above.
74. Marechal, De 'Arabische' Campagne, 240.
75. See note 49 above.
76. WHI, BMA, 43/55 XV/192, Edgard Cerckel Papers. Clipping from Le Soir, October 1952 entitled 'Un survivant de la campagne anitesclavagiste du Congo; and 43/55 XV/188 Florent Cassart Papers. Extract from Le Franc Tirreur, 7 December 1913.
77. Donny, L'art militaire au Congo, 41–2.
78. Callwell, Lessons to be Learnt, 370.

79. Donny, *L'art militaire au Congo*, 141–2.
80. Beckett, *The Victorians at War*, 182.
81. Callwell, *Lessons to be Learnt*, 397.
82. *Baron Dhanis*, 25; Wah, "Le Congo: le retour de Dhanis," 35.
83. *Baron Dhanis*, 27.
84. Hinde, *The Fall of the Congo Arabs*, 231.
85. *Baron Dhanis*, 28.
86. For more on this see Laqua, *The Age of Internationalism and Belgium*, 56–7; and Pakenham, *The Scramble for Africa*, 586.
87. WHI, BMA, 43/55 XV 284/1, Extract from *La Belgique Militaire*, 24 October 1897.
88. MRAC, Francis Dhanis Papers, HA.01.0003/138, Louis-Napoléon Chaltin to Francis Dhanis, 17 November 1893; WHI, BMA, 43/56 XV 426, Notes of Servive by Auguste Théophile Léon Rom [Undated]; and Wah, "Le Congo: le retour de Dhanis," 37.
89. Porch, *Counterinsurgency*, 16, 20; and, "Bugeaud, Galliéni, Lyautey," 378.
90. Callwell, *Lessons to be Learnt*, 407.
91. Vanderstraeten, "Léopold II et la Force Publique," 15.
92. Donny, *L'art militaire au Congo*, 42–5.
93. Vanderstraeten, "Léopold II et la Force Publique," 23; and Wah, "Le Congo: le retour de Dhanis," 36.
94. Gann and Duignan, *The Rulers of Belgian Africa*, 57.

Disclosure statement

No potential conflict of interest was reported by the author.

Bibliography

Primary & Contemporary Sources
  La Belgique Militaire
Musée Royal d'Afrique Centrale
  Francis Dhanis Papers
Royal Archives, Brussels
  Cabinet Léopold II, Expansion, 144.
War Heritage Institute
  Archives of Belgian Military Abroad, (Congo Free State and Belgian Congo (1885–1960)), 43/51, 43/55 & 43/56.
Newspapers
  Le Soir

Secondary Sources

Beckett, I.F.W. *Modern Insurgencies and Counter-Insurgencies: Guerrillas and Their Opponents since 1750*. London & New York: Routledge, 2001.

Beckett, I.F.W. *The Victorians at War*. London: Hambledon Continuum, 2006.

Boulger, D.C. *The Congo State: Or, the Growth of Civilisation in Central Africa*. Cambridge: Cambridge University Press, 2013.

Callwell, C.E. "Lessons to Be Learnt from the Campaigns in Which British Forces Have Been Employed since the Year 1865." *RUSI* 31, no. 139 (1887): 357–412.

Callwell, C.E. *Small Wars: Their Principles and Practice*. London: HMSO, 1896. Reprinted by Oregon: Watchmaker Publishing, 2010.

Club Africain d'Anvers. *Le Baron Dhanis Au Kwango Et Pendant La Campagne Arabe*. Antwerp: J.-B. Van Caneghem, 1910.

Cuelemans, R.P.P. *La Question Arabe Et Le Congo (1883–1892)*, 64–85, 98–121, 147–235 & 269–367. Brussels: Académie royale des Sciences colonials: Classe des Sciences Morales et Politiques, 1959.

De Boeck, G. *BAONI: Les Revoltes De La Force Publique Sous Léopold II 1895–1908*. Antwerp: Les Editions EPO, 1987.

Deuxième Section de l'État-Major de la Force Publique. *La Force Publique de sa Naissance à 1914: Participation des militaires à l'histoire des premières années du Congo*. Gebloux: J. Duculot, 1952.

Donny, Colonel. *L'art Militaire Au Congo*. Brussels: G Muquardt, 1897.

Draper, M. *The Belgian Army and Society from Independence to the Great War*. Basingstoke: Palgrave Macmillan, 2018.

Duchesne, A. *L'Expédition Des Volontaires Belges Au Mexique 1864–1867: Au Service De Maximilien Et De Charlotte*. Vol. 1 & 2. Brussels: Musée Royale de l'Armée et d'Histoire Militaire, 1967–68. doi:10.1016/0006-2944(75)90147-7.

French, D. *The British Way in Counter-Insurgency, 1945–1967*. Oxford: Oxford University Press, 2011.

Gann, L.H., and P. Duigan. *The Rulers of Belgian Africa 1884–1914*. Princeton, NJ: Princeton University Press, 1979.

Gordon, D.M. "Interpreting Documentary Sources on the Early History of the Congo Free State: The Case of Ngongo Luteta's Rise and Fall." *History in Africa* 41 (2014): 5–33. doi:10.1017/hia.2014.6.

Hinde, S.L. *The Fall of the Congo Arabs*. London: Methuen & Co, 1897.

Hull, I.V. *Absolute Destruction: Military Culture and the Practices of War in Imperial Germany*. Ithaca, NY: Cornell University Press, 2004.

Johnson, R. "General Roberts, the Occupation of Kabul, and the Problems of Transition, 1879–1880." *War in History* 20, no. 3 (2013): 300–322. doi:10.1177/0968344513483227.

Laqua, D. *The Age of Internationalism and Belgium, 1880–1930: Peace, Progress and Prestige*. Manchester: Manchester University Press, 2013.

Leconte, J.R. *Les Tentatives d'Expansion Coloniale Sous Le Regne De Léopold 1er*. Antwerp: V. Van Dieren & Co, 1946.

Lotar, R.P.L. *La Grande Chronique De l'Uele Suivant La Collection Du Mouvement Géographique, De La Belgique Coloniale, etc. Ainsi Que Des Documents Inédits En Possession De L'auteur Ou Lui Communiqués*. Brussels: Académie royale des Sciences colonials: Classe des Sciences Morales et Politiques, 1946.

Louis, W.R. "The Stokes Affair and the Origins of the Anti-Congo Campaign, 1895–1896." *Revue Belge De Philologie Et D'histoire* 43, no. 2 (1965): 572–584. doi:10.3406/rbph.1965.2576.

Lyautey, H. 1900. "Du Rôle Colonial De L'armée." *Revue des deux mondes*, January 15.

Macola, G. *The Gun in Central Africa: A History of Technology and Politics*. Athens, OH: Ohio University Press, 2016.

Marechal, P. *De 'arabische' Campagne in Het Maniema-Gebied (1892–1894) Situering Binnen Het Kolonisatieproces in De Onafhankelijke Kongostaat.* Tervuren: Musée Royal de l'Afrique Centrale, 1992.

Omissi, D. *The Sepoy and the Raj. The Indian Army 1860–1940.* Basingstoke: Palgrave Macmillan, 1994.

Pakenham, T. *The Scramble for Africa, 1876–1912.* London: Abacus, 1992.

Porch, D. "Bugeaud, Galliéni, Lyautey: The Development of French Colonial Warfare." In *The Makers of Modern Strategy: From Machiavelli to the Nuclear Age,* edited by P. Paret, 376–407. Oxford: Clarendon Press, 1986.

Porch, D. *The French Foreign Legion: A Complete History.* London: Macmillan, 1991.

Porch, D. *Counterinsurgency: Exposing the Myths of the New Way of War.* Cambridge: Cambridge University Press, 2013.

Reid, R.J. *Warfare in African History.* Cambridge: Cambridge University Press, 2012.

Rid, T. "The Nineteenth Century Origins of Counterinsurgency Doctrine." *The Journal of Strategic Studies* 33, no. 5 (2010): 727–758. doi:10.1080/01402390.2010.498259.

Roy, K. *War and Society in Colonial India, 1807–1945.* Oxford: Oxford University Press, 2006.

Roy, K. *The Army in British India: From Colonial Warfare to Total War, 1857–1947.* London: Bloomsbury, 2012.

Slade, R. *King Leopold's Congo: Aspects of the Development of Race Relations in the Congo Independent State.* London: Oxford University Press, 1962.

Spiers, E. *The Late Victorian Army, 1868–1902.* Manchester: Manchester University Press, 1992.

Stanard, M.G. "Learning to Love Leopold: Belgian Popular Imperialism, 1830–1960." In *European Empires and the People: Popular Responses to Imperialism in France, Britain, the Netherlands, Belgium, Germany and Italy,* edited by J.M. Mackenzie, 124–157. Manchester: Manchester University Press, 2011.

Sullivan, A.T. *Thomas-Robert Bugeaud: France and Algeria, 1784–1849: Politics, Power, and the Good Society.* Hamden, CT: Archon Books, 1983.

Vanderstraeten, L.F. "Léopold II Et La Force Publique." In *Léopold II Et La Force Publique Du Congo,* edited by P Lefèvre, L-F. Vanderstraeten, and P. De Gryse, 1–29. Brussels: Cercle royal des anciens officiers des campagnes d'Afrique, 1985.

Vandervort, B. *Wars of Imperial Conquest in Africa, 1830–1914.* London: University College London Press, 1998.

Vanthemsche, G. *Belgium and the Congo 1885–1980.* revised by Connelly, K. Translated by A. Cameron and S. Windross. Cambridge: Cambridge University Press, 2012.

Wah, Lord. "Le Congo: Le Retour De Dhanis – La Guerre Arabe." extract from *Revue de l'Armée Belge* (Tomes II & III, 1894–95), 1–46. Liège: Direction & Administration, 1894.

Walter, D. *Colonial Violence: European Empires and the Use of Force,* Translated by Peter Lewis. London: Hurst & Company, 2017.

Remembering and forgetting Mirambo: Histories of war in modern Africa

Richard Reid

ABSTRACT

While the study of organised violence is considered essential to understanding the history of the West, and accordingly imbued with various layers of meaning and remembrance, war is widely regarded as inimical to the modern nation in Africa and stable development more broadly. Using examples drawn from primarily from East Africa, this chapter considers the ways in which warfare in the deeper ('precolonial') past has been framed and envisioned in recent decades, in particular by governments whose own roots lie in revolutionary armed struggle and who began life as guerrilla movements. While in some cases particular elements of the deeper past were indeed mobilised in pursuit of contemporary political goals, in many other scenarios histories of precolonial violence were beheld as problematic and unworthy of remembrance. This chapter highlights the paradox and ambiguity which has attended the memory of key aspects of Africa's deeper past.

Introduction

For African political leaders in the 1950s and 1960s histories of violence in the deeper, supposedly the products of the colonial experience and the machinations of decolonisation, and for the postcolonial nations beyond, histories of violence in the deeper past had the potential to be mobilised in pursuit of modern political projects. But they were also problematic. This chapter offers some reflections on that ambiguity, using an East African focus in the main, but drawing on other examples where appropriate. Approaches to the deeper past were fragmentary and highly selective at best; at the other end of the spectrum, precolonial history was eschewed in favour of the presentist revolutionary fervour which underpinned political struggle.

For some, the precolonial past offered up heroes, who looked good if the lighting and the context were right. For backward-looking positivists, these were warriors of undeniable repute and achievement – men, and a few women too, who embodied the African genius through their deployment of violence in pursuit of glorious state-building projects, and the creation of innovative cultures. This was not violence for the sake of it: in addition to material outcomes – new political orders and the creation of wealth – this was war led by those who embodied the finest human qualifies. The experience of war was transformative and enlightening, and brought forth the best – and of course, for those defeated enemies, the worst – of human behaviour. Narratives of war were crafted to provide moral and didactic tales of goodness, valour, courage, embedded in oral tradition to offer guiding principles for governance and military command. The 'precolonial' past was, in short, seminal.

Yet for others, there was much which was best forgotten, or even vilified. In many ways, histories of war perpetuated the negative racial imagery which was rooted in a late nineteenth-century European interpretation of African war as savage, wantonly destructive, and lacking in any of the higher attributes of 'civilised' warfare; Africa itself was accordingly a place of insufferable and ceaseless violence which reflected fundamentally the chronically backward nature of African culture and the African personality. These were ideas which were prevalent from the late eighteenth century, when they formed the intellectual core of the anti-slave trade movement; later, Christian missionaries and sundry Victorian explorers expanded on the central themes – arguing, increasingly, for robust humanitarian intervention in order to save Africa from itself.[1] Those racial tropes proved remarkably pervasive, both in Europe and, more importantly, in Africa itself. The nationalists of the mid-twentieth century were loath to look back to a barbaric past, for they too wished to embrace what they saw as 'modernity'. Although there was a brief but intense period of interest in the deeper past as part of the nation-building projects of the 1960s, more generally there was a profound anxiety about 'Dark Continent' imagery among Africa's first generation of leaders. It was anxiety manifest in a reluctance to embrace a violent past and therefore a struggle to know how to engage with, let alone celebrate, that great temporal plain. Visions of a nineteenth century characterised by rampant violence – as it unquestionably was – were reminiscent of the 'Dark Continent' tropes common since the 1870s and 1880s.[2]

Moreover, the 'scramble' for Africa and the conquest involved had effectively extinguished the idea of 'usable' African military glory. More practically, few of the new nations corresponded with the polities underpinned by revolutionary war in the previous century; in fact most of them comprised several such polities, and so these were histories of violence which in fact underlined the *fragility* of those new nations, rather than providing them with precedents for political achievement and, more specifically, the constructive deployment of organised violence. The new nations of the 1950s and 1960s were thus cut across by violent fault lines and rested on a series of volatile precolonial frontiers. While in the precolonial era those frontiers might have represented the founts of African political creativity, as argued by Kopytoff in his treatise on the subject,[3] in the twentieth century they were seen only as destructive, as destabilising, and ultimately as inimical to postcolonial modernity. These were emphatically *not*, in Terence Ranger's famous formulation, 'usable' histories at the level of the nation.[4] The imposition of colonial rule had forced a violent rupture in terms of wider understandings of the historical relationship between political and military development over *la longue durée*.

There is nothing peculiarly *African* about this apparent paradox, of course: the remembrance of war is always contested, everywhere. The ways in which conflict are remembered globally are complex and numerous: there is no linear path, no easily chartable manner in which histories of violence can be mapped and understood. While some European states have long venerated older military traditions as part of the national heritage, revolutionary nationalisms have frequently involved similar kinds of rebirth as witnessed in Africa in the middle third of the twentieth century: renaissance and the 'rebooting' of History can be seen, for example, in the context of revolutionary France in the 1790s. And it is worth noting, too, that while national identity in Europe might occasionally be built around comparatively 'ancient' or more distant wars – the Hundred Years' War for England, for example – rather more commonly it was constructed around more *recent* conflicts, just as in twentieth century Africa: the Franco-Prussian War and the two World Wars are prime examples. And consider, too, in the recent past the agonising in the US over the Vietnam War, and the growing consensus around the 'foreign policy disaster' that was the Iraq War. The ambiguities around the role of war in the US are clear enough, and we see similar processes at work in Europe, with the attempts to create a usable past around the First World War on the occasion of the centenary since 2014. The relationship between war and nationalism is a contested, often unpredictable one, and the memory of violence can refract in dangerous ways.[5]

All that said, the ways in which history, and particularly histories of violent conflict, is a dichotomised realm in Africa is particularly stark. When we consider Africa from a Western military historical perspective, it might be argued that European war in the deeper past has long been the subject of celebration and commemoration, and has been the focus of scholarly literature of marked range and depth; such cultures of commemoration are fundamentally centred on the idea of the putatively transformative power of war itself, and certainly in the forging of national identities.

Europeans share common cultures of memory and commemoration, even if they don't always agree on the details of cause and effect. The contrast, in terms of the historical relationship between war and nation, between Africa and the West is striking. In Europe and North America, long histories of violent contest gave rise to sites and dates of mourning and celebration which became part of the essential equipment of consolidated nationhood.[6] And this is evident in the realms of popular military history, too: notably the work of military historians who return timelessly to Napoleon, Agincourt, Marlborough, Waterloo, Westphalia, Stalingrad, and who are keen always to reiterate those time-worn topics.[7] But in Africa, the military inheritance is all the more problematic, and so much more disconnected from the putatively 'modern' – although attempts at such connections were indeed made.

Mobilising the precolonial

The late precolonial era, including the transformative nineteenth century, was in many respects an era of remarkable political and military achievement – tantamount to a military revolution across a swathe of sub-Saharan Africa.[8] The transformative nature of the era itself presented a dilemma to late colonial nationalists, who were therefore not able to draw on authentic 'traditions' from the nineteenth century; the armed polities of the era, as Hodgkin observed, 'were in many cases either the products of revolutionary upheavals or attempting to carry out politics of internal reform and modernization'.[9] The irony lay in the fact that the nineteenth century was replete with warriors of global stature – if not of global fame – as builders of new states and reformers of old ones; it was an era in which the scale and objectives of warfare expanded considerably, involving new weaponry, and new methods of command and organisation. The military revolution, driven by a markedly dynamic armed entrepreneurialism, underpinned a parallel expansion of political scale and reach. In the economic sphere, these new polities deployed violence as a means to economic growth, the expansion of commercial operations, and the ever-more efficient mobilisation of factor endowments: people, metals, animals, textiles, soil. Africa's nineteenth-century military transformation could be conceived of as laying the groundwork for political modernity, in much the same way as Europe's own putative military revolution between the sixteenth and the eighteenth centuries created the conditions for modern state-formation and the construction of nations with those state structures at their centres.[10]

For our purposes, Uganda exemplifies these broader trends. Within the territory of the modern republic, the era between the fifteenth and the eighteenth centuries was the epoch of the warrior-king, associated with salvation and reconstruction.[11] Rich oral traditions, first recorded in the late nineteenth century, provided stories of military glory, dastardly rebellion, individual heroism, and the sagacity of founding fathers for several centuries before that. War was vital. Indeed, it was life itself, and was viewed in Manichaean terms: legitimate and righteous, evil and brutal, drawing out the extreme qualities in the human character – the cowardly, the heroic, the accomplished, the tragic, the richly rewarded, the bereft. In so many ways, war and the deployment of arms was posited as the purest form of human endeavour and leadership; heroes were celebrated, and transgressors used as moral indices for the individual as well as the wider community. Across the region – in kingdoms such as Buganda, Bunyoro, Nkore, and Toro – political centralisation meant tightened political control of increasingly pro-fessional armies for the purpose of external defence and domestic solidity.[12] The trend was clear enough among the forest kingdoms of southern Uganda, but it is in evidence, too, in the turbulent north: among the Acholi, and the Langi, and others, the momentum was toward the enlarge-ment of political and military scale, the emergence of charismatic military leadership, and the consolidation of ever larger field forces underpinned by distinctive warrior identities and cultures of militarism.[13] War, moreover, facilitated social mobility, as encapsulated in the story of a man named Kasindula, as told to the explorer Henry Morton Stanley in 1875. A humble local sub-chief of Buganda with little worldly wealth, Kasindula nonetheless brought himself to the attention of the mid-nineteenth century *kabaka* Suna by singlehandedly waging war on the Soga across the Nile and gathering more booty (an improbable amount of it) than any of the great chiefs, the men of privilege, at the royal court. He was duly rewarded with rank and riches, and in a stirring public address – doubtless embellished by the journalist Stanley – Suna pointedly contrasted his achievements with the inert privilege of his senior chiefs.[14] War was politics, and politics war: as the king of Bunyoro explained to the explorer John Hanning Speke, 'The brothers fought for [the throne], and the best man gained the crown.'[15] The early twentieth-century Ganda author, the Rev. Bartolomayo Zimbe, agreed: 'a Kingdom is always conquered, not succeeded to.'[16]

It was hardly surprising, then, that war was foremost in the minds of the first generation of Ugandan writers and historians who went into print in the 1900s and 1910s. Most of them had experienced the violent vicissitudes of the late nineteenth century, and – in the case of Sir Apolo Kagwa, most famously, prime minister of Buganda from 1889 to 1926 – had been prota-gonists in those upheavals. They had also inherited versions of those long-standing oral texts which commemorated and glorified violence as a means to individual and collective accomplishment. They wrote their histories with a view to building the new colonial society, and none doubted that violence had been at the centre of that process of remaking Uganda. Thus were Kagwa's now-classic texts on Buganda's history and customs – inspired in no small measure by his reading of Holy Scripture in Luganda as a young convert, or *musomi* ('reader') at the royal court – replete with military narratives, details on great wars and specific campaigns, the character and behaviour of commanders, and the rise and fall of men based on their qualities in battle: here there were stories of hubris, over-ambition, guile, cowardice, raw courage, humility, loyalty.[17] The importance of war and military culture is reflected, too, in the writings of Kagwa's close collaborator, the missionary John Roscoe, who clearly derived much of his information from Kagwa.[18]

By the 1930s, the generation which had 'managed' Buganda's transition to colonial modernity was going into print. Some – such as Samusoni Mazinga and the reigning king himself, Daudi Chwa – remembered the precolonial past in romantic hues, and depicted a pristine society of impeccable manners and remarkably little interpersonal violence.[19] But for others, their advancing years was a time to recall the righteous religious wars of the late nineteenth century, when they had wrestled with the forces of darkness to rescue the kingdom from civil strife and usher in spiritual truth. Produced in the course of the 1930s and early 1940s, the war memoirs of Albert Lugolobi, Zimbe, James Miti, and Ham Mukasa – formerly Kagwa's secretary – differed somewhat in the detail, but all shared a desire to describe the heroics of their generation in fighting monarchical tyranny in the name of Christian truth.[20] In so doing, they served to remind a younger generation – distracted by Westernisation and the fripperies of modernity – not to lose sight of their achievements in waging war on behalf of an increasingly forgetful kingdom. Their military recollections were infused with a powerful sense of nationalist sentiment.[21] What they tended to omit, of course, was the violence inflicted by the Anglo-Ganda armies on a range of neighbours – not least Bunyoro – in the course of the 1890s, wars of conquest which laid the parameters, and the contours, of the nation-to-be.

Early access to literacy and the printed word had meant that Ganda narratives of war had pioneered the genre, but where Kagwa had first trod, others followed. Y.K. Lubogo's history of Busoga, written between 1921 and 1938, may have reflected a desire to demonstrate that the Soga had been 'a complete tribe for many generations now with firmly established boundaries', but its intricate narrative described the significance of inter-clan warfare as well as the struggle of communities to protect themselves from outside incursions in which martial heroes abound.[22] Bunyoro, Buganda's old foe and close neighbour, was also by the 1930s producing counter-narratives to the Ganda texts; but the form and thematic emphasis was much the same, with due attention paid to military trials and tribulations, glorious feats of arms, and the endeavours of great leaders of note – in the Nyoro case, those of the great warrior Kabalega. The reigning *mukama* or king of Bunyoro Tito Winyi (1924–1967) wrote a set of articles under a pseudonym in the mid-1930s in the *Uganda Journal* in which he highlighted Nyoro's own claims to be the great military power of the region's recent past,[23] while John Nyakatura's extended text was likewise heavily martial in content and indeed, in some respects, purpose.[24] Armed endeavour was prominent, too, in Lazaro Kamugungunu and A.G.Katate's history of the kings of Ankole, written once more in the style of Kagwa and Nyakatura.[25] In each case, the text reflected unambiguously and unashamedly the importance of martial prowess to 'traditional' leadership, and to political and cultural identity. War, the authors were saying in each text, had *made* the kingdom. Perhaps, too, there was – at least in the case of Bunyoro and Ankole, but conceivably also in the case of Buganda – an element of wistful nostalgia for an age in which the agency of political change and social progress had been the man-at-arms; some primordial yearning for an epoch which must have seemed heroic to a generation of bureaucrats in colonial employment, as Nyakatura, Katate, Kamugungunu, and ultimately Kagwa himself – survivor of late-nineteenth-century traumas though he was – had all been.

Nationalists coming after the Kagwa generation at first followed suit. It seemed that the precolonial remained fertile ground and that it provided the kinds of heroes who could be relied upon to support nationalist struggle. In Tanzania, the Tanganyika African National Union (TANU) of the 1950s at first embraced continuity and demonstrated a self-conscious desire to historicise the African struggle. In this case, the relevant hero was a remarkable figure named Mirambo, who led a veritable military revolution among the Nyamwezi people of central northern Tanzania between the 1860s and the 1880s. He embodied the nineteenth-century transformation of warfare in Africa more broadly, with his energetic reform and leadership, his reorganisation of the practice and culture of war, his creation of a wholly new – if short-lived – polity. Famed for his military innovations, and his domination of coast-bound trade routes which linked the region to the global economy,[26] Mirambo looked to the nationalists of the 1950s like a ferocious defender of African sovereignty and a champion for their times. President Julius Nyerere stuck to broad themes in his opening address to the famous meeting of Africanist historians in Dar es Salaam in 1965,[27] but the nineteenth century loomed large in TANU's historical imagination. This ostensibly most progressive of nationalist movements adapted a song regularly performed by Mirambo's victorious soldiers as they returned from a successful campaign – *Ohoo! Chuma chabela mitwe* [Oh! Iron has broken heads] – to *Ohoo! TANU yajenga chi*; or 'Oh! TANU builds the country'. As local oral historian J.B. Kabeya explained:

> It has become like the second National Anthem ... [Nyerere], at a meeting of the TANU Youth League held in Tabora in 1962, said that this old battlesong should have new words to be sung in the building of our nation ... Immediately the people took this up and since then this song has

spread like wildfire all over Tanzania. These two great leaders, King Mirambo and Mwalimu Nyerere, are far apart in time. But they are both of one mind. They•both wanted to build one nation and to make the country peaceful.

A street was named after this 'hero of Unyamwezi and the whole of Tanzania' in Dar es Salaam.[28]

For others, indeed, there were tangible lineages between Africans' violent resistance to European imperialism in the late nineteenth and early twentieth centuries, and the 'modern' nationalist movements of later years.[29] In Tanzania, again, TANU appealed directly to the spirit of the Maji Maji rebellion against German rule in 1905–7, self-consciously mobilising ideas around radical nationalism and awarding these deeper historical roots: according to a nationalist newsletter in 1967, 'On the ashes of Maji-Maji our new nation was founded.'[30] Similarly, Zimbabwean nationalists in the late 1950s and early 1960s evoked the spirit of the 'first Chimurenga' – the uprising against European settlers by Ndebele and Shona peoples in 1896–97 – in mobilising resistance against the white minority regime in Southern Rhodesia.[31] Later, in the 1980s and 1990s, Eritrean nationalists likewise occasionally sought to deepen the historical roots of their armed struggle by pointing to centuries of violent subjugation by 'Ethiopians' and long histories of resistance to that subjugation.[32] Even the rapid emergence of army officers as leaders of the postcolony – the recurrent theme of African politics in the 1960s and 1970s – and the prevalence of the *coup d'etat* was, some proposed, emblematic of a much older pattern in the African body politic: as Ali Mazrui argued in the 1970s, perhaps Amin represented the '*Re-Africanization*' of Africa, and was the manifestation of deeper, older forces at work in the postcolony.[33] Certainly, there is an argument to be made about the extent to which Amin was symptomatic of the nineteenth-century militarisation of political culture, and its renaissance in the middle decades of the twentieth.[34]

So, too, arguably, was the emergence of revolutionary guerrilla movements – such as the National Resistance Movement in Uganda – which could be seen as representative of a longstanding pattern of rupture and rebirth, in which the armed frontier was mobilised to march on the centre, reorganising violence in order to effect political and social change. Despite the NRM's ambivalence about the past – more on which below – a cursory glance into Uganda's precolonial history revealed at least some elements worthy of admiration, even if grudging. NRM leader and, from 1986, President of Uganda Yoweri Museveni approved of some of the more expansionist 'tribal' chiefs of the precolonial era:

Some of the chiefs in our past history, however, tried to consolidate their rule, but today the neo-traditionalists are actually divisive ... The old kings ... were expansionists ... [However] [t]he precolonial indigenous states were weak because they had not yet succeeded in uniting all the tribes and consolidating the resources in this region. Some had made more progress than others, but they had not yet evolved enough power and unity to immunize themselves against foreign encroachment: that is why they were conquered.[35]

Wars of expansion were implicitly awarded the presidential seal of approval here – the violence of 'the old kings' was the means to an end, namely unity, and Museveni and the Movement, on this reading, were simply finishing the job.

Yet more than that, there was heroism to be mobilised and celebrated. Sometime in the early 1980s, Museveni – the student of revolutionary politics and sworn opponent of backward tribalism – was said to have visited, on at least one occasion, the shrine of the Ganda war god Kibuka to pay his 'respects' and perhaps to make sure all bases were covered in his campaign against incumbent Ugandan president Milton Obote.[36] Around the same time, the National Resistance Army named two of its units 'Kabalega' and – a little more controversially, perhaps – 'Mwanga', 'in honour', as Museveni himself described it, 'of African heroic figures [who were] noted Ugandan opponents of British imperialism in the late 19th century'.[37] These were the remarkable evocations of a national history at a time when the nation itself was being torn apart by the bloody violence of the second Obote regime. Museveni was clearly particularly fixated with Kabalega, the Nyoro king who was unambiguously a resister of foreign colonialism, as well as a gifted military reformer, and has long been recalled by the downtrodden and marginalised Nyoro as one of their great historical figures.[38] In 2009, Kabalega was declared a national hero, which entitled him to a medal and a 21-gun salute.[39] Things were a little more problematic with respect to Mwanga, whose heroism, after all, was of a rather more ambivalent hue: he had actively worked with the British for several years before rebelling, briefly, in 1897, when he threw his lot in with Kabalega in resisting the consolidation of British rule. Somewhat more problematically, Mwanga was probably homosexual, which in the context of escalating homophobia in Uganda in recent years was probably more iniquitous than collaboration with the colonial power – although, for Museveni, Mwanga was flawed and susceptible rather than downright evil.[40] Either way, there would be no gun salute for him, despite some scholarly revisionism.[41]

The NRM in Uganda was not the only revolutionary liberation front to select usable champions from the precolonial past. In South Africa, the Zulu 'warrior tradition' – ostensibly at least – furnished the ANC with a model of armed resistance. At his trial in 1962, Nelson Mandela invoked the name of Dingane, Shaka's half-brother and Zulu king from 1828 to 1840, and the Zulu uprising of 1906–8, as the progenitors of the heroic struggle of which the ANC and others were now a part. He also made reference to the resistance offered against white expansion by his own ethnic group, the Xhosa.[42] Mandela later recalled that date chosen in 1961 for the launch of the armed wing of the ANC, Umkhonto we Sizwe ('Spear of the Nation') – 16 December – was 'Dingane's Day'. 'That day', wrote Mandela, 'white South Africans celebrate the defeat of the great Zulu leader Dingane as the Battle of Blood River in 1838 ... We chose 16 December to show that the African had only begun to fight, and that we had righteousness – and dynamite – on our side.'[43] Yet such invocations were not unproblematic. By the 1980s, ideas about the Zulu military tradition were serving to fuel a distinctive Zulu nationalism in Natal which was pitted against all outsiders and non-Zulu speakers – including the ANC and its supporters.[44] Shaka himself might exemplify a heroic black militarism and tradition of resistance; but he was a divisive, contested figure, and there was no unified vision – or even ownership – of him among black South Africans.[45]

What is clear from the foregoing is that particular historical figures, and the circumstances within which they supposedly operated, served the purposes of nationalist struggle, whether violent or otherwise. Various actors – including the leaders of movements, and their supporters – were able to draw lines of descent from precolonial warfare to their own political projects. Yet this historicity needs to be set alongside an equally evident discomfiture with the violent past, and the ways in which an increasingly radical politics had little time for the heroes of precolonial yore. The nationalist movements which emerged across Africa from the 1960s onward increasingly wrestled with war in the deep past, and with the acquisition of usable warrior heroes from the precolonial era.

The blinding violence of the present

The utilisation of the deeper past was a complicated business. In general terms 'nationalism' in Africa was seen as the product of modernity, of sociopolitical change induced by colonial rule, change which political leaders sought to harness and mobilise in the pursuit of narrow-defined provincial, ethnic, or religious agendas.[46] And of course most – if indeed not all – African nationalists thought in terms of those sectional or regional interests: the modern 'tribalism' which was such a focus for political scientists, historians and anthropologists in the 1960s.[47] The pioneering scholar of African nationalism, and a contemporary observer of it, Thomas Hodgkin, pointed out in the mid-1950s that 'African nationalists have been compelled to develop their own counterattack; to answer the myth of African barbarism and backwardness with the counter-myth of African civilisation and achievement.'[48] Leaving aside the unfortunate implied scepticism that Africans might actually have access to anything like 'civilisation and achievement', the argument remains of interest: which historic figures, particularly in the mould of great warrior, could nationalists refashion to their own ends? In Francophone West Africa, Malians might marshal the great four-teenth-century king Mansa Musa,[49] and Mande-speakers could harness the epic of their folk-hero Sunjata.[50] But in truth, while these characters might be repackaged as state-builders, ideologues, or political entrepreneurs, they were not always ideal *vis-à-vis* the needs of modern nationalists and the exigencies of modern nationalism. The latter wished to stress *not* histories of violence, but pacific achievement – in economy, technology, culture.[51] Histories relating to war, militarism, force of arms, were generally regarded as negative stereotypes, redolent of missionary and colonial narratives of Africa's past in the late nineteenth and early twentieth centuries.

In Nyerere's Tanzania in the 1960s and 1970s, the Department of Antiquities neglected the upkeep of Mirambo's grave near Tabora: weeds grew up around the memorial, and the story of his heroics became con-sumed by the bush in the years after independence.[52] In fact, as Kabeya noted, local ambivalence toward Mirambo could be traced back some years: 'Even the colonial Government tried to place a memorial ... on his grave, but evil-minded people went and broke it down and the District Council of Tabora has not up to this day paid any attention to this grave.'[53] The forgetting of the great Nyamwezi leader was emblematic of the ways in which histories of violence had begun to be troublesome to those with one eye on the enduring, nefarious idea of a Dark Continent crippled by mind-less tribalism, and the other on the grand new vistas of African modernity. By the 1970s, Mirambo had come to represent the violent and traumatic upheaval of the precolonial era, and was of dubious utility to would-be nation-builders and modernisers. Nor did it help that as much as anything else Mirambo was a slave-dealer – one of the biggest in the East African interior – and that did not sit well with Nyerere's own romantic visions of a beatific African past, characterised as a communal idyll undisturbed by the ravages of industrial revolution.[54] And as Kabeya conceded, 'There are people who say that Mirambo was only a robber leader who attacked people on the roads and in the bush ... ' So much for the Mars of Africa, as Henry Morton Stanley – with perhaps just a touch of ironic hyperbole – had anointed him.[55]

In Uganda, meanwhile, the early twentieth-century authors discussed above were indeed keen on nineteenth-century history; but naturally they framed the late precolonial past in particular ways. War was prominent, clearly, and of ineffable significance. In many respects it was seen as leading inexorably to peace and salvation, while at the same time they struggled with the meanings of the 'modern', and especially its meanings vis-à-vis the hardening conceptualisation of the nineteenth century as a murky, savage epoch which paved the way for enlightened civilisation. This was especially true of the generation that emerged in the 1930s and 1940s – the progeny of Kagwa and others – which was increasingly involved in what are broadly termed 'nationalist politics'. In Uganda, from the outset, the brittle nature of the nation meant that there could be no use for *longue durée* histories of war at the level of state-building. On the eve of independence, while the Ruwenzururu insurgency erupted in the far west,[56] Buganda, the great martial power of the nineteenth century, was threatening to the derail the nation itself by threatening to secede. In 1966, in one of the first significant acts of political violence against the past in sovereign Uganda, Milton Obote sent troops against the Ganda royal enclosure, forcing the *kabaka* into exile[57]; within months, the main kingdoms – Buganda, Bunyoro, Toro, Ankole, and Busoga – had been constitutionally abolished in what seemed like an act of violent revenge against the privileged southern states by a disgruntled northern politician, an echo of precolonial violence in what was a putatively modernist attempt to refashion the nation in unitary terms.[58]

By the 1970s, the army was intervening in politics across the continent, a phenomenon which attracted much analysis at the time.[59] The military *coup d'etat* was a typology of rupture exemplified by Idi Amin in Uganda, and Mengistu Haile Mariam in Ethiopia. These may well have been representative, as Ali Mazrui and others argued at the time, of precolonial dynamics: 'Man on Horseback', the armed saviour, or even a reassertion of some timeless African masculinity briefly suppressed by the emasculation that was colonial rule.[60] But these soldiers did not see themselves as the products, or the exponents, of some deeper culture of military involvement in the polity: they believed themselves to be emblematic of military modernity, the saviours of the nation, and professionals who could be entrusted with the task.[61] Where civilians had failed, embroiled as *they* were in the tawdry business of ancient squabbles and pecuniary ambition, professional soldiers would hold steady the ship of state. In Ethiopia – perhaps more forcibly and dramatically than elsewhere – the forces of military modernity swept through the musty corridors, anterooms and myriad shadowed corners of the Imperial Palace when in 1974 the army moved against Haile Selassie. It was a story told in stark, if highly stylised, terms by Ryzsard Kapuscinski, whose tale of the Emperor's overthrow was one of how swiftly the past might be destroyed if ever there was one.[62] The military command – known as the *Derg* ('committee') – which led Ethiopia from the mid-1970s to 1991 espoused an aggressive, authoritarian Marxist-Leninism with all its modernist rhetoric and paraphernalia. The social and political transformation was embodied, too, in the emergence of the *Derg*'s eventual leader, Mengistu Haile Mariam, for he was a professional soldier from the long-marginal Konso ethnic group in the southwest of Ethiopia, and probably of slave origins. The modern army had facilitated his rise, and his displacement of a royal house which had stood for seven hundred years.[63]

Mengistu's counterpart in Uganda was Idi Amin, who dispatched Milton Obote into exile in 1971 and who consolidated the control of the army over Uganda for the next decade. Amin was a northerner – from the Kakwa, in the far northwest corner of the country – and indeed independent Uganda had its roots in the colonial-era creation of a north-south dichotomy: the centre of political gravity and economic 'development' was in the south, especially Buganda, while the north was a source of labour, including military labour. The north had long been the favoured hunting ground for recruits into the colonial army.[64] The image of the northern recruit is captured in Idi Amin's own early history, as the loyal NCO in the King's African Rifles – hardly bursting with intelligence, but took orders well, and was a fine boxer[65]; while the role of the soldier in Acholi society, another northern source of recruitment, is attested by the figure of the burly sergeant, apparently a veteran of the Second World War, in Okot p'Bitek's classic Luo-language novel *Lak Tar*, first published in 1953.[66] Idi Amin's professional credentials were beyond question, as a former NCO in the King's African Rifles and then one of East Africa's first commissioned officers. Paradoxically, in some ways, it was his exposure to *both* British military modernity *and* to British military tradition which burnished him with that credibility. The sense of progress engendered by the British imperial mission further emphasised the striking disconnect between African past and African present. Yet Amin sought to present himself, too, as the vanquisher of the British, using war or the threat of it, and certainly the rhetoric of violence, to consolidate an idiosyncratic vision of the brittle nation. His expulsion of the Ugandan Asian community was the central plank in what he described as an 'economic war'.[67] He turned on its head the narrative suggested by his previous service in the KAR (he had fought Mau Mau rebels in Kenya) by promoting himself to Field Marshal with a splendiferous uniform to match, and designating himself 'Conqueror of the British Empire'.[68]

These regimes were soon to be challenged by a new phenomenon, the armed revolutionary front – and these identified themselves as the harbingers of a new social and political order which eschewed the failures of the past and emphasised the necessity of rebirth. In many ways this distinctive revolutionary historicism first emerged in Algeria. Frantz Fanon's exhortation to revolutionary anticolonial violence prompted a shift toward a more presentist, strikingly ahistorical culture of arms within which the role of war in Africa's deeper past was regarded with suspicion, even anxiety. His classic 1961 treatise – *Les Damnés de la Terre*, published in English as *The Wretched of the Earth* – quickly established an intellectual and moral framework within which a new generation of African political warriors could organise their armed struggles.[69] Violence, he argued in the context of the Algerian *Front de Libération Nationale*'s struggle against France, was necessary: not only to utterly purge the insidious and pervasive influences of

European colonial rule, but to address the psychological disorders which had arisen in the African mind as a result of foreign domination. Only violent struggle could truly liberate Africa; it would lay the groundwork for new nations, and point the way to socialist futures which were, after all, the only futures worth contemplating. By sheer force of arms would Africans eschew the awful past and march into postcolonial modernity. Fanon himself, of course, is the subject of a veritable canon of literature which has sought to clarify his meanings and intents, and at least some of which suggests that he was not actually exhorting Africans to violence.[70] But that is how he was read and received by a generation of activists and nationalists across the continent – a generation born in the 1940s, which clutched copies of *Wretched of the Earth* to its breast at school and at university and ultimately in 'the bush', where he was taken into battle in pursuit of the new nation, and in flight from the evil past. In the late 1960s, Yoweri Museveni carried Fanon into the FRELIMO-liberated areas of northern Mozambique, and believed he had seen Fanon's theory of revolutionary violence 'verified'.[71] Fanon created the conceptual baseline for liberation movements and provided the legitimacy for revolutionary violence – a kind of purification through fire of the political kingdom. Self-consciously modernising revolutionary movements sought to build nations as entities of the (socialist) future, rather than as things of the past; they would be arenas of political modernity.

Few mid-twentieth century movements were able to tap into, or reinvent, the kind of nationalist history of anticolonial violence available in Algeria. Here, importantly, there were the benefits of Arabic literacy and literary culture, as well as the 'advantage' of long-term resistance to foreign invasion which could be moulded to serve contemporary needs.[72] Yet Algeria and Fanon served to inspire new kinds of political movements and ultimately new ideas about the utilisation of violence across the continent, whether consciously or otherwise. These movements, with their roots in the political turbulence and excitement of the 1960s, espoused revolutionary violence and aimed at the salvation – nothing less – of the postcolonial nation. Revolutionary people's war, in other words, was at least implicitly *dehistoricising* and evangelical, and would bring about much-needed rupture between dark past and bright future. In Ethiopia, Eritrea, Zimbabwe, Mozambique, South Sudan, Rwanda and Uganda, there emerged movements which – though each distinct in specific objectives and local exigencies – espoused a form of violent political rebirth: armed liberation fronts organised around the need for fundamental social revolution through warfare.[73] Of course, as we have already noted in some cases, these kinds of movements *did* sometimes seek to connect with the deeper past. But at the same time they espoused, and aspired to, a self-consciously modernist patriotism, and increasingly expressed an ambivalence about History as a discipline, and the past in general. Revolutionaries might occasionally make forays into military history in order to make a contemporary point; but revolution itself was not served by retrievals of the deep past. These movements advocated gender equality, land reform, education for 'the masses' and in particular the awakening and politicisation of slumbering peasantries. Histories of war in the deep past were largely irrelevant, and the only history that really mattered was the recent past of political oppression and violent marginalisation at the hands of occupying regimes or internal tyrannies.

One of the starkest cases is Eritrea, where the Eritrean People's Liberation Front (EPLF) waged one of Africa's longest wars, for independence from Ethiopia, from the early 1970s until 1991. While a handful of activists sought to claim continuity for the armed struggle from the deeper past, as noted earlier, the EPLF itself was scarcely interested in anything much before the 1950s, although of course Italian colonial rule (c.1890–1941) putatively provided a legal basis for independent statehood. The EPLF's objective was a socialist revolution, to be achieved through the blood sacrifice of its fighters.[74] In developing its political programme,[75] the Front sought a modernist, revolutionary antidote to the oppressive antiquity of its neighbour (and occupier) Ethiopia. After achieving independence in 1991, the EPLF reified the public memory of the armed struggle and the resultant sacrifice on which the nation was built, notably through the use of national days of mourning (Martyrs' Day, 20 June) and commemoration (Independence Day, 24 May).[76] A similar culture of memorialisation was evident in Rwanda, where the horrors of the 1994 genocide – and the salvation of the nation by the Rwandan Patriotic Front – bore witness to catastrophic deeper histories of tribalism and division. Those histories needed careful managing as a result, and quarantined or even banned altogether.[77]

A process of historical quarantine was also at work in Ethiopia, though here the situation was complicated by the long recorded history of which Ethiopians – or some of them, at least – were justly proud. The Tigray People's Liberation Front (TPLF) was able, to some extent, to draw on the memory of the last Tigrayan emperor, Yohannes IV (1872–1889)[78]; but as the vanguard movement in the Ethiopian People's Revolutionary Democratic Front (EPRDF), a coalition of liberation movements ranged against the *Derg*, the TPLF was increasingly ambivalent about Ethiopia's deeper past. Tragic heroes such as Emperor Tewodros – caught in the pre-dawn between savage primitivism and modernising reform, and who committed suicide in the face of British invasion in 1868 – might continue to haunt the Ethiopian imaginary.[79] And of course the battle of Adwa in 1896 gave rise to one of the most important annual national days in the calendar, 2 March, commemorating the moment when Ethiopia successfully defeated an invading Italian army, thus preserving its independence at the height of the

European scramble for Africa.[80] But even Adwa Day, while inviolable as a public holiday, was not unproblematic to those ethnic groups – Tigrayans and Oromo, most obviously – who felt a little uneasy about lauding the Amhara emperor, Menelik, who had won the battle itself, for Menelik was also the creator of an imperial hegemony which oppressed many non-Amhara. The Tigrayan-led EPRDF, in power from 1991, sought to neutralise the violent past and detach the age of kings and their wars from the brave new world of social development and economic growth.[81]

Uganda further elucidates the paradox and ambiguity which often attended the memory and mobilisation of violent conflict in pursuit of national identity, and of national history; the rejection of the idea that historical violence had had any meaningful utility, and, despite the NRM's own immediate history, the embracing of normative modern ideas about violence as inimical to 'development'. Come the NRM, come another new beginning; except this one, said Museveni, was *for real*. *This* new beginning really was the Rupture from the Past to end all Ruptures; their war, indeed, was the war to end all wars. And so, for all the grudging respect to those expansionist precolonial chiefs noted earlier, Museveni was doggedly, at least publicly, Anti-History. During the public discourse on history during the jubilee celebrations celebrating Uganda's fiftieth birthday as an independent state in 2012, the deeper past, whether represented by personality or by process, was conspicuous by its absence – notwithstanding a handful of exceptions, such as Nyoro king Kabalega. The cause of so much apparent amnesia was violence, or more precisely the perception that the past was rather too full of it, or at least the kind that had gotten Africa precisely nowhere. Museveni offered his own interpretation of those histories in a speech at a political seminar in 1989:

> Before colonial rule, we had backward tribal states here. If anyone tries to glamorise them, he is telling you a lie. If these tribal entities were equal to the tasks of a viable nation-state, why did they surrender their sovereignty to foreigners? Can any foreigner come here now and take away our sovereignty? It is impossible! This is because our present state is superior to the ancient tribal ones.[82]

Ultimately, the nation is defined as a thing of modernity, while the deeper past is characterised by savage tribalism and internecine warfare – there could be no 'Uganda' there.[83] Movements such as the NRM needed to be highly selective in their approach to war in the deeper past – a process of foreshortening, indeed – in order to consolidate the nation, which was not served by histories of violence in the precolonial era.

All nations need wars for internal consumption; they require heroes, and martyrs, in order to legitimise incumbent regimes. But those wars, and their fallen, tended to be of more recent vintage. In Uganda, the reluctance to endow too many precolonial processes with constructive significance in the context of war and militarism contrasts sharply with the government's willingness to acknowledge the talent and patriotism of two very recent adversaries, General Tito Okello Lutwa, Museveni's immediate predecessor, and General Oyite Ojok, Obote's army chief staff in the early 1980s. Ojok was declared a national hero in 2010, and was awarded due respect as an honourable and worthy foe in the panoply of historic patriots during the 2012 jubilee celebrations.[84] Tito Okello's record in power (July 1985-January 1986) was regarded as a mixed bag: a professional soldier, he was remembered alternately as a man with some 'good ideas' but with little time to implement them, as bringing some economic stability to Uganda, and as presiding over violent political instability and widespread extrajudicial killing. Nonetheless, in 2010, Museveni posthumously awarded him the Kagera National Medal of Honour for his role in fighting Amin in the 1970s.[85] Their efforts were, at least, somehow patriotic, and legitimate, if wrongheaded.

An entirely different interpretation was offered of the war in the north. The military rupture engendered by the NRA's seizure of power in 1986 was followed by Museveni's men ravaging and pillaging north of the Nile in what was in effect punishment for northern tyranny and militarism; it was awful brutality which itself gave rise to local resistance, in the form, first, of the Holy Spirit Movement under the prophetess Alice Lakwena, and then the Lord's Resistance Army (LRA) under her erstwhile lieutenant Joseph Kony. The terrible war in the north which resulted enabled the NRM to militarise and in effect further marginalise the north as a troublesome frontier zone, where wayward and backward people had brought dreadful violence on themselves; northern Uganda became a place where the UPDF – its inability to actually defeat the LRA or capture Kony notwithstanding – could demonstrate military prowess, and the government could, through repeatedly reheated but largely ineffectual 'development' schemes, demonstrate its political benevolence.[86] It was a bitter irony indeed for a swathe of territory which served as a useful badland to a government desirous of maintaining a degree of militarisation, and a head of state who still liked to periodically don the khakis. For the same reason, the neglected and fly-blown district of Karamoja in the far northeast corner of Uganda – another place where desperate herdsmen toted AK-47s in apparent defiance of modernity and civility – had its own dedicated government minister, for a time none other than the President's own wife, Janet Museveni. All nations have cores and peripheries, centres and margins. And it is in those borderlands that a military regime such as the NRM demonstrates its power and its prowess; the 'Otherisation' of troublesome people in turbulent peripheries has long enabled regimes to achieve justification and set up the juxtaposition of internal legitimacy versus the chaotic brutality of the badlands. The latter are also critical externally, for peripheries can be useful buffers and militarised zones, platforms for external adventurism and the armed entre-preneurialism on which the region's political culture is largely based. Uganda is the product of a complex network of military frontiers and rough edges. Those frontiers operate in much the same way as they always have: facilitating armed creativity, facilitating essentialised senses of 'self' for all involved, and permitting hegemonic regimes to project benevolence and to develop security agendas. More recently, Museveni has directed military adventurism outward, with war becoming an increasingly central part of Ugandan foreign policy across the region.[87] The Movement has used regio-nal threats to strengthen its internal security agenda, while positioning Uganda as a pivot of global security. Museveni has to date given no indica-tion that he believes any of these wars might be to Ugandans what the Trojan War was (according to Thucydides) to the Greeks: the unity in armed purpose of formerly disparate tribes.[88] So far, external adventures have had, if anything, the opposite effect; but, in any case, Uganda's militarised foreign policy is in keeping with a nation whose deep roots are in war.

Conclusions

War, history, and the political kingdom

In the West, oftentimes, the relationship between recent war and nation-building is more significant than anything the deeper past has to offer. No doubt, Raphael Samuel had a powerful point when he thoughtfully tackled the relationship between nation and military memory, memorably observing:

> The disasters of the Gallipoli campaign, by common consent, marked Australia's coming-of-age; are we to say less of the battle of Thermopylae … ? By what right do we claim the Maginot line … for the history of modern nationalism, while leaving the fortifications of Anatolia … to the archaeologists?[89]

Hobsbawm was more sceptical.[90] But there is no doubt that, in much of sub-Saharan Africa from the mid-twentieth century onward, political fragility rendered the deeper past an unhealthy environment. This was true even in Ethiopia, a modern nation which boasts an unusually richly recorded past but where the EPRDF is uneasy about the ascription of glorious military victories to an *ancien regime* rooted in ethnic imperialism. According to Charles Tilly in the European context, states made war, and war made states, and by extension, nations.[91] Not so in Africa, or at least not in quite such linear terms. As René Lemarchand has demonstrated for Central Africa, nationalism has not been a unifying force when mobilised for war, but rather the opposite, exacerbating deep fissures between ethnic and cultural communities, and in some cases leading to some of the worst violence on the planet since the end of the Cold War.[92] War, according to most analyses over the last half-century or so, has been at the very least inimical to national cohesion and construction, to the consolidation of the postcolonial political order so necessary to achieve broadly-defined socio-economic 'development'.[93] It has even been the destroyer of all those things held dearest in the modern humanitarian lexicon, and in many ways this anxiety, or at least uneasy ambivalence, has been extrapolated backwards into the deeper past.

Visions of violence danced before the eyes of would-be nation-builders, who were haunted by the ghastly spectacle of bloodthirsty despots and implacable tribes, tropes beloved of colonial officials. Imperial partition cast long shadows, but produced short memories; at the same time, the evangelical spell created visions of a past that was damned, and Christianity distorted ideas about how time worked, cursed history, led to new emphases on rupture and rebirth. How to use (or not use) these turbulent, creative histories in the modern era? Just as people built places to belong to, so they built time to which to belong, too. As the twentieth century drew on, there emerged a different attitude to those martial pasts: at best, one of uneasy ambivalence, among those who wrestled with cultural pride in 'traditional history' on the one hand and with the hunger for modernity on the other; at worst, downright hostile, marked by a sense that the deeper past could serve no purpose other than to remind everyone – Africans and foreigners alike – just how barbaric their history was, and how backward things were until Europe showed up to nudge progress along. It was a nuanced and elongated process, not a sudden moment of revelation and transformation; but the upshot was a creeping presentism, a gradual foreshortening of historical trajectories, and a cauterisation of Time itself. To a very large extent it reflected an abiding notion embedded within developmental discourse that war was 'bad' and peace 'good' – especially in Africa, even though Europe's historical experience demonstrated something a little more complicated. 'Peace' itself, of course, is something of a novel concept.-[94] But the idea has taken hold that in Africa's deepest darkest past internecine tribal wars were inimical to the realisation of political and economic 'modernity'. The reality is rather more convoluted, and altogether more nuanced.

In the end, the past became securitised, in order that the nation might survive against the existential threat supposedly posed by history itself: history was a matter of national security, a coalescence of violently sectarian issues which were inimical to modern development and which constituted an existential threat to the political community.[95] Earlier we discussed Mirambo, who, despite his remarkable political vision and his ambition – not so much *ruga ruga* as hopeful builder of territorial statehood, no less than Tewodros was mere *shifta* – has not really endured as an historic figure. Mirambo ultimately had no place in a Tanzania in which concepts of modern nationhood were rooted not in a muddy and violent precolonial past – although he had a brief resurgence in the 1950s and 1960s as a proto-national hero – but in ideas about modernity, development, the future. What, after all, had Mirambo *actually* achieved? A violent, short-lived polity which fed off the nefarious trade in people, and which collapsed soon after Mirambo's own death in 1884, was hardly the stuff on which great nations could be built. And thus, for all Nyerere's early enthusiasm for the African past – which was in truth as much a matter of decolonising the curriculum as of any intrinsic interest in precolonial history *per se* – there was a subtle shift toward the African present, and the future, in the course of the 1970s and 1980s. In Uganda, too, Yoweri Museveni was loath to celebrate the achievements of the precolonial past which was, for the NRM, the source of savage sectarianism, and the positive aims and outcomes of which had been greatly exaggerated. There was nothing glorious about tribal chieftains running around stabbing one another but unable to get to grips with encroaching global modernity.

Ultimately, while certain elements of the deeper past might be mobilised in the service of political and military struggle, longer-range histories were increasingly viewed with ambivalence. At the same time, the wars of the late twentieth and early twenty-first centuries – and the military cultures they produced – have tended to eclipse deeper histories of warfare, not least because of the desire to recover from recent violence and construct nations around 'peace and development' agendas. Still, deeper histories of violence were never very far from the surface, despite the claims to radical violence, and to violent rebirth, in the second half of the twentieth century.

Notes

1. A sample of writings from East Africa which capture the dark imaginings of Europeans would include Burton, *The Lake Regions of Central Africa*; Speke, *Journal of the Discovery of the Source of the Nile*; Stanley, *Through the Dark Continent*; and Mackay, *Pioneer Missionary of the Church Missionary Society*. For analysis, see Reid, "Revisiting Primitive War"; and Porter, *Military Orientalism*.
2. Brantlinger, "Victorians and Africans".
3. Kopytoff, *The African Frontier.* especially Kopytoff's own introduction.
4. Ranger, "Towards a Usable African Past".
5. Hall and Malesevic, *Nationalism and War*.
6. Notably Winter, *Sites of Memory, Sites of Mourning*. For a compelling account of histories of national catastrophe drawing on the examples of the American South after 1865, France after 1871, and Germany after 1918, see Schivelbusch, *The Culture of Defeat*.
7. I am thinking here of Anthony Beevor, Andrew Roberts, and numerous others.
8. For example, Reid, "The Fragile Revolution". See also Uzoigwe, "The warrior and the state in precolonial Africa", in Mazrui, *The Warrior Tradition in Modern Africa*.

9. Hodgkin, "The relevance of "Western" ideas," 66; also, Coleman, "Tradition and Nationalism in Tropical Africa".

10. For useful summations of the European experience, see for example Howard, *War in European History*, 13–14; and more detailed examinations in Parker, *The Military Revolution*; and Tilly, *Coercion*.

11. For example, Oliver, "Discernible developments in the interior"; Chretien, *The Great Lakes of Africa*. See also Schoenbrun, "Conjuring the Modern in Africa".

12. Kagwa, *The Kings of Buganda*, 124; Reid, *Political Power in Pre-Colonial Buganda*; Médard, *La Royaume du Buganda au XIXe siècle*. For Bunyoro, see Nyakatura, *Anatomy of an African Kingdom*; Doyle, *Crisis and Decline in Bunyoro*. On Busoga: see Cohen, *The Historical Tradition of Busoga*; and for Toro and the western kingdoms, Steinhart, *Conflict and Collaboration*.

13. Atkinson, *The Roots of Ethnicity*, 271–2; Lamphear, "The evolution of Ateker "New Model" Armies"; Lawrance, *The Iteso*, 13–16.

14. Stanley, *Through the Dark Continent*, 289–94.

15. Speke, *Journal*, 547; see also Nyakatura, *Anatomy*, 100; Roscoe, *Twenty-Five Years in East Africa*, 254; *The Bakitara or Banyoro*, 314.

16. Zimbe, "Buganda ne Kabaka [Buganda and the King]," 83, 107.

17. Kagwa's early writings – many of which were produced on his own printing press – include *Basekabaka be Buganda* – 'the Kings of Buganda' (1901); *Empisa za Baganda*, 'Customs of the Baganda' (1907); and *Ebika bya Baganda* (1912), 'Clans of the Baganda'.

18. In Roscoe's *The Baganda: an account of their native customs and beliefs* (London, 1911), Chapter 10 is devoted to 'Warfare' which also appears intermittently throughout the lengthy tome. A useful overview of the relationship between the two men appears in Rowe, 'Roscoe's and Kagwa's Buganda'.

19. Rowe, "Myth, Memoir and Moral Admonition," 24–5; *Kabaka* Daudi Chwa, *Education, Civilisation, and Foreignisation in Buganda*, 104–8.

20. Zimbe, "Buganda"; Rowe, "Myth", 24; and Miti, "A History of Buganda".

21. Rowe, "Myth," 24–5.

22. Lubogo, "A History of Busoga," 3.

23. K.W, "The Kings of Bunyoro-Kitara"; K.W, "The Kings of Bunyoro-Kitara Part II' and 'Part III'," respectively.

24. Nyakatura, *Anatomy*. This was first published as *Abakama ba Bunyoro* in 1947.

25. Kamugungunu and Katate, *Abagabe b'Ankole*. ('The Kings of Ankole'); much of this material was translated into English and included in Morris, *The Heroic Recitations of the Bahima of Ankole*.

26. Bennett, *Mirambo of Tanzania, 1840–1884* For contemporary accounts, see Broyon-Mirambo, "Description of Unyamwesi," London Missionary Society Archives (SOAS Special Collections), Central Africa, Incoming, Box 3: Southon to Thomson, 28 March 1880, encl. 'History, Country, and People of Unyamwezi'.

27. See 'Speech by the President of Tanzania, Mwalimu Julius Nyerere', in Ranger (ed.), *Emerging Themes of African History*.

28. Kabeya, *King Mirambo*, ix–xi.

29. Ranger, "Connexions between "Primary Resistance" movements and modern mass nationalism".

30. Quoted in Ranger, "Connexions," 636.

31. Ibid., 635–6.

32. Pateman, *Eritrea*; Haile, "Historical Background to the Ethiopia-Eritrea conflict,"; Cliffe and Davidson, *The Long Struggle of Eritrean for Independence and Constructive Peace*, 12–5; and Gebre-Medhin, "Eritrea (Mereb-Melash) and Yohannes IV of Abyssinia".

33. Mazrui, "Soldiers as Traditionalizers".

34. I myself argue for this in my *Warfare in African Histoy*.

35. 'Political substance and political form', speech at the opening of a political seminar for National Resistance Council members, 6 September 1989, in Museveni, *What is Africa's Problem?*, 163–4.

36. Author's field notes and informal interviews, Kampala, 6 August 2010. Kibuka was a warrior from the Sesse Islands on Lake Victoria whose help was recruited by the *kabaka* of Buganda during a particularly bruising war with Bunyoro, possibly in the sixteenth century. He was killed in the fighting, despite being able to fly, and at some point afterwards was elevated even higher to become the main *lubaale* or national spirit of Ganda warfare.

37. Museveni, *Mustard Seed*, 137.

38. Beattie, *The Nyoro State*, 31–2, 58; Ingrams, *Uganda*, 241–2; and more recently, 'Kabalega: the symbol of colonial resistance', *NV*, 9 January 2012.

39. 'Kabalega named national hero', *New Vision* (Kampala), 10 June 2009; author's field notes and informal interviews, Kampala, 16 August 2012. For rather more sceptical interpretation, based on the President's need for Nyoro political support and economic co-operation, see 'Gen. Museveni woos Banyoro with medals', *The Observer* (Kampala), 16 June 2009.

40. See for example 'Uganda's president admits gays part of Africa's heritage, *Changing Attitude*, 3 April 2012; Rao, "Re-membering Mwanga".

41. Lunyiigo, *Mwanga II*.

42. Hallencreutz, "Thomas Mofolo and Nelson Mandela on King Shaka and Dingane,"; and Granqvist, *Culture in Africa*, 185.

43. Mandela, *Long Walk to Freedom*, 274–5.

44. Hamilton, *Terrific Majesty*, 202–3.

45. See for example Golan, *Inventing Shaka*; Wylie, *Myth of Iron*. Of course, Shaka was one of those characters who transcended national boundaries. A Ugandan informant, asked to reflect on the significance of the precolonial past, enthused not about a figure from Uganda's history, but about how he was inspired by 'the history of ... past great people and their contributions to the communities that they came from e.g. Shaka Zulu in South Africa among the Zulu people; hence the spirit of fighting and protection of your people that may be oppressed and exploited by superior groups.' Indeed, the same informant was inspired by 'the philosopher of nationalism and patriotism like Nyungu ya Mawe among the Ndebele' [sic], and more generally there was succour from the deeper past in terms of 'how societies protected themselves from external aggressors', which was a central motif for this particular interviewee: Interview with Khaukha Musungu Paul, Bubutu, Uganda, January 2014.

46. Lonsdale, "Some Origins of Nationalism in East Africa"; and Davidson, *Which Way Africa?*.

47. Gulliver, *Tradition and Transition in East Africa*.

48. Hodgkin, *Nationalism in Colonial Africa*, 173.

49. Ibid., 174.

50. See for example 'Introduction' in Suso and Kanute, *Sunjata*.

51. Hodgkin, *Nationalism*, 174.

52. Kabeya, *King Mirambo*, x.
53. Ibid., x–xi.
54. Nyerere, *Ujamaa*.
55. Stanley, *Dark Continent*, 384.
56. Syahuka-Muhindo, *The Rwenzururu Movement and the Democratic Struggle*; Kyaminyawandi, *The Faces of the Rwenzururu Movement*. See also Peterson, "States of Mind".
57. This is described in exactly these terms in The Kabaka of Buganda, *The Desecration of My Kingdom*.
58. Low, *Buganda in Modern History*, 245–6; 'Address to the Nation by the President Dr. A. Milton Obote on the occasion of the sixth anniversary of independence on 9th October 1968', Kabale District Archives COM18/CM157/ Independence and Republic Celebrations and Labour Day.
59. Lee, *African Armies and Civil Order*; Gutteridge, *Military Regimes in Africa*.
60. See various contributions in Mazrui, *The Warrior Tradition in Modern Africa*.
61. For example, Decalo, *Coups and Army Rule in Africa*, 33–8.
62. Kapuscinski, *The Emperor*.
63. For fascinating contemporary analysis, see Halliday and Molyneaux, *The Ethiopian Revolution*; and Schwab, *Ethiopia*. See also Tiruneh, *The Ethiopian Revolution, 1974–1987*.
64. Thompson, *Governing Uganda*, 104–5.
65. For an eloquent examination of the imagery, see Leopold, *Inside West Nile*, 57–67. See also Kyemba, *State of Blood*.
66. It was later translated into English as *White Teeth*.
67. 'Message to the Nation by His Excellency the President General Idi Amin Dada, on British citizens of Asian origin and citizens of India, Pakistan and Bangla Desh living in Uganda … .12th/13th August, 1972', *Speeches by His Excellency the President General Idi Amin Dada*.
68. For a selection of contemporary, often sensationalist, assessments, see Melady and Melady, *Idi Amin Dada*; Martin, *General Amin*; Donald, *Confessions of Idi Amin*; Kamau and Cameron, *Lust to Kill*; and Richardson, *After Amin*.
69. Fanon, *The Wretched of the Earth*; and *Toward the African Revolution*.
70. See for example Alessandrini, *Frantz Fanon*.
71. Museveni, "Fanon's theory on violence"; Ngoga, "Uganda," 92. See also Museveni, *Sowing the Mustard Seed*, 24–5.
72. McDougall, *History and the Culture of Nationalism in Algeria*.
73. See for example Clapham, *African Guerrillas*.
74. Connell, *Against All Odds*.
75. 'National Democratic Programme, Eritrean People's Liberation Front, March 1987', in Cliffe and Davidson, *Long Struggle*, 207–8.
76. Author's field notes and informal interviews, Eritrea, 1997–2008. See also Reid, "Writing Eritrea".
77. Reid, "States of Anxiety," 256–7. Warnings of the dangers of history have been incorporated into the very constitution of Rwanda, revised on 26 May 2003: see http://www.rwandahope.com/constitution.pdf.
78. Young, *Peasant Revolution in Ethiopia*, 94, 99.
79. Assefa, "Tewodros in Ethiopian historical fiction".
80. See for example Milkias and Metaferia, *The Battle of Adwa*.

81. Author's field notes and informal interviews, Addis Ababa, 2005–2014; Orlowska, "Forging a nation".
82. 'Political substance and political form', speech at the opening of a political seminar for National Resistance Council members, 6 September 1989, in Museveni, *What is Africa's Problem?* 163–4.
83. Author's field notes and informal interviews, Uganda, 2010–2015; see also Reid, "Ghosts in the Academy".
84. 'Oyike Ojok, one of Uganda's best soldiers', *New Vision*.
85. See for example 'Tito Okello: the president who was kept on his toes', *New Vision*, 25 January 2012.
86. By Human Rights Watch, *The scars of death*; *Abducted and Abused*; and *Uprooted and Forgotten*. See also Allen and Vlassenroot, *The LRA*.
87. Reyntjens, *The Great African War*; and International Crisis Group, *South Sudan*.
88. The Dale, *The History of the Peloponnesian War*, 2–3; and Samuel, "Epical History," 6.
89. 'Epical History: the idea of nation', in *Island Stories: Unravelling Britain. Theatres of Memory, Vol II*, 8.
90. Hobsbawm, *On History*, 6–7.
91. Tilly, *Coercion, Capital and European States, AD 990–1992*. It is worth noting, however, that Tilly – a historical sociologist – has attracted criticism from historians, some of whom see the catchy formulation as overly materialistic: see Hall and Malesevic, "Introduction"; in *Nationalism and War*, 5, 11.
92. Lemarchand, "War and nationalism: the view from Central Africa".
93. For example, Williams, *War and Conflict in Africa*; Chabal, Engel and Gentili, *Is Violence Inevitable in Africa?*; Kaarsholm, *Violence, Political Culture and Development in Africa*; Nhema and Zeleza, *The Roots of African Conflicts*; and Nhema and Zeleza, *The Resolution of African Conflicts*.
94. Howard, *The Invention of Peace and the Reinvention of War*.
95. Buzan and de Wilde, *Security*.

Disclosure statement

No potential conflict of interest was reported by the author.

Bibliography

Alessandrini, Anthony C., ed. *Frantz Fanon: Critical Perspectives*. London: Routledge, 1999.

Allen, T., and K. Vlassenroot, eds. *The LRA: Myth and Reality*. London: Zed Books, 2010.

Assefa, Taye. "Tewodros in Ethiopian Historical Fiction." *Journal of Ethiopian Studies* 16 (1983): 115–128.

Atkinson, Ronald R. *The Roots of Ethnicity: The Origins of the Acholi of Uganda before 1800*. Philadelphia: University of Pennsylvania Press, 1994.

Beattie, John. *The Nyoro State*. Oxford: Oxford University Press, 1971.

Bennett, N.R. *Mirambo of Tanzania, 1840–1884*. New York: Oxford University Press, 1971.

Brantlinger, Patrick. "Victorians and Africans: The Genealogy of the Myth of the Dark Continent." *Critical Enquiry* 12, no. 1 (1985): 166–203. doi:10.1086/448326.

Broyon-Mirambo, P. "Description of Unyamwesi, the Territory of King Mirambo, and the Best Route Thither from the East Coast." *Proceedings of the Royal Geographical Society* 22, no. 1 (1877–78): 28–38.

Burton, Sir, and Richard Francis. *The Lake Regions of Central Africa*. 2 vols. London: Longman, Green, Longman & Roberts, 1860.

Buzan, Barry, Ole Wæver, and Jaap de Wilde. *Security: A New Framework for Analysis*. Boulder, CO: Lynne Rienner Publishers, 1998.

Chabal, Patrick, Ulf Engel, and Anna-Maria Gentili, eds. *Is Violence Inevitable in Africa? Theories of Conflict and Approaches to Conflict Prevention*. Leiden & Boston: Brill, 2005.

Chretien, J.-P. *The Great Lakes of Africa: Two Thousand Years of History*. New York: Zone Books, 2003.

Chwa, Kabaka Daudi. "Education, Civilisation, and Foreignisation in Buganda. (orig. published 1935), reproduced." In *The Mind of Buganda: Documents in The Modern History of a Kingdom,* edited by D.A. Low. London: Heinemann, 1971.

Clapham, Christopher, ed. *African Guerrillas*. Oxford: James Currey, 1998.

Cohen, D.W. *The Historical Tradition of Busoga: Mukama and Kintu*. Oxford: Clarendon Press, 1972.

Coleman, James Smoot. "Tradition and Nationalism in Tropical Africa." In *Nationalism and Development in Africa: Selected Essays*, edited by J.S. Coleman and Richard L. Sklar. Berkeley: University of California Press, 1994.

Connell, Dan. *Against All Odds: A Chronicle of the Eritrean Revolution*. Lawrenceville, NJ: Red Sea Press, 1997.

Dale, Rev Henry. *The History of the Peloponnesian War*. London: Henry G. Bohn, 1849.

Davidson, Basil. *Which Way Africa? The Search for a New Society*. Harmondsworth: Penguin, 1964.

Decalo, Samuel. *Coups and Army Rule in Africa: Motivations and Constraints*. New Haven & London: Yale University Press, 1990.

Donald, T. *Confessions of Idi Amin*. London: W.H. Allen & Co, 1978.

Doyle, Shane. *Crisis and Decline in Bunyoro: Population Ad Environment in Western Uganda, 1860–1955*. Oxford: James Currey, 2006.

Eritrean People's Liberation Front. "National Democratic Programme, Eritrean People's Liberation Front, March 1987." In *The Long Struggle of Eritrean for Independence and Constructive Peace*, edited by L. Cliffe and B. Davidson, 205–213. Trenton, NJ: Red Sea Pres, 1988.

Fanon, Frantz. *Toward the African Revolution*. New York: Grove Press, 1964, 1967.

Fanon, Frantz. *The Wretched of the Earth*. (tr. Constance Farrington). London: Penguin, 1967, 1990.

Gebre-Medhin, Jordan. "Eritrea (mereb-melash) and Yohannes IV of Abyssinia." *Eritrean Studies Review* 3, no. 2 (1999): 1–42.

Golan, Daphna. *Inventing Shaka: Using History in the Construction of Zulu Nationalism*. Boulder, CO: Lynne Rienner Publishers, 1994.

Gulliver, P.H., ed. *Tradition and Transition in East Africa: Studies of the Tribal Element in the Modern Era*. London: Routledge & Kegan Paul, 1969.

Gutteridge, W.F. *Military Regimes in Africa*. London: Methuen, 1975.

Haile, Semere. "Historical Background to the Ethiopia-Eritrea Conflict." In *The Long Struggle of Eritrean for Independence and Constructive Peace*, edited by Lionel Cliffe and Basil Davidson, 11–31. Trenton, NJ: Red Sea Press, 1988.

Hall, John A., and Sinisa Malesevic, eds. *Nationalism and War*. Cambridge: Cambridge University Press, 2013.

Hallencreutz, Carl F. "Thomas Mofolo and Nelson Mandela on King Shaka and Dingane." In *Culture in Africa: An Appeal for Pluralism*, edited by Raoul Granqvist. Uppsala: Nordiska Afrikainstitutet, 1993.

Halliday, Fred, and Maxine Molyneaux. *The Ethiopian Revolution*. London: Verso, 1981.

Hamilton, Carolyn. *Terrific Majesty: The Powers of Shaka Zulu and the Limits of Historical Invention*. Cape Town and Johannesburg: David Philip Publishers, 1998.

Hobsbawm, Eric. *On History*. London: Abacus, 1997.

Hodgkin, Thomas. *Nationalism in Colonial Africa*. London: Frederick Muller, 1956.

Hodgkin, Thomas. "The Relevance of "western" Ideas for the New African States." In *Self-Government in Modernizing Nations*, edited by J. Roland Pennock. Englewood Cliffs, NJ, 1964.

Howard, Michael. *The Invention of Peace and the Reinvention of War*. London: Profile Books, 2001.

Howard, Michael. *War in European History*. Oxford: Oxford University Press, 2002.

Human Rights Watch. *The Scars of Death: Children Abducted by the Lord's Resistance Army in Uganda*. New York: Human Rights Watch, 1997.

Human Rights Watch. *Abducted and Abused: Renewed Conflict in Northern Uganda*. New York: Human Rights Watch, 2003.

Human Rights Watch. *Uprooted and Forgotten: Impunity and Human Rights Abuses in Northern Uganda*. New York: Human Rights Watch, 2005.

Ingrams, Harold. *Uganda: A Crisis of Nationhood*. London: HMSO, 1960.

International Crisis Group. *South Sudan: Keeping Faith with the IGAD Peace Process*. Africa Report No.228. 27 July 2015.

K.W. "The Kings of Bunyoro-Kitara." *Uganda Journal* 3, no. 2 (1935): 155–160.

K.W. "The Kings of Bunyoro-Kitara Part II." *Uganda Journal* 4, no. 1 (1936): 75–83.

K.W. "The Kings of Bunyoro-Kitara Part III." *Uganda Journal* 5, no. 2 (1937): 53–69.

Kaarsholm, Preben, ed. *Violence, Political Culture and Development in Africa*. Oxford: James Currey, 2006.

Kabaka of Buganda. *The Desecration of My Kingdom*. London: Constable, 1967.

Kabeya, J.B. *King Mirambo: One of the Heroes of Tanzania*. Nairobi: East African Literature Bureau, 1976.

Kagwa, A. *The Kings of Buganda*. (tr. & ed. M.S.M.Kiwanuka). Nairobi: East African Publishing House, 1971.

Kamau, J., and A. Cameron. *Lust to Kill: The Rise and Fall of Idi Amin*. London: Corgi Books, 1979.

Kamugungunu, Lazaro, and A.G. Katate. *Abagabe b'Ankole*. Dar es Salaam: Eagle Press, 1955.

Kapuscinski, Ryszard. *The Emperor: Downfall of an Autocrat*. London: Penguin, 2006. 1st1978.

Kopytoff, Igor, ed. *The African Frontier: The Reproduction of Traditional African Societies*. Bloomington & Indianapolis: Indiana University Press, 1987.

Kyaminyawandi, Augustine. *The Faces of the Rwenzururu Movement*. n.p., 2001.

Kyemba, Henry. *State of Blood: The inside Story of Idi Amin's Reign of Fear*. London: Corgi Books, 1977.

Lamphear, John. "The Evolution of Ateker "new Model" Armies: Jie and Turkana." In *Ethnicity and Conflict in the Horn of Africa*, edited by K. Fukui and J. Markakis, 63–92. London: James Currey, 1994.

Lawrance, J.C.D. *The Iteso: Fifty Years of Change in a Nilo-Hamitic Tribe of Uganda*. London: Oxford University Press, 1957.

Lee, J.M. *African Armies and Civil Order*. London: Chatto & Windus, 1969.

Lemarchand, René. "War and Nationalism: The View from Central Africa." In *Nationalism and War*, edited by John A. Hall and Sinisa Malesevic. Cambridge, 2013.

Leopold, Mark. *Inside West Nile: Violence, History and Representation on an African Frontier*. Oxford: James Currey, 2005.

Lonsdale, John. "Some Origins of Nationalism in East Africa." *Journal of African History* 9, no. 1 (1968): 119–146. doi:10.1017/S0021853700008380.

Low, D.A. *Buganda in Modern History*. London: Weidenfeld & Nicolson, 1971.

Lubogo, Y.K. *A History of Busoga*. Kampala: translated & unpublished manuscript in Makerere University Library, 1960.

Lunyiigo, Samwiri Lwanga. *Mwanga II: Resistance to Imposition of British Colonial Rule in Buganda, 1884–1899*. Kampala: Wavah Books, 2011.

Mackay, A.M. *Pioneer Missionary of the Church Missionary Society to Uganda*. London: Hodder & Stoughton, 1890.

Mandela, Nelson. *Long Walk to Freedom*. London: BCA, 1995.

Martin, D. *General Amin*. London: Faber, 1974.

Mazrui, Ali. "The Social Origins of Ugandan Presidents: From King to Peasant Warrior." *Canadian Journal of African Studies* 8 (1974): 3–23.

Mazrui, Ali. "Soldiers as Traditionalizers: Military Rule and the Reafricanisation of Africa." In *The Warrior Tradition in Modern Africa*, edited by Ali Mazrui, 236–258. Leiden: Brill, 1977.

McDougall, James. *History and the Culture of Nationalism in Algeria*. Cambridge: Cambridge University Press, 2006.

Médard, Henri. *La Royaume Du Buganda Au XIXe Siècle*. Paris: Karthala, 2007.

Melady, T., and M. Melady. *Idi Amin Dada: Hitler in Africa*. Kansas City: Sheed Andrews & McMeel, 1977.

Milkias, Paulos, and Getachew Metaferia, eds. *The Battle of Adwa: Reflections on Ethiopia's Historic Victory against European Colonialism*. New York: Algora Publishing, 2005.

Miti, James. "A History of Buganda." unpublished manuscript, SOAS Library, London, 1938.

Morris, H.F. *The Heroic Recitations of the Bahima of Ankole*. Oxford: Clarendon Press, 1964.

Museveni, Yoweri. "Fanon's Theory on Violence: Its Verification in Liberated Mozambique." In *Studies in Political Science*, edited by N.M. Shamuyarira, 3 vols, 1–24. Dar es Salaam: Tanzania Publishing House, 1974.

Museveni, Yoweri. *Sowing the Mustard Seed: The Struggle for Freedom and Democracy in Uganda*. Oxford: Macmillan, 1997.

Museveni, Yoweri. *What Is Africa's Problem?*. Minneapolis: University of Minnesota Press, 2000.

Ngoga, Pascal. "Uganda: The National Resistance Army." In *African Guerrillas*, edited by C. Clapham, 91–106. Oxford: James Currey, 1998.

Nhema, Alfred, and Paul Tiyambe Zeleza, eds. *The Roots of African Conflicts: The Causes & Costs*. Oxford: James Currey, 2008.

Nhema, Alfred, and Paul Tiyambe Zeleza, eds. *The Resolution of African Conflicts: The Management of Conflict Resolution & Post-conflict Reconstruction*. Oxford: James Currey, 2008.

Nyakatura, John. *Anatomy of an African Kingdom: A History of Bunyoro-Kitara*. (ed. G.N. Uzoigwe). New York: Doubleday/Anchor Press, 1973.

Nyerere, Julius K. "' Ujamaa: The Basis of African Socialism'." In *Government and Politics in Africa: A Reader*, edited by Okwudiba Nnoli, 51–58. Harare: AAPS Books, 2000.

Oliver, R. "Discernible Developments in the Interior, C.1500-1840." In *History of East Africa*, edited by R. Oliver and G. Mathew, I vols, 169–211. Oxford: Clarendon Press, 1963.

Orlowska, Izabela. "Forging a Nation: The Ethiopian Millennium Celebration and the Multi-ethnic State." *Nations and Nationalism* 19, no. 2 (2013): 296–316. doi:10.1111/nana.12021.

p'Bitek, Okot. *White Teeth*. Nairobi: Heinemann Kenya, 1989.

Parker, Geoffrey. *The Military Revolution: Military Innovation and the Rise of the West, 1500–1800* [1996]. Cambridge: Cambridge University Press, 1988.

Pateman, Roy. *Eritrea: Even the Stones are Burning*. Lawrenceville, NJ: Red Sea Press, 1990. 1998.

Peterson, D.R. "States of Mind: Political History and the Rwenzururu Kingdom in Western Uganda." In *Recasting the Past: History Writing and Political Work in Modern Africa*, edited by D.R. Peterson and G. Macola, 171–190. Athens, OH: Ohio University Press, 2009.

Porter, Patrick. *Military Orientalism: Eastern War through Western Eyes*. London: Hurst, 2009.

Ranger, T.O., ed. *Emerging Themes of African History*. Nairobi: Heinemann Educational Books, 1968.

Ranger, T.O., ed "Connexions between "primary Resistance" Movements and Modern Mass Nationalism in East and Central Africa', Parts I & II." *Journal of African History* 9, no. 3 & no. 4 (1968): 437–453 & 631–641.

Ranger, T.O. "Towards a Usable African Past." In *African Studies since 1945: A Tribute to Basil Davidson*, edited by Christopher Fyfe, 17–30. London: Longman, 1976.

Rao, R. "Re-membering Mwanga: Same-sex Intimacy, Memory, and Belonging in Postcolonial Uganda." *Journal of Eastern African Studies* 9, no. 1 (2015): 1–19. doi:10.1080/17531055.2014.970600.

Reid, Richard. *Political Power in Pre-Colonial Buganda: Economy, Society and Warfare in the Nineteenth Century*. Oxford: James Currey, 2002.

Reid, Richard. "Revisiting Primitive War: Perceptions of Violence and Race in History." *War and Society* 26, no. 2 (2007): 1–25. doi:10.1179/072924707791591677.

Reid, Richard. *Warfare in African History*. New York: Cambridge University Press, 2012.

Reid, Richard. "The Fragile Revolution: Rethinking War and Development in Africa's Violent Nineteenth Century." In *Africa's Development in Historical Perspective*, edited by E. Akeampong, R.H. Bates, N. Nunn, and J. Robinson, 83–115. Cambridge: Cambridge University Press, 2014.

Reid, Richard. "Writing Eritrea: History and Representation in a Bad Neighbourhood." *History in Africa* 41 (2014): 239–269.

Reid, Richard. "Ghosts in the Academy: Historians and Historical Consciousness in the Making of Modern Uganda." *Comparative Studies in Society and History* 56, no. 2 (2014): 351–380. doi:10.1017/S0010417514000073.

Reid, Richard. "States of Anxiety: History and Nation in Modern Africa." *Past and Present* 229 (2015).

Reyntjens, Filip. *The Great African War: Congo and Regional Geopolitics, 1996–2006.* Cambridge: Cambridge University Press, 2009.

Richardson, M.L. *After Amin: The Bloody Pearl.* Atlanta: Majestic Books, 1980.

Roscoe, John. *The Baganda: An Account of Their Native Customs and Beliefs.* London: Macmillan, 1911.

Roscoe, John. *Twenty-Five Years in East Africa.* Cambridge: Cambridge University Press, 1921.

Roscoe, John. *The Bakitara or Banyoro.* Cambridge: Cambridge University Press, 1923.

Rowe, John. "Roscoe's and Kagwa's Buganda." *Journal of African History* 8, no. 1 (1967): 163–166.

Rowe, John. "Myth, Memoir and Moral Admonition: Luganda Historical Writing, 1893-1969." *Uganda Journal* 33, no. 1 (1969): 17–40.

Samuel, Raphael. *Island Stories: Unravelling Britain. Theatres of Memory.* Vol. II. London & New York: Verso, 1998.

Schivelbusch, Wolfgang. *The Culture of Defeat: On National Trauma, Mourning, and Recovery.* London: Granta Books, 2003.

Schoenbrun, David. "Conjuring the Modern in Africa: Durability and Rupture in Histories of Public Healing between the Great Lakes of East Africa." *American Historical Review* 111, no. 5 (2006): 1403–1439. doi:10.1086/ahr.111.5.1403.

Schwab, Peter. *Ethiopia: Politics, Economics, and Society.* London: Frances Pinter, 1985.

Speke, J.H. *Journal of the Discovery of the Source of the Nile.* Edinburgh & London: William Blackwood & Sons, 1863.

Stanley, H.M. *Through the Dark Continent.* Vol. 2 London: George Newnes, 1878. 1899.

Steinhart, E.I. *Conflict and Collaboration: The Kingdoms of Western Uganda, 1890–1907.* Princeton, NJ: Princeton University Press, 1977.

Suso, Bamba, and Banna Kanute. *Sunjata.* (ed. Lucy Duran & Graham Furniss), London: Penguin, 1974, 1999.

Syahuka-Muhindo, A. *The Rwenzururu Movement and the Democratic Struggle.* Kampala: Centre for Basic Research, 1991.

Thompson, Gardner. *Governing Uganda: British Colonial Rule and Its Legacy.* Kampala: Fountain, 2003.

Tilly, C. *Coercion, Capital, and European States, AD 990–1992.* Malden MA & Oxford: Blackwell, 1992.

Tiruneh, Andargachew. *The Ethiopian Revolution, 1974–1987: A Transformation from an Aristocratic to a Totalitarian Society.* Cambridge: Cambridge University Press, 1993.

Uzoigwe, G.N. "The Warrior and the State in Precolonial Africa." In *The Warrior Tradition in Modern Africa,* edited by A.A. Mazrui, 20–47. Leiden: Brill, 1977.

Williams, Michael C. "Words, Images, Enemies: Securitization and International Politics." *International Studies Quarterly* 47 (2003): 511–531.

Williams, Paul D. *War and Conflict in Africa.* Cambridge: Polity, 2011.

Winter, Jay. *Sites of Memory, Sites of Mourning: The Great War in European Cultural History*. Cambridge: Cambridge University Press, 1995.

Wylie, Dan. *Myth of Iron: Shaka in History*. Scottsville: University of KwaZulu-Natal Press, 2006.

Young, John. *Peasant Revolution in Ethiopia: The Tigray People's Liberation Front, 1975–1991*. Cambridge: Cambridge University Press, 1997.

Zimbe, B.M. "Buganda Ne Kabaka [Buganda and the King]." unpublished ms., Makerere University Library, Kampala, 1939.

Conclusion

Mark Lawrence

Irregular warfare in the nineteenth century, in contrast to some aspects of the American and French Revolutions at the end of the eighteenth century and Marxist insurgencies of the twentieth century, evokes political and military failure. The political chaos of the Ibero-American and Chinese worlds, the failed insurrections in Ireland and Poland, and tribal disunity vis-à-vis European encroachment in Africa and Asia all seemed to confirm the general disdain felt by revolutionaries and governments alike. In industrialised North America and Europe, irregular aspects to campaigns were either overlooked against a backdrop of military revolution or distorted for political ends. The global nineteenth century's irregular warfare cannot escape from relegation of nationalistic and militaristic grand narratives imbued with a Whiggish 'presentism'. It falls between the stools of the French and Russian revolutions and fades out of focus in a century busy with imperialism and industrialised warfare.

Yet this study has challenged commonly made assumptions about nationalism, imperialism and militarism. Irregular warfare transcended the usual categories. Foreign veterans of the Napoleonic Wars fought as volunteers in asymmetrical wars in southern Europe and the Americas, lending their military talents to nation-building projects which were not their own. And civil rather than foreign wars, like the Spanish, Chinese, Latin American and American examples explored in this volume, accounted for more nation-building than the wars of Napoleon and Bismarck combined. These civil wars made greater emotional strains on combatants and civilians, owing to the greater intimacy of violence and the extent of insurgent warfare. Guerrilla forces collaborated with invaders and counterinsurgents borrowed from the tactics of the enemy. Responses to irregular warfare account for most of the continuity and hybridity in a century otherwise marked by military revolution.

Insurgent warfare in the global nineteenth century also reveals non-Western agency. Indigenous populations from Mexico to Africa were not passive objects of state-building and imperialism but adaptive in resisting and collaborating where necessary. Layers of imperialism overlapped, turning the Mexican Cora into temporary allies of French invaders against the long-standing imperialism of mestizo settlers, and the Yeke into collaborators of European imperialism in their struggles with rival tribes in Katanga. Euro-African cooperation was, in the words of Giacomo Macola, 'forged in blood, with African men of violence making a decisive contribution to the colonial takeover'.[1] Throughout the twentieth century, this reality was selectively forgotten or distorted by revolutionary and anti-imperial regimes on the one hand, and overlooked by Western historians impressed by the 'tools of empire' on the other.

The tools of the West were blunted once counterinsurgents moved out of range of river and rail supplies and into jungle and mountains. Here, insurgents retained the traditional advantages of cold arms, intelligence and metaphysics to explain the upheaval caused by outsiders. Outsiders, especially in Africa, often encountered highly militarised societies able to inflict temporary defeats on unsuspecting European forces.[2] European counterinsurgents also continued to succumb to disease. The last third of the century witnessed the advance of antiseptics, of better logistics and hygiene, all of which reduced non-combat casualties in armies waging regular campaigns. But colonial counterinsurgencies in such places as Burma and Cuba continued to exact a far higher proportion of epidemiological casualties from counterinsurgents until the century's end. As Edward Spiers explained, colonial campaigns were first and foremost 'wars against nature'.[3]

The peculiar challenges posed by counterinsurgency were understood by contemporary practitioners, but their political impact seldom was. General Zuo's conquest of Xinjiang locked in tension between Han Chinese and Muslims, which has persisted until today. The Spanish army's defeat of the Carlists saddled Spain with a praetorian tradition in politics that would not be expunged until the death of Franco.[4] French defeat in the Franco-Prussian War was blamed on the 'African' military establishment, which had grown complacent since the start of the Algerian adventure in 1830. The swarming tactics of desperadoes serving in Algeria, Mexico and elsewhere against ill-armed and non-European enemies seemed to indict the French army, which in 1870 suddenly found itself poorly prepared against a modern, European foe. Counterinsurgency was messy and its legacy unwelcome. It seemed to attract unlikely characters like the humble-born General Zuo, who developed his doctrine in middle age, and the hard-living General Skobelev. The British mercenary officer, 'Chinese' Gordon, was condemned as a prima donna by Prime Minister Gladstone, and even armies which developed no distinct colonial identity, such as the German army, seemed to attract impecunious and otherwise eccentric officers in African campaigns. Only the drive to partition Africa, which became a proxy for Great Power rivalry, restored respect for the colonial soldier and interest in small wars in European academies.[5]

Thus, the closing of the nineteenth century witnessed the convergence of imperialism and counterinsurgency. The pioneering French 'oil-slick' method associated with such men as Joseph Gallieni and Hubert Lyautey offered civilians economic incentives alongside garrisons.[6] General Zuo's pacification of rebel areas of China had marked civic and economic dimensions. British officers in the north-west frontier of India from the late nineteenth century operated effectively as politicians, using charm, bribes and diplomacy in meetings with tribal leaders.[7] French counter-insurgent authorities from the late nineteenth century started to distinguish 'civilised' from 'savage' colonial subjects. Public infrastructure projects, including railways, roads and ports, served to impose a socially hierarchized system. The system placed those who had adopted Eurocentric values and culture (*les évolués*) above those unwilling or incapable of doing so (*les indigènes*).[8]

Yet before political counterinsurgency could begin, the brute force of military counterinsurgency had to take effect. The violent *razzia* made famous by Thomas Bugeaud's campaigning in Algeria from the 1830s was accepted towards the end of the century by the small wars expert, Charles Callwell, who relished decisive action and risk-taking over the logistics and heavy firepower dominating calculations for regular warfare during his time.[9] Callwell thought that supposed primitive peoples were impressionable, and other Western armies thought the same. The German army was well known for its emphasis on decisive action, doubly so when it fought non-Western opposition. In the words of Isabel Hull, 'tactical retreat vis-à-vis blacks was simply unacceptable.'[10] The savage results of concentrating civilians in China, Cuba, South Africa and elsewhere were sometimes the product of racism and sometimes the product of overwhelmed logistics.

What, then, of the insurgent? It suited occupation armies to describe him as a bandit, in Spain, China, Mexico or Burma, and elsewhere. The rules of war mostly excluded him from the right to receive quarter and there were few scruples about using collective punishment against villagers suspected of harbouring him. Although insurgents often acquired Romantic status, especially when their cause was wrapped in religion, only advanced constitutional states, like Britain's in Ireland or the Union in its war with the Confederacy, afforded them any legal protection. In this regard, the insurgent nineteenth century was very long indeed. Only after 1945 did the international legal order improve from an insurgent's point of view. The Hague Conventions gradually extended *jus in bello* rights to insurgents fighting racist regimes, enemy occupations and (from 1977) anti-colonial struggles. The intellectual environment praised the insurgent and condemned the counterinsurgent, a complete *volte-face* from the nineteenth century. Regular armies, especially those retreating from empires, felt obliged at least to pay lip-service to 'hearts and minds'. Leftist intellectuals like Robert Taber celebrated the 'war of the flea' and the *guerrillero heróico* photograph of Ernesto 'Che' Guevara adorned the student bedrooms of imperialist oppressors.

However, if the laws of war (*jus in bello*) now protect the insurgent, the question of whether he has legitimate grounds to wage war in the first place (*jus ad bellum*) is not much clearer now, after a century of self-determination and over half a century of embargoes on formal declarations of war, than it was in the nineteenth century. *Jus ad bellum* prescribes under what circumstances states as formally constituted powers may wage war. Insurgents, by contrast, have no such formally constituted power at the outset, and thus must succeed with violence in order to prove their 'right' to wage war in the first place.[11]

Notes

1 Macola, *The Gun in Central Africa*, 94.
2 Reid, *Warfare in African History*, 135.
3 Spiers, *The Late Victorian Army*, 275.
4 Lawrence, *The Spanish Civil Wars*, 10.
5 Porch, *Counterinsurgency*, 40–8.
6 Rid, "Nineteenth Century Origins of Counterinsurgency Doctrine," 727–31.
7 Tripodi, "Peace-making Through Bribes or Cultural Empathy," 148.
8 Bigon and Njoh, "Power and Social Control in Settler and Exploitation Colonies," 942.
9 Callwell, *Small Wars*, 50–80.
10 Hull, *Absolute Destruction*, 22.
11 Gross, *The Ethics of Insurgency*, 1–38.

Bibliography

Bigon, Liora and Njoh, Ambe J. "Power and Social Control in Settler and Exploitation Colonies: The Experience of New France and French Colonial Africa." *Journal of Asian and African Studies* 53, no. 6 (2018): 932–951, p. 942.
Callwell, Charles. *Small Wars: Their Principles and Practice*. London: Her Majesty's Stationery Office, 1899.
Gross, Michael L. *The Ethics of Insurgency*. Cambridge: Cambridge University Press, 2014.
Hull, Isabel. *Absolute Destruction: Military Culture and the Practices of War in Imperial Germany*. London: Cornell University Press, 2006.
Lawrence, Mark. *The Spanish Civil Wars*. London: Bloomsbury, 2017.
Macola, Giacomo. *The Gun in Central Africa*. Athens: Ohio University Press, 2016.
Porch, Douglas. *Counterinsurgency: Exposing the Myths*. Cambridge: Cambridge University Press, 2013.
Reid, Richard. *Warfare in African History*. Cambridge: Cambridge University Press, 2012.
Rid, Thomas. "The Nineteenth Century Origins of Counterinsurgency Doctrine." *The Journal of Strategic Studies* 33, no. 5 (2010): 727–758.
Spiers, Edward. *The Late Victorian Army*. Manchester: Manchester University Press, 1992.
Tripodi, Christian. "Peace-making Through Bribes or Cultural Empathy: The Political Officer and Britain's Strategy towards the North West Frontier, 1901–1945." *Journal of Strategic Studies* 31 (2008): 123–151.

Index